Wear-Resistant Coatings and Film Materials

Wear-Resistant Coatings and Film Materials

Guest Editors

Xuan Yin
Jianxun Xu

MDPI

Basel • Beijing • Wuhan • Barcelona • Belgrade • Novi Sad • Cluj • Manchester

Guest Editors
Xuan Yin
School of Mechanical and Electrical Engineering
Beijing University of Chemical Technology
Beijing
China

Jianxun Xu
School of Energy, Power and Mechanical Engineering
North China Electric Power University
Beijing
China

Editorial Office
MDPI AG
Grosspeteranlage 5
4052 Basel, Switzerland

This is a reprint of the Special Issue, published open access by the journal *Lubricants* (ISSN 2075-4442), freely accessible at: https://www.mdpi.com/journal/lubricants/special_issues/MY13PI3CMF.

For citation purposes, cite each article independently as indicated on the article page online and as indicated below:

Lastname, A.A.; Lastname, B.B. Article Title. *Journal Name* **Year**, *Volume Number*, Page Range.

ISBN 978-3-7258-5179-9 (Hbk)
ISBN 978-3-7258-5180-5 (PDF)
https://doi.org/10.3390/books978-3-7258-5180-5

Contents

lubricants

MDPI

Review

Nanostructure of Superlubricating Tribofilm Based on Friction-Induced a-C:H Films under Various Working Conditions: A Review of Solid Lubrication

Xuan Yin [1,*], Linyuan Mu [1], Zihang Jia [1], Haosheng Pang [2,*], Chunpeng Chai [3,*], Huan Liu [4], Chang Liang [1], Bing Zhang [1] and Dameng Liu [4]

1 College of Mechanical and Electrical Engineering, Beijing University of Chemical Technology, Beijing 100029, China; 2021030059@buct.edu.cn (L.M.); 2021030475@buct.edu.cn (Z.J.); liangchang@mail.buct.edu.cn (C.L.); zhangbing@mail.buct.edu.cn (B.Z.)
2 Chinese Aeronautical Establishment, Beijing 100012, China
3 School of Materials Science and Engineering, Beijing Institute of Technology, Beijing 100081, China
4 State Key Laboratory of Tribology in Advanced Equipment, Department of Mechanical Engineering, Tsinghua University, Beijing 100084, China; liuhuansklt@163.com (H.L.); ldm@tsinghua.edu.cn (D.L.)
* Correspondence: yinxuan@buct.edu.cn (X.Y.); panghs.2020@tsinghua.org.cn (H.P.); chaicp@bit.edu.cn (C.C.)

Abstract: Diamond-like carbon (DLC) film has gained widespread popularity as a versatile and important solid lubricant material in the field of tribology. Among various types of DLC films, hydrogen-rich DLC (a-C:H) film as a high-performance material has greatly enhanced anti-friction and anti-wear. However, despite its remarkable capabilities, the surface chemical properties and tribological performance of a-C:H film are significantly influenced by the surrounding environment, in special atmospheric conditions. Its super-slip mechanism involves the participation of hydrogen atoms, which can weaken the normal electron number of the outermost layer of a-C:H film. What is more, it is essential to investigate tribofilms in a vacuum or inert gas environment to ascertain the appropriate tribological properties of a-C:H film, which helps in mitigating oxidation effects. When non-doped DLC films are subjected to friction in a dry nitrogen or argon environment, they create sp^3-C-rich transfer films on the contact surface, resulting in macroscopic super-slip effects. This paper aims to introduce and discuss the diverse nanostructures of in situ tribofilms in a-C:H film, focusing on the working environment, and explore the prospective application directions of a-C:H film.

Keywords: diamond-like carbon film; tribofilm; two-dimensional materials; lubrication mechanism

1. Introduction

Diamond-like carbon (DLC) is a synthetic material that has garnered significant attention due to its unique properties, making it an excellent coating material for various mechanical devices. As a widely engineered and industrial application, DLC film has already been utilized in engines and bearings resulting in a friction coefficient lower than 0.01 (a superlubricity state [1,2]). Superlubricity helps reduce friction and wear, thereby extending the life of materials. The superlubricity of DLC is due to the formation of a highly ordered tribofilm with a thickness of only a few nanometers. This highly ordered structure reduces the friction between surfaces, resulting in superlubricity. The growing focus on superlubric DLC has caught the attention of many. Besides its excellent tribological performance, carbon thin film also possesses low surface energy, which has been explored in diversified superlubricity pathways. Super-low friction and wear are common characteristics of DLC films, which exhibit significant tribological behaviors and negligible energy consumption during rubbing. As a representative of amorphous carbon (a-C) films in relevant tribological fields [3–6], DLC film has bonding structures and properties resembling both diamond (sp^3-C phase) and graphite (sp^2-C phase), as shown in Figure 1. Among various types of DLC films, hydrogen-rich DLC (a-C:H) films have been successfully designed to achieve ultra-low (friction coefficient < 0.1) or even super-low (friction coefficient < 0.01) friction in dry inert atmospheres like dry nitrogen atmospheres. This is mainly attributed to the buildup of a nanostructured graphitic tribofilm (a thin film that forms on the surface of an induced-friction transfer substance, reducing its abrasion) and the associated surface passivation from –H or –OH free groups along the sliding interface [7–9]. This kind of induced-friction transfer substance is called a tribofilm, which does not have any hardness and cannot be tested independently. Essentially, a tribofilm is a substance generated through physical action or a chemical reaction induced by shear force during the friction process. The tribological performances are obtained by characterizing the contact structure and friction coefficient of the friction pair that carries the transfer film. However, when friction occurs in the atmosphere, the structure of the a-C:H film surface is easily destroyed due to O and moisture-induced strong tribochemical reactions [10–12], leading to the failed protection of materials or key mechanical moving parts. The tribofilm generated by the DLC coating possesses a distinct nanoscopic structure, smaller than a micrometer. The superlubricating property of DLC can be ascribed to the aforementioned nanoscopic structure of the tribofilm. The tribofilm generated by DLC exhibits a highly organized arrangement and is only a few nanometers thick. Therefore, it is crucial to comprehensively investigate the nanostructures of in situ tribofilms on the contacts for broadened environments.

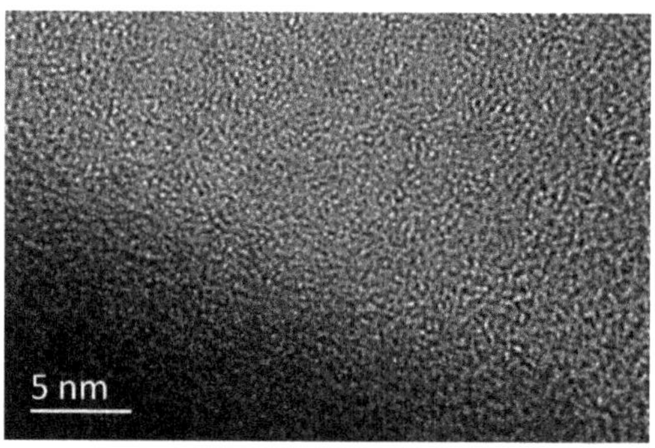

Figure 1. HRTEM images showing the atomic microstructures of DLC film.

In the past decade, low-dimensional (2D) materials and zero-dimensional (0D) materials have been extensively studied as efficient alternatives for surface modification, to exploit novel superlubricity systems. Graphene, 2D titanium carbide (Ti_3C_2, one type of MXenes), graphene quantum dots (GQDs), and nanodiamonds, owing to their special atomic-scale structure, are gradually being utilized in fundamental applied research [13–17]. In general, the combination of DLC films and low-dimensional materials has become the core of modified solid lubrication films. The major characterizations of in situ tribofilm nanostructures are high-resolution transmission electron microscopy (HRTEM), the focused ion beam–scanning electron microscopy dual-beam system (SEM-FIB), and dual-aberration-corrected scanning transmission electron microscopy and electron energy-loss microscopy (STEM-EELS) [15,18–21]. These characterization methods have already been utilized in the interfacial structure and surface morphology of materials for basic research and mechanical engineering applications. In a few research groups aspiring to investigate the nanostructures of in situ tribofilms, our group has carried out a large amount of work based on the structures and characterizations of macro-lubrication. Due to the drawbacks of DLC films, their surface properties need to be modified or matched with proper tribo-couples to meet the increasingly stringent requirements of superlubricity in complicated working conditions. In previous experiments [15,16,22], we found that nanocoating, which is utilized as a highly effective solid lubricant, accelerated the formation of the tribofilm as a protective surface layer, especially in a dry nitrogen atmosphere. In addition, the selection of friction counterparts has a significant impact on tribological performances and in situ tribofilm structures in mechanical engineering [19,23–28]. Mechanical performances such as hardness and elastic modulus exert the most pivotal influence on the tribological behaviors of the tribo-couples. Commonly, among the counterpart materials, steel [29,30] and ceramic [26,31] are the most frequently used tribo-couples through intensive adhesive or abrasive interactions between sliding surfaces. DLC-based steel or silicon nitride sliding against the same type of film material substrate (namely a self-mated system) has the superiority to establish the superlubricity state (friction coefficient namely $\mu < 0.01$). During the sliding process, the contact surfaces of the counterfaces are usually covered by forming an in situ tribofilm to reduce friction and wear [9]. The nanostructures of in situ tribofilms are governed by tribophysics and tribochemistry on the interfacial contacts. Weak interactions along the sliding interface are accordingly realized by low-shear-strength lubrication phases such as the sp^2-C phase or the highly passivated contact surfaces induced by van der Waals forces such as the sp^3-C phase [6].

It is crucial to choose the appropriate friction counterparts to optimize the overall performance and efficiency of mechanical systems. The selection process involves considering various factors such as material compatibility, load capacity, and environmental conditions [23,25]. The design of tribo-couples also plays a crucial role in reducing wear and increasing the lifespan of moving components. The geometry, surface roughness, and lubrication strategies are carefully evaluated and implemented to minimize friction and enhance tribological performances. For instance, steels [29,30] and ceramics [31,32] are the most commonly used substrates among the counterpart materials. Generally, the mechanical properties, such as hardness and elastic modulus, have the most significant influence on the tribological behaviors of tribo-couples. Additionally, friction and wear are also greatly affected by the surface morphologies and bonding chemistry of the counterpairs [24,33]. To improve the sliding accommodation to the working conditions, the surfaces of mechanical components are typically coated with a layer of solid film or lubricated with various lubricants [34]. With breakthroughs in technological methods and algorithm innovations, there have been significant advancements in the study of DLC friction behavior, specifically the microstructure of transfer films. These approaches have provided valuable insights into the interactions, conformational changes, and dynamic processes between molecules within transfer films, shedding light on their formation mechanisms, stability, and functionality [35,36]. Simulating the interactions between molecules in transfer films allows for the investigation of changes in parameters such as binding forces, bond lengths, and angles, thereby exploring the mechanisms behind transfer film formation. Furthermore, molecular dynamics (MD) can simulate the movement of molecules within transfer films, including torsion, rotation, and displacement, providing insights into conformational changes and dynamic processes.

To date, there has been a lack of comprehensive summaries regarding the nanostructures of in situ tribofilms that induce superlubricity for DLC films. Therefore, this paper aims to introduce and discuss the diverse nanostructures of in situ tribofilms that induce superlubricity for DLC films. The influencing factors of superlubricating tribofilm nanostructures are also considered, with a focus on the working environment atmosphere (such as atmospheric, dry inert, harsh, and humid environments), which is supplemented with computational simulation techniques. Furthermore, the lubrication mechanisms of obtained tribofilms based on various working conditions are summarized at the end of the paper. This paper provides perspectives on the nanostructures of in situ tribofilms that induce superlubricity, particularly in the field of surface engineering. Our goal is to establish a system that provides useful guidance for understanding and analyzing the nanostructures of in situ tribofilms for DLC films and solid lubrication films. Meanwhile, we hope that interested researchers will offer valuable suggestions for characterizations and the analysis of nanostructures of in situ tribofilms.

2. Atmospheric Environment

2.1. Nanostructure of Transfer Film

Atmospheric conditions at room temperature are commonly used operating environments for aircraft, cars, high-speed trains, ships, and other vehicles. In such environments, it is crucial to investigate the structure of the super-slippery surface of a-C to guide the improvement of surface lubrication for large critical motion mechanisms operating in atmospheric conditions at room temperature. In the atmospheric environment, the lowest μ of hydrogen-free DLC is 0.016, as shown in Table 1. The reason for the low μ is that under strong shear, there is unstable sp^3-C bond break and bond length and direction change, causing adjacent atoms to rearrange carbon bonds along the shear direction to reduce lattice strain energy. Ultimately, transfer materials containing discontinuous graphite nanocrystalline (cluster clusters) are formed on the contact surface, and much complete amorphous carbon is generated along the direction of maximum shear stress [37]. After hydrogen doping, the μ of DLC can be further reduced to 0.005, resulting in a super-slippery state, as shown in Figure 2. The mechanism behind this effect is that the participation of hydrogen weakens the normal electron number of the outermost layer of the a-C:H film's electric dipole. Due to the lower van der Waals force, the terminal hydrogen bonds randomly distribute and passivate the outermost electrons, further weakening the induced electric dipole moment and obtaining a lower polarization rate. Finally, a transfer film composed of curved multilayer graphene strips or onion-shaped carbon nanoparticles is formed in situ on the sliding contact surface, effectively reducing the μ and wear rate [27,38,39]. When other elements, such as Cr, are doped into amorphous carbon film, excluding hydrogen, it is found that the μ increases. This is because Cr is a hard metallic element that enhances the hardness of the contact surface, resulting in the structure of the transfer film being mainly composed of sp^2-C and Cr [40]. Other carbon-based amorphous films, such as onion-like carbon (OLC) and hydrogenated fullerene-like carbon (FL-C:H), also exhibit extremely low μ. The μ of OLC can reach 0.01, and the wear rate is also extremely low at 6.41×10^{-18} mm^3/Nm. Onion-like carbon can slide as a nanoscale "rolling bearing" during the friction process, providing an amorphous contact interface between the friction contact surface and the OLC film. In situ this generates an interface transfer film combining onion-like carbon and graphitized carbon, thereby achieving ultra-low friction–wear, and greatly reducing energy loss [41]. The μ of FL-C:H film is also less than 0.02, and a transfer film is generated in situ on the friction-sliding surface induced by friction. The composition of the transfer film is composed of pentagonal and heptagonal carbon rings and hexagonal graphene rings caused by the thermal and strain effects of repeated friction. These polygonal ring-shaped carbon structures form a stable fullerene-like structure, thereby achieving ultra-low friction and wear of FL-C:H in the atmospheric environment.

Table 1. Tribological data of DLC film under different working conditions.

Environment	Sample	Stable μ	Wear Rate (mm^3/Nm)	Reference
Atmosphere	a-C	0.016		[37]
	a-C:H	0.005		[27,38,39]
	a-C:Cr			[40]
	OLC	0.01	6.41×10^{-18}	[41]
	FL-C:H	<0.02		[41]
Dry inert	a-C:H	0.001~0.003	4.6×10^{-10}	[42]
	GLC	0.005		[43]
	FLC	0.009		[43]
	B_4C/a-C	0.035		[44]
	TaC/a-C	0.002	8.80×10^{-10}	[45]
	GQDs/a-C:H	0.01		[15,16]
Combustible gas-H_2	a-C:H	0.0001		[46,47]
–CH_4	a-C:H	0.0093	8.0×10^{-8}	[48–50]
Ultra-high vacuum	a-C:H	0.003–0.02	4.6×10^{-10}	[51,52]
Moisture	GO/a-C:H:Si	0.002		[53]

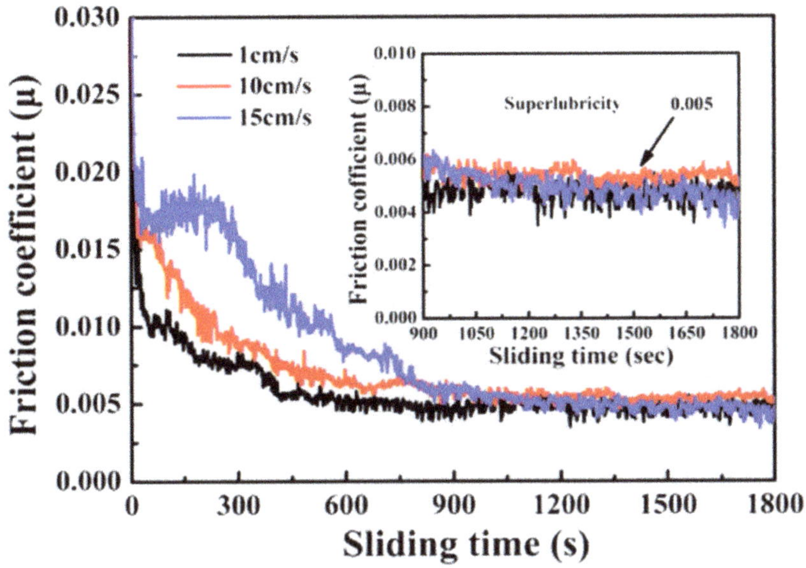

Figure 2. Experimental demonstration of the superlubricity regime for a-C:H film under 1, 10, and 15 cm/s linear velocity (air environment, room temperature, 2.89 GPa contact pressure) Rprinted with permission from Ref. [38]. Copyright 2017, WILEY-VCH Verlag GmbH & Co. KGaA, Weinheim.

In addition to regulating the composition ratio of C, H, and F in DLC and the hybrid structure of C, excellent lubrication effects can also be achieved by coating the surface with a coating. By coating graphene or modified MXenes on the polymer-like carbon (PLC) surface, it is found that the μ is less than 0.1, and the wear rate is at the level of 10^{-8} mm^3/Nm. The structure of the transfer film is mainly composed of low molecular weight polyacrylic acid forming a more ordered sp^2-C phase. Additionally, the surface-active groups of MXenes participate in the formation of multilayer lubricating transfer films [16,54].

2.2. *Effects on Surface Structure*

Through the study of the structure of transfer films, it can be observed that the air environment has a significant impact on the termination groups of the DLC surface, especially a-C:H film. Firstly, oxygen in the air reacts with the hydrogen atoms on the surface of a-C:H film, resulting in the formation of oxide surfaces. This will change the physicochemical characteristics of the a-C:H surface, such as the surface energy and friction coefficient. Additionally, oxide surfaces can reduce the hydrophobicity of a-C:H, reducing its lubricating performances [55,56]. Therefore, in applications where high lubrication requirements or key friction performance are needed, oxygen in the air may cause a decrease in the surface performances of a-C:H film. Secondly, in addition to oxygen, there may be other pollutants in the air, such as moisture and organic compounds [57,58]. These pollutants can interact with the surface of a-C:H film through adsorption or reaction. For example, moisture can cause hydrolysis reactions on the surface of a-C:H film, leading to the detachment of hydrogen atoms from the surface and a decrease in the hydrogen content of the DLC surface [59–61]. This will affect the hardness and conductivity of DLC films. For instance, organic pollutants may form adsorption layers on the surface of a-C:H film, affecting its optical transparency and optical performance [62,63]. To address these issues, measures can be taken to reduce the impact of the air environment on the surface terminal groups of a-C:H film. For example, special surface treatment methods, such as surface modification or coating, can be used to protect the surface terminal groups of a-C:H film and maintain its effective lubricating properties when subjected to friction in the air [64].

3. Dry Inert Environment

3.1. Nanostructure of Transfer Film

The room temperature atmospheric environment is currently the most variable service condition environment, including unstable humidity, temperature, and oxygen content. Therefore, by controlling the variables and singling out the service environment, studying the ultra-smooth transfer film structure and lubrication mechanism of amorphous carbon films is beneficial. Among the commonly used methods for controlling service conditions, dry inert environments (such as dry nitrogen and dry argon) are the easiest to implement and operate in the laboratory. The advantage of dry inert gas environments is that they do not react with the upper and lower friction pairs. Their surface coatings remain intact during the friction process, thereby they do not affect the composition of the ultra-smooth transfer film structure.

Several studies by researchers [1,65,66] have found that under dry nitrogen conditions, silicon wafers coated with high-hydrogen DLC (namely 40at.%H-DLC, which is made from 25% methane and 75% hydrogen) films have extremely low μ. During the running-in period, the μ is 0.15 [67]. During the stable period, the μ is 0.001~0.003, and the wear rate is 4.6×10^{-10} mm^3/Nm, as shown in Table 1. The transfer film structure on the surface is mainly composed of amorphous carbon layers. The most directly contacted surface contains a large amount of released hydrogen elements and hydrogenated sp^3-C, which passivate the surface chemically and increase the local density, as shown in Figure 3 [42]. This passivated surface cannot form strong chemical bonds and covalent σ bonds (which can lead to high adhesion/friction) on the sliding interface. The variation results in the enhanced hydrophobicity of the film and reduced possibility of frictional adhesion by reducing the number of carbon dangling bonds, thus achieving ultra-low friction and wear. The friction-induced hydrogenated amorphous carbon layer is crucial to the wear of diamond films. In addition to hydrogenated DLC, graphite-like carbon (GLC) and fullerene-like carbon (FLC) DLC also exhibit ultra-smooth effects. Wang's study [43] found that steel balls and plates coated with GLC or FLC on the surface (i.e., self-mated friction pairs) have extremely low μ during wear. The self-mated GLC has a μ of 0.005 and forms some ordered graphene layer structures on the sliding interface. These layered structures are formed by GLC through adaptive sliding-direction hybridization and rearrangement along the sliding direction, resulting in a larger and more ordered graphene structure in the transfer film on the surface. The self-mated FLC has a slightly higher μ of 0.009. The reason for this is that the spherical nanoparticles with a graphite shell formed on the sliding interface of FLC reduce the contact area. FLC forms spherical nanoparticles with a graphite shell through hybridization and rearrangement from the surface to the center of the particles but partially resists the rearrangement of its graphene layer. Compared with GLC, FLC exhibits a lower degree of hybridization under low and high loads and can only achieve ultra-low friction at higher loads (>0.77 GPa). Different initial structures of sp^2-C films are transformed into different friction products through different hybridization pathways and achieve ultra-low friction through different friction reduction pathways such as reducing the contact area or interlayer sliding. In other words, the structural characteristics of sp^2-C films determine the transformation pathway in advance, thus producing a friction reduction mechanism in macroscopic contact.

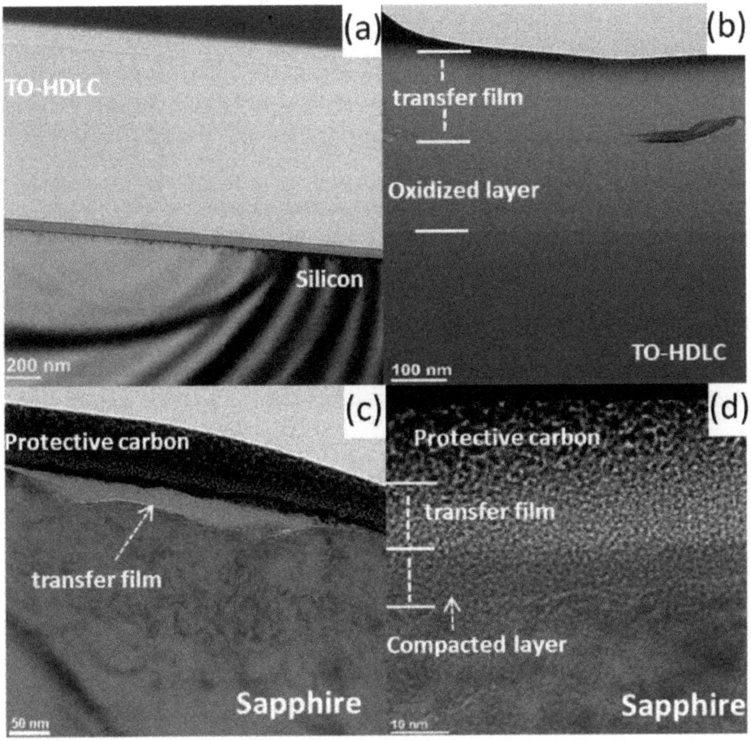

Figure 3. HRTEM cross-section images from a-C:H wear track and sapphire hemisphere contact region. (**a**) Wear track center. (**b**) Edge of the wear track. (**c**) Cross-section HRTEM micrograph of the sapphire hemisphere. (**d**) The transfer film on the sapphire surface. Rprinted with permission from Ref. [42]. Copyright 2017, WILEY-VCH Verlag GmbH & Co. KGaA, Weinheim.

3.2. *Effects on Surface Structure*

The μ of amorphous carbon films is between 0.015 and 0.15 [68,69]. In addition to pure amorphous carbon films, doped amorphous carbon films also exhibit excellent tribological properties. Borides are effective graphite catalysts that can accelerate the graphitization process of different carbon materials. The μ of B_4C/a-C films doped with boron carbide (B_4C) during wear with steel balls is $\mu \approx 0.035$. The reason for this is that under the action of shear force, B_4C forms a compact and ordered graphite-like carbonaceous tribolayer on the sliding interface of the upper and lower friction pairs, ultimately leading to the low friction and wear of B_4C/a-C films [44]. In addition, introducing atomically dispersed gold (Au) atoms into the nanocomposite structure of tantalum carbide (TaC)/amorphous carbon (a-C) achieves a μ of 0.002 and a wear rate of 8.80×10^{-10} mm^3/Nm [45]. In the initial stage of running-in, high shear forces strip TaC nanoparticles, causing them to roll and merge into larger TaC nanoparticles to reduce surface energy. Meanwhile, under the assistance of shear forces and nano-Au atoms, a-C is graphitized in situ, generating graphite-like carbon (GLC) flakes that wrap TaC particles in the shear layer to form TaC/GLC nanocoils. Finally, a multicontact configuration and non-conforming contact are achieved by obtaining a robust macroscopic ultra-smooth state. Recently, low-dimensional materials such as GQDs, MXenes, and graphene have been preliminarily applied in research on surface engineering. These low-dimensional materials have excellent layered nanostructures, which are beneficial for causing interfacial sliding and accelerating effective lubrication during the frictional process. The sliding counterparts contain transfer layers, namely in situ tribofilms, which act as protective shielding to weaken friction and wear. Through comprehensive analysis of the interfacial structures of in situ tribofilms, it has been found that the in situ tribofilms always contain some sp^2-C phase such as nanoflakes. For instance, this is distinctly different from the as-formed nanostructured tribofilm on the contact counterfaces. Studies have shown that the contact surface of the upper counterfacing ball was covered by 2D-layered carbon and graphitic lubricants induced via the structural transformation of GQDs during the whole sliding [15,16]. Meanwhile, the tribofilm of the disk wear track was composed of a silica-like SiO_x boundary layer and a multicomponent mixed layer induced by tribochemistry. Compared to the self-mated DLC system, the structural boundary enriched with SiO_x compounds was not formed at the bottom region of the tribofilm for the bare steel system; however, the disk wear track was covered by a thicker tribofilm containing plenty of degraded GQDs. This inferred that the formation of a nanostructured sliding interface was the key to realizing superlubricity.

Dry nitrogen gas does not contain oxygen. Hence, the inert atmosphere does not react with the hydrogen atoms of the contact surface, avoiding the formation of oxide surfaces. The inert connection helps maintain the hydrophilicity of a-C:H and reduces the adsorption or reaction of other pollutants, such as organic compounds. This helps maintain the physical and chemical properties of the DLC surface, thereby maintaining its lubricity and wear resistance.

4. Harsh Environment

4.1. Combustible Gas Environment

4.1.1. Nanostructure of Transfer Film

Combustible gas environments, such as hydrogen and methane, are also complex service environments. The main concern is to ensure the safety of the friction pair, including the moving mechanism, during the movement process, while also ensuring the safety of personnel. From 1960 to 2000, fuel cells developed rapidly as an important tool for hydrogen utilization. The application practice in aerospace, power generation, and transportation fields fully proves the feasibility of hydrogen as a secondary energy source. The upstream link of hydrogen energy is hydrogen production, mainly including industrial by-product hydrogen, fossil fuel hydrogen production, and renewable energy hydrogen production. The midstream link is hydrogen storage, transportation, and refueling, mainly including low-temperature liquid, high-pressure gas, and solid-state hydrogen storage. Refueling stations are important infrastructure for hydrogen refueling. The downstream link is the fuel cell and application link, and the current main application areas are various fuel-cell vehicles. Therefore, achieving the stable and safe operation of the moving mechanism in a hydrogen environment is imperative. DLC, as a stable and inert film, can maintain great physical and chemical stability in hydrogen.

In a hydrogen environment, the μ of DLC is extremely low, reaching 0.0001, as shown in Figure 4. Its super-slip transfer film is composed of hydrocarbons generated by the polymerization of hydrogen and ethanol vapor during the friction process [46,47]. By controlling the surface tribochemistry in a hydrogen environment, DLC films can induce the formation of hydrogen-rich transfer materials on the contact surface, achieving a super-slip state. Methane, as a clean energy source, is one of the pillars of global energy. However, damage to moving parts operating in methane can put pressure on social security and economic development. Hydrogen-rich DLC films are believed to protect moving parts operating in natural gas, mainly composed of methane. At a pressure of 10 kPa, the steady-state μ is 0.0093, and the wear rate is 8.0×10^{-8} mm^3/mN [48–50], as shown in Table 1. The transfer film is composed of a graphitized structure in the central region of the contact interface. Meanwhile, by dissociating methane molecules into hydrogen and methane groups, and combining them with carbon dangling bonds, the smoothness of the sliding interface is increased. Through the chemical termination and graphitization process of the sliding interface (the re-hybridization process from sp^3-C to sp^2-C), the tribofilm induces a weak interaction between the friction interfaces to achieve low friction and wear.

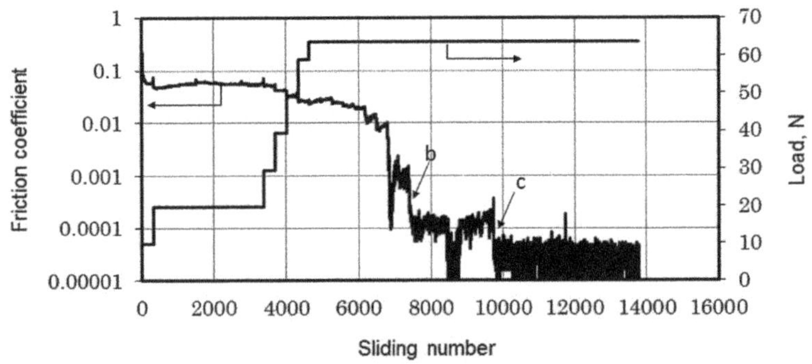

Figure 4. Friction coefficient trace. Rprinted with permission from Ref. [46]. Copyright 2017, WILEY-VCH Verlag GmbH & Co. KGaA, Weinheim.

4.1.2. Effects on Surface Structure

In a combustible gas environment, the termination groups on the a-C:H surface may be susceptible to erosion and reaction by gas molecules. For a-C:H film, the bond between hydrogen and carbon atoms is very strong, giving it high chemical stability. However, there may be unique reaction mechanisms in combustible gases that can lead to the decomposition or alteration of the termination groups on the a-C:H surface. The specific reaction mechanisms are as follows:

Firstly, gas molecule adsorption, where molecules in combustible gases can adsorb onto the terminal groups on the a-C:H surface. This adsorption can result in structural changes or the formation of new chemical bonds in the terminal groups. Secondly, gas molecule reaction, where molecules in combustible gases can chemically react with the terminal groups on the film surface [70]. This reaction can lead to the breaking, alteration, or formation of new chemical bonds in the terminal groups. Thirdly, gas molecule erosion, where molecules in combustible gases can physically or chemically erode the a-C:H surface [71]. This erosion can cause damage or detachment of the a-C:H film and even the matrix, thereby affecting its lubricity, wear resistance, and other physical and chemical properties.

4.2. Ultra-High Vacuum Environment

4.2.1. Nanostructure of Transfer Film

With the development of vacuum acquisition technology, vacuum applications are increasingly expanding into various aspects of industry and scientific research. Vacuum applications refer to the use of the physical environment of rarefied gases to complete specific tasks. The energy transfer in outer space outside the atmosphere is similar to that in an ultra-high vacuum. Therefore, an ultra-high vacuum can be used for space simulation. Under ultra-high vacuum conditions, the time for the formation of a single molecular layer is long (in hours), which can be used to study surface properties. Before being contaminated by gas, vacuum conditions take advantage of a sufficiently long time for the formation of a contact surface.

Atomic-scale friction behavior mainly depends on chemical reactions and interactions on the friction interface. Friction experiments with DLC in an ultra-high vacuum can reveal interface reactions and atomic-scale wear mechanisms. The results show that the μ of a-C:H is <0.02. Under the action of a frictional force, the hydrogen concentration and methane gas emission in the transfer film structure are high, and this interface emission induces DLC atomic-level chemical wear. With the increase in hydrogen concentration in a-C:H, chemical wear and mechanical wear increase and decrease, respectively. Total wear shows a concave hydrogen concentration dependence, with an optimal hydrogen concentration of about 20% [51,52]. Injecting low-temperature nitrogen into the ultra-high vacuum environment, DLC's μ can be further reduced to 0.003, with the wear rate of almost zero (4.6×10^{-10} mm^3/Nm), as shown in Table 1. A graphitized structure was also found in its transfer film structure, and a local friction-induced increase in the content of sp^2-C bonds (Figure 5), thereby achieving ultra-low friction and wear of the hard carbon film. Among a-C:H films against various ceramic-based balls, the significant reason for the failure in superlubricity is attributed to the van der Waals force [72]. The weak interface adhesion between the Al_2O_3/a-C:H interfaces helps to maintain a stable tribofilm on the surface of the friction ball and forms a smooth graphitized surface on the wear track, resulting in long-lasting superlubrication. The strong adhesion between the ZrO_2 ball, Si_3N_4 ball, and SiC ball and the contact interface of a-C:H film causes serious damage to the surface of the friction ball, making the superlubrication fail.

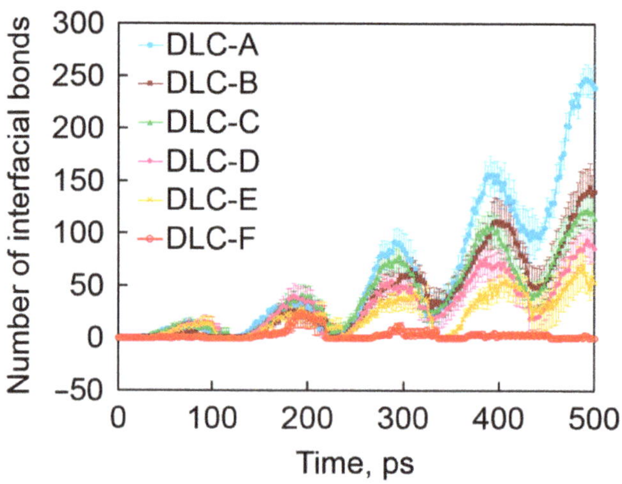

Figure 5. Time evolution of the number of interfacial C-C bonds. Rprinted with permission from Ref. [52]. Copyright 2017, WILEY-VCH Verlag GmbH & Co. KGaA, Weinheim.

4.2.2. Effects on Surface Structure

On one hand, one possible change that may occur in an ultra-high vacuum environment is that the dissociative hydrogen ions may dissociate from termination groups on the surface and react with ions or molecules in the surrounding air, thereby affecting the properties of the a-C:H film. On the other hand, in an ultra-high vacuum environment, due to the absence of gases or impurities, the termination groups on the surface of the a-C:H film may undergo rearrangement, altering the structure and properties of the a-C:H film.

5. Humid Environment

5.1. Nanostructure of Transfer Film

A water vapor environment is a common environment faced by humid areas near the equator, and DLC does not have low friction and wear characteristics in high-humidity environments. Therefore, lubricating oil needs to be added to achieve lubrication in high-humidity environments [73–78]. Therefore, achieving ultra-low friction and reducing the wear of DLC in high-humidity environments is a problem that needs to be solved. By using ceramic-based balls (such as Al_2O_3 and Si_3N_4 [78]), an altered tip showed a notably low adhesion on the original a-C:H film. It was additionally discovered that the rubbing factor of the a-C:H/Al_2O_3 duo steadily declined as the oxygen level on the surface of the a-C:H substrate decreased. This occurrence can be elucidated by a shift in the connective surface from an oxygen ending with potent hydrogen bond connections to a hydrogen ending with feeble van der Waals interactions. By using graphene oxide (GO) nanosheets as additives, the lubrication of silicon-doped DLC (a-C:H: Si) films was achieved, and the μ reached 0.002. On the sliding contact surface of the upper and lower friction pairs, it was found that the surfaces of both were covered by a layer of silicon gel induced by frictional chemistry. At the same time, through the physical adsorption between the upper and lower friction pairs of the friction surface, touched GO nanosheets were transferred from the Si_3N_4/a-C:H: Si interface to the GO/GO shear interface. This further reduces the shear stress and achieves macroscopic super-slip on a synergistic basis of GO and a-C:H: Si, as shown in Figure 6 [53]. This is similar to the friction of diamonds in humid environments. In high-humidity environments, the μ of nanodiamonds is ≤0.05, and the wear rate is in the order of 10^{-10} $mm^3/(Nm)$, as shown in Table 1. The passivation mechanism of dangling bonds on the contact surface depends on the fracture of bonds in the slip process (such as contact stress, sliding rate, and temperature settings). The passivation equilibrium is induced by dissociative adsorption of relative humidity, which is the re-hybridization process of graphitized carbon ($sp^3 \rightarrow sp^2$). The structure of the interface transfer film is mainly diamond nanoparticles embedded in the amorphous carbon layer. The collision between nanodiamond particles causes friction and the amorphization of chemically induced carbon in the contact area, and then breaks in the uneven contact point, resulting in nanodiamond particles adhering to the amorphous carbon phase [68,79].

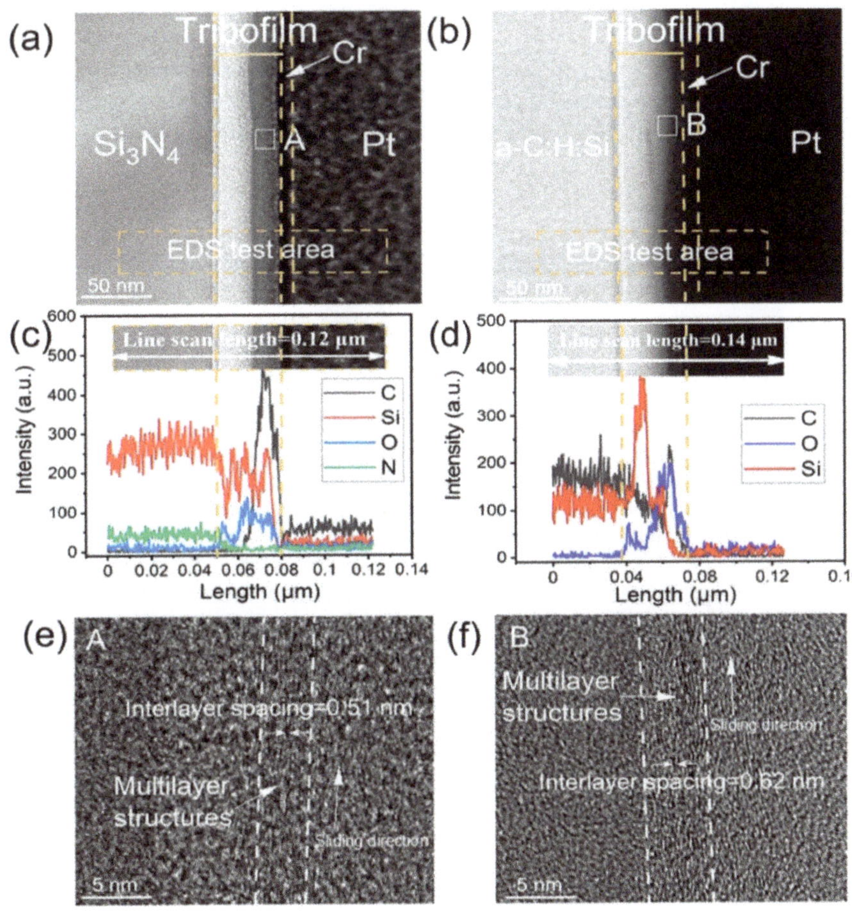

Figure 6. The cross-sectional structure of the tribofilm on the Si_3N_4 surface and a-C:H: Si film after a friction test. (**a**,**b**) The tribofilm formed on the Si_3N_4 surface and a-C:H: Si film, respectively.

(**c**) Elemental distributions along the cross-sectional profile in (**a**). (**d**) Elemental distributions along the cross-sectional profile in (**b**). (**e**,**f**) Enlarged views of the tribofilms (area: A and B) formed on the Si_3N_4 surface and a-C:H: Si film, respectively. Rprinted with permission from Ref. [53]. Copyright 2017, WILEY-VCH Verlag GmbH & Co. KGaA, Weinheim.

5.2. Effects on Surface Structure

In a humid environment, similar to the air environment, there is a high concentration of oxygen, water vapor, and organic compounds, which can cause some physical and chemical changes on the surface of DLC. Firstly, the interaction between water molecules and the hydrogenous surface may result in moisture absorption [80,81], leading to an increase in friction and subsequently an increase in the friction coefficient. Secondly, the positive terminal groups on the surface of a-C:H film are prone to react with water molecules and oxygen in a humid environment, forming hydroxyl groups [82,83]. Unlike the original hydrogenated groups, hydroxyl groups introduce new functional groups, thereby altering the chemical reactivity of the DLC surface and reducing its hydrophobicity. Thirdly, water molecules in a humid environment may form microscopic droplets with passivating groups on the surface [84,85]. These droplets can create an uneven liquid layer on the surface of a-C:H film, increasing the lubrication.

6. Lubrication Mechanism

6.1. Computational Simulation of the Formation of the Transfer Film

Computational simulation is an emerging research method. With the breakthrough of technology and algorithm innovation, the accuracy of the simulation calculation is increasing, resulting in accurate predictions of the structural properties of materials, avoiding material waste caused by experimental research, and saving costs. For the study of the microstructure and properties of DLC, there are mainly three types of research methods: density functional theory (DFT) simulation based on quantum mechanics, first-principles calculation based on classical mechanics, and MD simulation based on classical mechanics. Some good progress has been made among these methods. MD is a set of molecular simulation methods that mainly rely on Newtonian mechanics to simulate the motion of molecular systems, extract samples from systems composed of different states of molecular systems, and calculate the integral configuration of the system. MD can be used to calculate the thermodynamic properties and other macroscopic properties of the system based on the results of the integral configuration. MD studies atoms or groups of atoms, and the system size is relatively large compared to DFT calculations. The bond length and bond angle between molecules are determined by the force field, making it difficult to carry out detailed research. MD is mostly used for large molecular systems and is often used in conjunction with DFT calculations when studying microproperties, mainly for the construction of DLC models. Through the MD modeling of DLC films, it has been found that the super-slip properties of DLC, that is, surface passivation, mainly depend on two factors [86]: the degree of carbon hybridization and hydrogen content, and the test environment conditions.

For non-hydrogenated DLC, such as a-C or Ta-C films, the molecular modeling design of the friction process simulates interface sliding between graphene-like layers by adjusting the steady-state velocity. This aims to explain the formation of graphitized and carbon-rich transfer layers in the transfer film, as shown in Figure 7 [87–89]. The results show that the formation of well-separated graphene-like interface layers on the interface can be observed when the interlayer spacing is large. The interlayer bond fracture is reduced, and the final bond disappears. That is, triggered by plasticity between adjacent amorphous carbons, the sp^3 to sp^2 transition of the carbon phase occurs, greatly increasing the content of sp^2-C at the interface and forming a soft amorphous carbon friction transfer layer. Finally, the friction result is super-slip ($\mu \leq 0.01$), that is, a sliding interface with almost no friction and wear. These results are also consistent with the results of super-slip experiments.

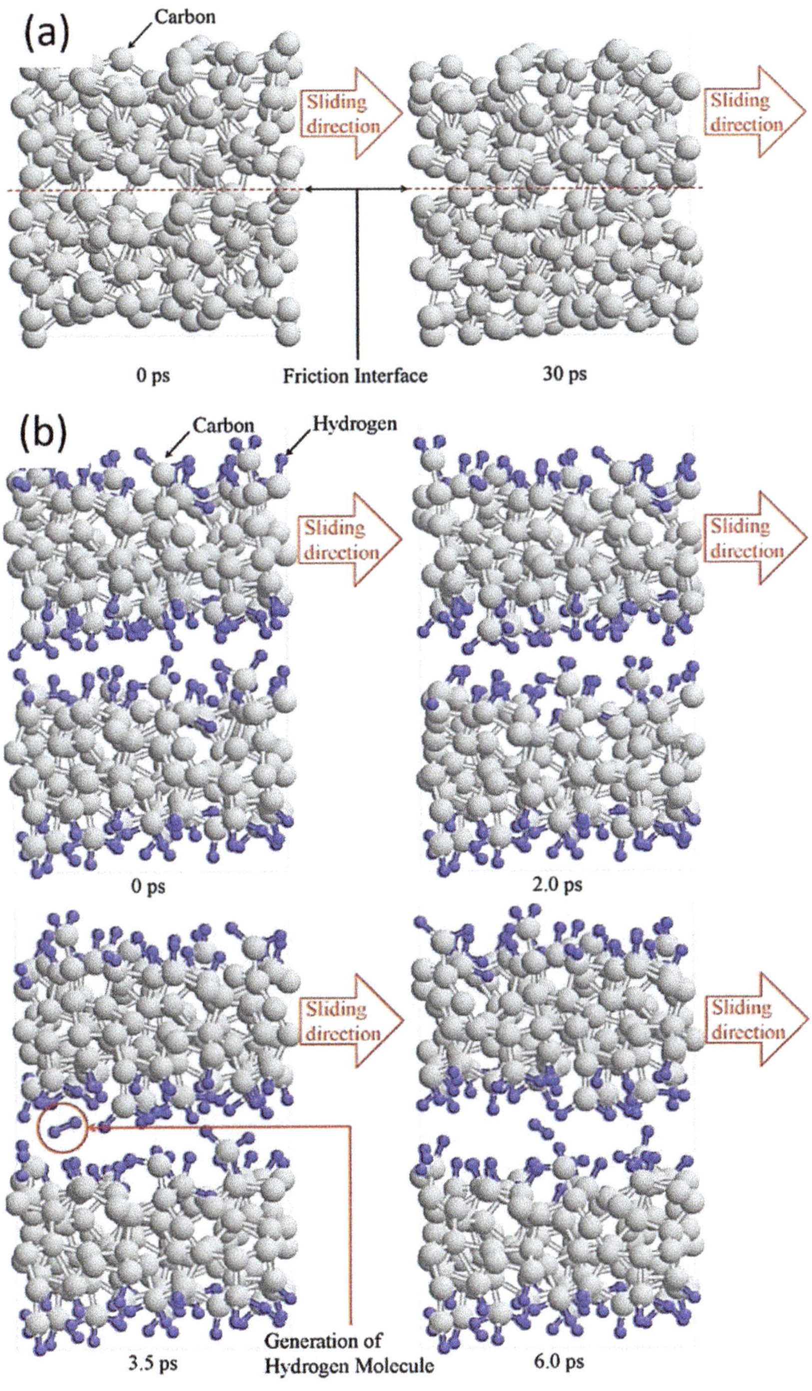

Figure 7. Constant height asperity collision simulation of (**a**) a-C and (**b**) a-C:H under a load pressure of 1.0 GPa. Rprinted with permission from Ref. [90]. Copyright 2017, WILEY-VCH Verlag GmbH & Co. KGaA, Weinheim.

Compared with non-hydrogenated DLC films, the addition of hydrogen in hydrogenated DLC films leads to a significant reduction in unsaturated carbon bonds at the film interface. These unsaturated carbon atoms serve as contact points for the formation of C-C covalent bonds between the friction pair surface and the a-C:H film [4,90–93]. The adhesive force between these contact points can increase the friction force during sliding. As the sliding time increases, the steric hindrance effect of hydrogen molecules generated can remove the load on the substrate and inhibit the formation of C-C bonds. At the same time, surface hydrogen atoms play a key role in promoting the cleavage of C-C bonds during the sliding process of a-C:H films. As C-C covalent bonds form and break, the number of unsaturated carbon atoms (including sp^3- and sp^2-hybridized carbon atoms) decreases, inducing the formation of a transfer film. When covalent bonds break, the friction effect weakens, and the local temperature of the interface increases, leading to the formation of a passivation layer. Compared with DLC films without H or F, a higher sp^3-C content can cause the DLC film to produce smaller deformation during sliding, thereby reducing the contact area and adhesion [92,93]. In addition, there are a large number of hydrogen or fluorine atoms doped in the DLC film, which form covalent bonds with unsaturated carbon atoms. Non-bonded interactions result in a lower μ for a-C:H or a-C:F than for a-C. The anti-bonding effect exists in the friction interface of a-C:F and the repulsive force between the interface atoms is strong. However, a-C:H has weak van der Waals forces on the interface, resulting in a higher μ (0.13) for a-C:H than for an a-C:F end DLC model ($\mu = 0.07$). Therefore, the number of unsaturated C-C bonds on the sliding interface is reduced, resulting in a lower adhesive force and a lower steady-state value of interface friction force.

When the tribo-couples are changed, the tribology behavior of a-C:H film changes accordingly, as shown in Figure 8. Mechanical action should also play an important role in the superlubricity of a-C:H films in a vacuum [94–96]. Compared with Al_2O_3 balls, ZrO_2, Si_3N_4, and SiC balls have stronger adhesion to the diamond surface. Weak interfacial adhesion contributes to the stable superlubricity of the a-C:H surface (such as the Al_2O_3/a-C:H interface) under vacuum conditions. However, excessive interfacial adhesion [97–99] can lead to the wear of a-C:H film or the friction pair (such as the ZrO_2/a-C:H, Si_3N_4/a-C:H, and SiC/a-C:H interfaces), both of which shorten the lifespan of superlubricity. The interfacial adhesion between the SiC/a-C:H interfaces is weaker than that of the Si_3N_4/a-C:H interfaces, but the hardness and contact stress of SiC balls are higher, resulting in more severe wear of the a-C:H film and thus reducing the durability of superlubricity.

By surface doping a-C films with materials such as diamond or methanol, a super-slip model was established, and the simulated friction force was approximately zero [100,101]. Through calculation, it was found that an ultra-thin graphene-like friction layer was formed on the friction interface, and the specific formation reason was shear localization rather than uniform deformation. The main manifestation is the redirection (orientation angle $< 10^\circ$) and phase transition (sp^2-C ratio > 90%) of C-C covalent bonds that occur preferentially in the friction contact area (graphene-like friction layer) and structural ordering. As the graphene-like sheets gradually aggregate and layer in the friction layer, under the induction of normal stress, the graphene-like sheets transition from adhesive sliding to continuous sliding with ultra-low friction characteristics. This ultimately leads to shear weakening and achieves ultra-low friction. The proposed shear localization mechanism helps to reveal the mechanism of superlubrication.

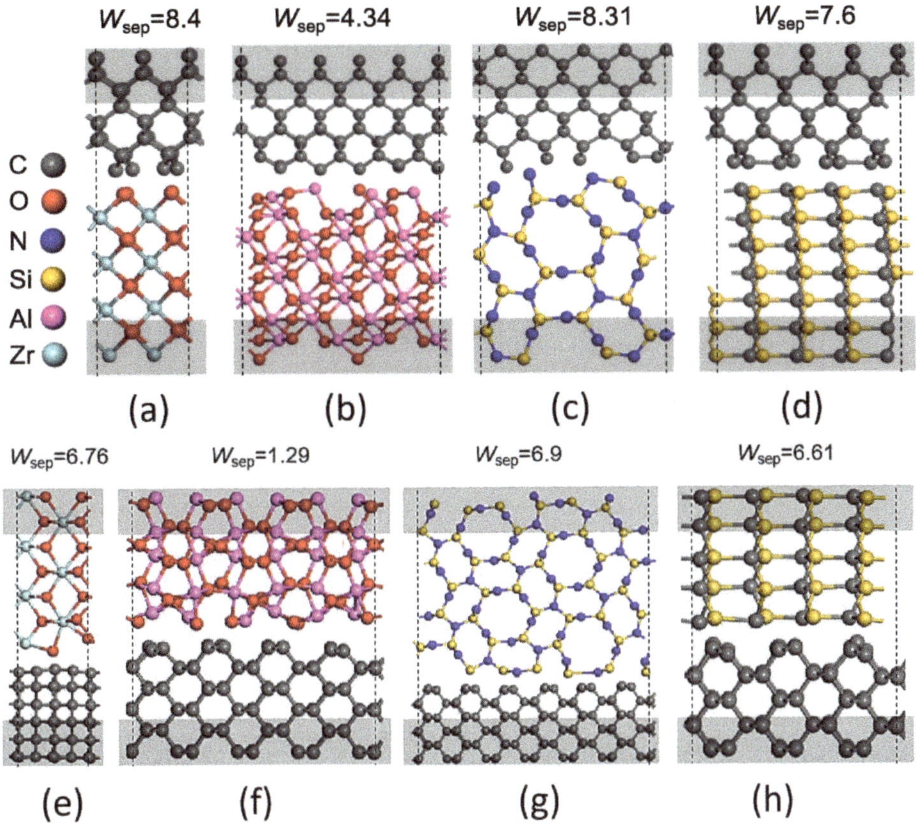

Figure 8. Atomic structures: (**a**) ZrO_2(001)/diamond(001), (**b**) Al_2O_3(010)/diamond(001), (**c**) Si_3N_4(110)/diamond(001), (**d**) SiC(110)/diamond(001), (**e**) ZrO_2(101)/diamond(110), (**f**) Al_2O_3(10–2)/diamond(110), (**g**) Si_3N_4(1–30)/diamond(110), (**h**) SiC(1–20)/diamond(110). The calculated works of separation (W_{sep}) with J/m^2 in units are marked upon each structure. Rprinted with permission from Ref. [72]. Copyright 2017, WILEY-VCH Verlag GmbH & Co. KGaA, Weinheim.

6.2. Material Composition of Tribo-Couples

The appropriate material composition of a-C films plays a crucial role in determining their comprehensive performances. When used as a tribo-couple material, a-C films offer excellent anti-friction and -wear, leading to improved performances and the extended lifespan of a-C films [102,103]. Additionally, their good adhesion to substrates contributes to reduced friction and wear. When selecting counterparts for a mechanical system, one must prioritize their ability to adhere to the substrate, such as metal and ceramic [104–108]. This consideration is vital as it directly impacts on the lubrication properties. The adhesion between the friction pair and the substrate ensures stability and durability, which can avoid slipping or even detachment that would result in system failure. One of the primary factors that affects the tribological performances of a-C film-based counterpairs is the film thickness [109–112]. Thicker films are generally associated with higher wear resistance and improved surface hardness to enhance the durability of the frictional system. However, excessive film thickness may lead to increased friction and hinder the smooth sliding motion. Another crucial aspect that affects the tribological performances is the deposition technique (such as plasma-enhanced chemical vapor deposition (PECVD)) used to fabricate the a-C films [40,113–120]. For instance, PECVD films tend to have a higher sp^3/sp^2

carbon ratio, which enhances their hardness and wear resistance. On the other hand, sputtered films exhibit a more amorphous structure, leading to lower friction coefficients. In addition, proper mating materials can help optimize the tribological properties [121–126]. For example, surface materials with low surface roughness can further reduce friction and wear with excellent chemical compatibility. The key to achieving desired tribological performances depends on the suitable thickness of the a-C film, deposition technique, and counterface materials. By considering these factors, researchers can enhance the lubrication performances of a-C film-based systems, enabling their widespread application across various industries.

6.3. Comparative Summary of the Formation of the Transfer Film

The formation of an in situ tribofilm is crucial in achieving ultra-wear resistance and low friction coefficients. The structure of the tribofilm plays a key role in its effectiveness, as demonstrated in Figure 9, which summarizes the tribofilms formed under various conditions. These tribofilms exhibit unique formations that enhance their mechanical performance and provide protection against wear and tear. In addition, the presence of certain elements in the environmental conditions or lubricants used can influence the composition and structure of the tribofilms, leading to even greater improvements in wear resistance and friction reduction. Overall, the careful management and optimization of in situ tribofilm formation have the potential to greatly benefit a wide range of industrial applications by providing long-lasting and effective protection against wear and tear.

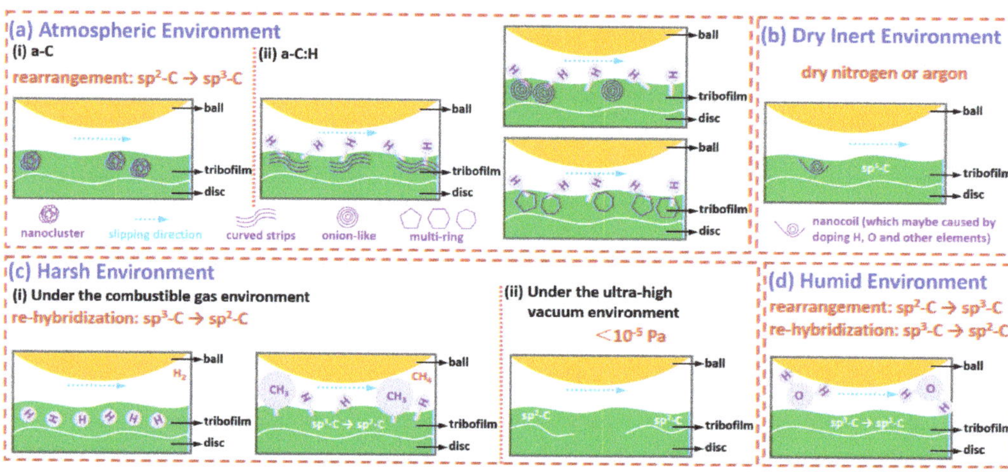

Figure 9. Lubrication mechanism diagram of DLC under different service conditions.

(a) Atmospheric environment

Under strong shear, unstable sp^3 carbon bonds in a-C break, and their length and direction change, causing adjacent atoms to rearrange along the shear direction to form stable sp^2 carbon bonds and reduce lattice strain energy. Eventually, discontinuous graphite-like nanocrystalline (nanoclusters) transfer materials are formed on the frictional contact surface, and complete graphite-like amorphous carbon is generated along the direction of maximum shear stress. The super-slip mechanism involves the participation of hydrogen atoms, which can weaken the normal electron number of the outermost layer of a-C:H film. Due to lower van der Waals forces, hydrogen end bonds will randomly distribute and cause the outermost electrons to be passivated, further weakening the induced dipole moment and obtaining a lower polarization rate. Finally, a transfer film composed of curved multilayer graphene strips or special structured carbon (such as onion-like or multi-ring) is formed in situ on the frictional sliding contact surface.

(b) Dry inert environment

When non-doped amorphous carbon films that contain elements other than carbon, hydrogen, and oxygen are subjected to friction in a dry nitrogen or argon environment, they create graphitized transfer films on the contact surface, which are rich in sp^3-C. These transfer films possess chemically inert passivation surfaces, producing macroscopic super-slip effects. However, when the films are doped with other elements, their transfer film structures not only contain graphitized structures but also exhibit new nanostructures due to friction-induced processes. These newly formed structures may exist as nanocoils and can further enhance the friction-reducing properties of the films. Adding these elements greatly enhances the films' super-slip effects, providing optimal conditions for various industrial and mechanical applications.

(c) Harsh environment

(c-i) Under the combustible gas environment

Controlling the target material during the preparation process and controlling the surface or friction chemistry during friction in a hydrogen environment can induce DLC films to form hydrogen-rich transfer materials on the contact surface, achieving a super-slip state. Methane molecules are dissociated into hydrogen and methane groups, which combine with carbon dangling bonds. Through the chemical termination effect of the sliding interface and the re-hybridization process from sp^3-C to sp^2-C, the smoothness of the sliding interface is increased, thereby weak interactions are induced between the friction interfaces and low friction and wear are achieved.

(c-ii) Under an ultra-high vacuum environment

Maintaining a suitable level of hydrogen can be advantageous in mitigating methane gas emissions and decreasing atomic-level chemical wear in the transfer film. Additionally, the transfer film structure reveals the presence of graphitized structures, and friction-induced augmentation of the sp^2 bond content in certain locations has a positive effect on the ultra-low friction and wear of hard carbon films.

(d) Humid environment

In increasingly humid environments, the application of DLC can undergo a transformative process, resulting in a combination of surface coatings and DLC that reduces the effects of abrasion and attrition. This unique feature is achieved by these two elements collaborating, working together to form a protective layer on the surface. The interface transfer membrane responsible for this process is created by combining unspecified carbon with graphitized carbon, which undergoes re-hybridization. The result is a layer that is resistant to wear and tear, providing enhanced durability to the material's surface. This attribute is particularly useful in applications where exposure to moisture and other elements can lead to corrosion and degradation, making this type of surface treatment ideal for use in a wide range of settings.

7. Conclusions and Perspectives

In the presence of oxygen, the DLC film's surface can oxidize, leading to alterations in its chemical properties and a decline in lubrication performance. However, when in a vacuum or inert gas environments, the DLC film surface experiences a minimal reaction with its surroundings, resulting in outstanding lubrication performance. Alternatively, hydrogen or methane gas can cause gas molecules to dissociate and combine with the carbon dangling bonds on the DLC surface. This process facilitates strong lubrication using chemical termination and graphitization at the sliding interface. Controlling the target material during the preparation process and the surface or friction chemistry during friction in a hydrogen environment can induce a-C:H film to form hydrogen-rich transfer materials on the contact surface, achieving a super-slip state. Methane molecules are dissociated into hydrogen and methane groups, which combine with carbon dangling bonds. Through the chemical termination effect of the sliding interface and the re-hybridization process from sp^3-C to sp^2-C, the smoothness of the sliding interface is increased, thereby weak interactions are induced between the friction interfaces and low friction and wear are achieved. For the tribological properties of a-C films, the composition of the material plays a crucial role. The introduction of elements such as silicon can enhance the hardness and anti-wear properties of a-C films, making them perform well in harsh environments such as high temperatures and high pressure. The introduction of oxygen, fluorine, and other elements can improve the adhesion and corrosion resistance of a-C film and improve its tribological properties in a specific environment. In addition to the addition of elements, the tribological properties of a-C films can also be affected by adjusting the preparation process (such as deposition temperature, deposition rate, and adding metal elements).

Overall, significant advancements have been made with superlubricating a-C film in terms of low friction coefficient and wear resistance. However, several areas require further improvement. Firstly, the preparation technology of superlubricating a-C film needs to be simplified and made more efficient. The current methods, such as PECVD and ion-beam deposition, are complex and costly. Developing simpler and more cost-effective preparation methods would enhance production efficiency and reduce costs. Secondly, the performance stability of superlubricating a-C film needs to be enhanced. While it exhibits excellent wear resistance, its performance may be compromised under harsh conditions like a combustible gas environment. Further research is necessary to improve the stability of the film, ensuring it maintains its excellent performance under various working conditions. Additionally, the application range of superlubricating a-C film should be expanded. Currently, it is primarily utilized in high-end fields like aerospace and electronic equipment. The correspondingly mature products have been applied in high-load-bearing wear-resistant fasteners and self-lubricating bushing in aircrafts. However, applications in other fields are limited in the depositing device and service conditions. Exploring and developing the potential applications of super-slippery carbon film in other industries would broaden its usage and increase its market demand. Although research on the tribofilm of a-C:H film has achieved certain results, there are still some challenges and issues for large-scale utilization. Firstly, enhancing the adhesion of a-C:H films is essential to prevent detachment and ensure their long-term effectiveness. By improving the sp^2/sp^3 ratio or doping of soft metal elements, a-C:H films can offer durable and reliable industrial manufacturing, where friction and wear are prominent concerns. Secondly, the production cost of a-C:H film is high, and the preparation process is complex, which limits the reliability of renewable energy technologies. By combining the coating technique with the deposition technique to enhance the tribological properties, a-C:H film can minimize energy losses and prolong the lifespan of critical parts upon high-load capacity and high wear resistance. Additionally, in the biomedical field, a-C:H film can offer innovative solutions in prosthetics and medical devices. Its low friction characteristics can improve the efficiency and functionality of joint replacements and implants, as it reduces the risk of adverse reactions and enhances patient safety. Overall, with further research, optimization, and cost reduction, a-C:H film holds tremendous potential to revolutionize a wide range of industries in industrial manufacturing, as well as biomedical fields.

Author Contributions: The manuscript was written via the contributions of all authors. X.Y. wrote the original draft. Z.J., L.M. and C.L. searched for literature resources. X.Y., H.P. and C.C. confirmed the final draft. H.L., D.L. and B.Z. conceived the conceptualization. All authors have read and agreed to the published version of the manuscript.

Funding: This work is supported by the Fundamental Research Funds for the Central Universities (buctrc 202101) and the Tribology Science Fund of the State Key Laboratory of Tribology in Advanced Equipment (SKLTKF23B02).

Data Availability Statement: Data available in a publicly accessible repository.

Conflicts of Interest: The authors declare no conflicts of interest.

References

1. Erdemir, U.A.; Eryilmaz, O.L.; Nilufer, I.B.; Fenske, G.R. Synthesis of superlow-friction carbon films from highly hydrogenated methane plasmas. *Surf. Coat. Technol.* **2000**, *133–134*, 448–454. [CrossRef]
2. Yue, Z.F.; Wang, H.; Fan, X.Q.; Li, H.; Zhang, J.Y.; Zhu, M.H. Regulating the fretting behavior of diamond-like carbon films by changing the composition and structure. *Carbon* **2023**, *212*, 118097. [CrossRef]
3. Johnson, J.A.; Woodford, J.B.; Erdemir, A.; Fenske, G.R. Near-surface characterization of amorphous carbon films by neutron reflectivity. *Appl. Phys. Lett.* **2003**, *83*, 452–454. [CrossRef]
4. Stoyanov, P.; Romero, P.A.; Merz, R.; Kopnarski, M.; Stricker, M.; Stemmer, P.; Dienwiebel, M.; Moseler, M. Nanoscale sliding friction phenomena at the interface of diamond-like carbon and tungsten. *Acta Mater.* **2014**, *67*, 395–408. [CrossRef]
5. Donnet, C. Recent progress on the tribology of doped diamond like and carbon alloy coatings a review. *Surf. Coat. Technol.* **1998**, *100–101*, 180–186. [CrossRef]
6. Johnson, J.A.; Holland, D.; Woodford, J.B.; Zinovev, A.; Gee, I.A.; Eryilmaz, O.L.; Erdemir, A. Top-surface characterization of a near frictionless carbon film. *Diam. Relat. Mater.* **2007**, *16*, 209–215. [CrossRef]
7. Chen, X.; Li, J. Superlubricity of carbon nanostructures. *Carbon* **2020**, *158*, 1–23. [CrossRef]
8. Yu, Q.; Chen, X.; Zhang, C.; Luo, J. Influence Factors on Mechanisms of Superlubricity in DLC Films: A Review. *Front. Mech. Eng.* **2020**, *6*, 65. [CrossRef]
9. Liu, A.C.Y.; Arenal, R.; Miller, D.J. Structural order in near-frictionless hydrogenated diamondlike carbon films probed at three length scales via transmission electron microscopy. *Phys. Rev. B* **2007**, *75*, 205402. [CrossRef]
10. Zhang, Z.; Chen, P.; Duan, X.; Zang, K.; Luo, J.; Duan, X. Robust epitaxial growth of two-dimensional heterostructures, multiheterostructures, and superlattices. *Science* **2017**, *357*, 788–792. [CrossRef]
11. Saravanan, P.; Selyanchyn, R.; Tanaka, H.; Fujikawa, S.; Lyth, S.M.; Sugimura, J. Ultra-low Friction between Polymers and Graphene Oxide Multilayers in Nitrogen Atmosphere, Mediated by Stable Transfer Film Formation. *Carbon* **2017**, *122*, 395–403. [CrossRef]
12. Eryilmaz, O.L.; Erdemir, A. Surface analytical investigation of nearly-frictionless carbon films after tests in dry and humid nitrogen. *Surf. Coat. Technol.* **2007**, *201*, 7401–7407. [CrossRef]
13. Heimberg, J.A.; Wahl, K.J.; Singer, I.L.; Erdemir, A. Superlow friction behavior of diamond-like carbon coatings: Time and speed effects. *Appl. Phys. Lett.* **2001**, *78*, 2449–2451. [CrossRef]
14. Turcheniuk, K.; Trecazzi, C.; Deeleepojananan, C.; Mochalin, V.N. Salt-assisted ultrasonic deaggregation of nanodiamond. *ACS Appl. Mater. Interfaces* **2016**, *8*, 25461–25468. [CrossRef]
15. Yin, X.; Zhang, J.; Luo, T.; Cao, B.; Xu, J.; Chen, X.; Luo, J. Tribochemical mechanism of superlubricity in graphene quantum dots modified DLC films under high contact pressure. *Carbon* **2021**, *173*, 329–338. [CrossRef]
16. Yin, X.; Wu, F.; Chen, X.; Xu, J.; Wu, P.; Li, J.; Zhang, C.; Luo, J. Graphene-induced reconstruction of the sliding interface assisting the improved lubricity of various tribo-couples. *Mater. Des.* **2020**, *191*, 108661. [CrossRef]
17. Berman, D.; Erdemir, A.; Sumant, A.V. Approaches for achieving superlubricity in two-dimensional materials. *ACS Nano* **2018**, *12*, 2122–2137. [CrossRef] [PubMed]
18. Chen, X.; Yin, X.; Qi, W.; Zhang, C.; Choi, J.; Wu, S.; Wang, R.; Luo, J. Atomic-scale insights into the interfacial instability of superlubricity in hydrogenated amorphous carbon films. *Sci. Adv.* **2020**, *6*, eaay1272. [CrossRef] [PubMed]
19. Chen, X.; Zhang, C.; Kato, T.; Yang, X.; Wu, S.; Wang, R.; Nosaka, M.; Luo, J. Evolution of tribo-induced interfacial nanostructures governing superlubricity in a-C:H and a-C:H:Si films. *Nat. Commun.* **2017**, *8*, 1675. [CrossRef] [PubMed]
20. Merkle, A.P.; Erdemir, A.; Eryilmaz, O.L.; Johnson, J.A.; Marks, L.D. In situ TEM studies of tribo-induced bonding modifications in near-frictionless carbon films. *Carbon* **2010**, *48*, 587–591. [CrossRef]
21. Joly-Pottuz, L.; Matta, C.; de Barros Bouchet, M.I.; Vacher, B.; Martin, J.M.; Sagawa, T. Superlow friction of ta-C lubricated by glycerol: An electron energy loss spectroscopy study. *J. Appl. Phys.* **2007**, *102*, 064912. [CrossRef]
22. Xu, J.; Luo, T.; Chen, X.; Zhang, C.; Luo, J. Nanostructured tribolayer-dependent lubricity of graphene and modified graphene nanoflakes on sliding steel surfaces in humid air. *Tribol. Int.* **2020**, *145*, 106203. [CrossRef]
23. Fouts, J.A.; Shiller, P.J.; Mistry, K.K.; Evans, R.D.; Doll, G.L. Additive effects on the tribological performance of WC/a-C:H and TiC/a-C:H coatings in boundary lubrication. *Wear* **2017**, *372–373*, 104–115. [CrossRef]
24. Benamor, A.; Kota, S.; Chiker, N.; Haddad, A.; Hadji, Y.; Natu, V.; Abdi, S.; Yahi, M.; Benamar, M.E.A.; Sahraoui, T.; et al. Friction and wear properties of MoAlB against Al2O3 and 100Cr6 steel counterparts. *J. Eur. Ceram. Soc.* **2019**, *39*, 868–877. [CrossRef]
25. Wang, Q.; Zhou, F.; Zhu, L.; Zhang, M.; Kong, J. Mechanical and tribological evaluation of CrSiCN, CrBCN and CrSiBCN coatings. *Tribol. Int.* **2019**, *130*, 146–154. [CrossRef]
26. Yin, X.; Jin, J.; Chen, X.; Rosenkranz, A.; Luo, J. Ultra-Wear-Resistant MXene-Based Composite Coating via in Situ Formed Nanostructured Tribofilm. *ACS Appl. Mater. Interfaces* **2019**, *11*, 32569–32576. [CrossRef]
27. Echeverrigaray, F.G.; SR, S.d.M.; Leidens, L.M.; ME, H.M.d.C.; Alvarez, F.; Burgo, T.A.L.; Michels, A.F.; Figueroa, C.A. Towards superlubricity in nanostructured surfaces: The role of van der Waals forces. *Phys. Chem. Chem. Phys.* **2018**, *20*, 21949–21959. [CrossRef] [PubMed]
28. Liu, X.; Yang, J.; Hao, J.; Zheng, J.; Gong, Q.; Liu, W. A near-frictionless and extremely elastic hydrogenated amorphous carbon film with self-assembled dual nanostructure. *Adv. Mater.* **2012**, *24*, 4614–4617. [CrossRef] [PubMed]

29. Restuccia, P.; Righi, M.C. Tribochemistry of graphene on iron and its possible role in lubrication of steel. *Carbon* **2016**, *106*, 118–124. [CrossRef]
30. Chen, X.; Yin, X.; Jin, J. A Study on the Wettability of Ion-Implanted Stainless and Bearing Steels. *Metals* **2019**, *9*, 208. [CrossRef]
31. Wu, P.; Li, X.; Zhang, C.; Chen, X.; Lin, S.; Sun, H.; Lin, C.T.; Zhu, H.; Luo, J. Self-Assembled Graphene Film as Low Friction Solid Lubricant in Macroscale Contact. *ACS Appl. Mater. Interfaces* **2017**, *9*, 21554–21562. [CrossRef]
32. Murashima, M.; Hojo, K.; Ito, S.; Umehara, N.; Tokoroyama, T.; Takahashi, T.; Imaeda, M. Nanotextured Mold Surface with DLC Coating for Reduction in Residual Ceramic Particles. *Langmuir* **2021**, *37*, 3563–3574. [CrossRef]
33. Xiang, Z.Y.; Zhang, J.; Xie, S.; Mo, J.; Zhu, S.; Zhai, C. Friction-induced vibration and noise characteristics, and interface tribological behavior during high-speed train braking: The effect of the residual height of the brake pad friction block. *Wear* **2023**, *516*, 204619. [CrossRef]
34. Zhang, H.B.; Wu, X.H.; Wang, W.Z. Effect of fluid pressure on adhesive wear of spherical contact. *Tribol. Int.* **2023**, *187*, 108723. [CrossRef]
35. Zhu, X.; Wang, X.; Liu, Y.; Luo, Y.; Liu, Y.; Zhang, H.; Zhao, X. Effect of the Graphitization Mechanism on the Friction and Wear Behavior of DLC Films Based on Molecular Dynamics Simulations. *Langmuir* **2023**, *39*, 1905–1913. [CrossRef] [PubMed]
36. Li, Z.; Ma, G.Z.; Xing, Z.G.; Yong, Q.S.; Zhao, H.C.; Huang, Y.F.; Guo, W.L.; Zhang, Z.A.; Wang, H.D. The effects of Cr and B doping on the mechanical properties and tribological behavior of multi-layered hydrogenated diamond-like carbon films. *Surf. Coat. Technol.* **2022**, *431*, 127977. [CrossRef]
37. Wang, D.-S.; Chang, S.-Y.; Huang, Y.-C.; Wu, J.-B.; Lai, H.-J.; Leu, M.-S. Nanoscopic observations of stress-induced formation of graphitic nanocrystallites at amorphous carbon surfaces. *Carbon* **2014**, *74*, 302–311. [CrossRef]
38. Cao, Z.; Zhao, W.; Liang, A.; Zhang, J. A General Engineering Applicable Superlubricity: Hydrogenated Amorphous Carbon Film Containing Nano Diamond Particles. *Adv. Mater. Interfaces* **2017**, *4*, 1601224. [CrossRef]
39. Harada, D.; Tunmee, S.; Euaruksakul, C.; Rittihong, U.; Nakajima, H.; Aono, Y.; Hirata, Y.; Ohtake, N.; Akasaka, H. Investigation of the structure and tribological properties of laser-irradiated hydrogenated amorphous carbon films. *Diam. Relat. Mater.* **2023**, *131*, 109573. [CrossRef]
40. Xiang, D.; Tan, X.; Sui, X.; He, J.; Chen, C.; Hao, J.; Liao, Z.; Liu, W. Comparative study on microstructure, bio-tribological behavior and cytocompatibility of Cr-doped amorphous carbon films for Co–Cr–Mo artificial lumbar disc. *Tribol. Int.* **2021**, *155*, 106760. [CrossRef]
41. Gong, Z.; Bai, C.; Qiang, L.; Gao, K.; Zhang, J.; Zhang, B. Onion-like carbon films endow macro-scale superlubricity. *Diam. Relat. Mater.* **2018**, *87*, 172–176. [CrossRef]
42. Manimunda, P.; Al-Azizi, A.; Kim, S.H.; Chromik, R.R. Shear-Induced Structural Changes and Origin of Ultralow Friction of Hydrogenated Diamond-like Carbon (DLC) in Dry Environment. *ACS Appl. Mater. Interfaces* **2017**, *9*, 16704–16714. [CrossRef]
43. Wang, Y.; Gao, K.; Zhang, B.; Wang, Q.; Zhang, J. Structure effects of sp2-rich carbon films under super-low friction contact. *Carbon* **2018**, *137*, 49–56. [CrossRef]
44. Wang, J.; Cao, X.; Lu, Z.; Zhang, G.; Xue, Q. The improved mechanical and tribological properties of amorphous carbon film by doping boron carbide. *Ceram. Int.* **2020**, *46*, 9878–9884. [CrossRef]
45. Zhang, Y.; He, X.; Liu, M.; Zhang, K.; Singh, D.J.; Fan, X.; Wen, M.; Zheng, W. Enabling macroscopic superlubricity in TaC/a-C nanocomposite film by atomic-level Au. *Scr. Mater.* **2023**, *228*, 115329. [CrossRef]
46. Nosaka, M.; Morisaki, Y.; Fujiwara, T.; Tokai, H.; Kawaguchi, M.; Kato, T. The Run-in Process for Stable Friction Fade-Out and Tribofilm Analyses by SEM and Nano-Indenter. *Tribol. Online* **2017**, *12*, 274–280. [CrossRef]
47. Erdemir, A.; Eryilmaz, O. Achieving superlubricity in DLC films by controlling bulk, surface, and tribochemistry. *Friction* **2014**, *2*, 140–155. [CrossRef]
48. Chen, L.; Wu, J.; Lu, Z.; Shang, L.; Zhang, G.; Xue, Q. Probing tribological performances of hydrogenated amorphous carbon film applied in methane by structural modification with boron. *Wear* **2021**, *470–471*, 203610. [CrossRef]
49. Chen, L.; Wei, X.; Zhang, G.; Shang, L.; Lu, Z.; Nie, X.; Xue, Q. Probing the tribological performances of hydrogenated amorphous carbon film in methane atmosphere based on Hertzian elastic contact model. *Tribol. Int.* **2021**, *155*, 106790. [CrossRef]
50. Erdemir, A.; Martin, J.M. Superior wear resistance of diamond and DLC coatings. *Curr. Opin. Solid State Mater. Sci.* **2018**, *22*, 243–254. [CrossRef]
51. Rusanov, A.; Nevshupa, R.; Fontaine, J.; Martin, J.-M.; Le Mogne, T.; Elinson, V.; Lyamin, A.; Roman, E. Probing the tribochemical degradation of hydrogenated amorphous carbon using mechanically stimulated gas emission spectroscopy. *Carbon* **2015**, *81*, 788–799. [CrossRef]
52. Wang, Y.; Yamada, N.; Xu, J.; Zhang, J.; Chen, Q.; Ootani, Y.; Higuchi, Y.; Ozawa, N.; De Barros Bouchet, M.-I.; Martin, J.M.; et al. Triboemission of hydrocarbon molecules from diamond-like carbon friction interface induces atomic-scale wear. *Sci. Adv.* **2019**, *5*, eaax9301. [CrossRef] [PubMed]
53. Yi, S.; Chen, X.; Li, J.; Liu, Y.; Ding, S.; Luo, J. Macroscale superlubricity of Si-doped diamond-like carbon film enabled by graphene oxide as additives. *Carbon* **2021**, *176*, 358–366. [CrossRef]
54. Yin, X.; Chen, H.; Jiang, L.; Liang, C.; Pang, H.; Liu, D.; Zhang, B. Effects of polyacrylic acid molecular weights on V_2C-MXene nanocoatings for obtaining ultralow friction and ultralow wear in an ambient working environment. *Phys. Chem. Chem. Phys.* **2022**, *24*, 27406–27412. [CrossRef] [PubMed]

55. Khan, A.M.; He, X.; Wu, H.; Desanker, M.; Erdemir, A.; Chung, Y.-W.; Wang, Q.J. Acid Treatment of Diamond-Like Carbon Surfaces for Enhanced Adsorption of Friction Modifiers and Friction Performance. *Tribol. Lett.* **2018**, *66*, 128. [CrossRef]
56. Kalin, M.; Polajnar, M. The Effect of Wetting and Surface Energy on the Friction and Slip in Oil-Lubricated Contacts. *Tribol. Lett.* **2013**, *52*, 185–194. [CrossRef]
57. Bellón Vallinot, I.; de la Guerra Ochoa, E.; Echávarri Otero, J.; Chacón Tanarro, E.; Fernández Martínez, I.; Santiago Varela, J.A. Individual and combined effects of introducing DLC coating and textured surfaces in lubricated contacts. *Tribol. Int.* **2020**, *151*, 106440. [CrossRef]
58. Sun, H.; Yang, L.; Wu, H.; Zhao, L.; Ji, B. Pd nanoparticles lubricant additive catalyze the construction of carbon-based tribofilm to reduce graphitization-induced wear of DLC films under boundary lubrication. *Appl. Surf. Sci.* **2023**, *641*, 158545. [CrossRef]
59. Ruiz-Lopez, M.F.; Francisco, J.S.; Martins-Costa, M.T.C.; Anglada, J.M. Molecular reactions at aqueous interfaces. *Nat. Rev. Chem.* **2020**, *4*, 459–475. [CrossRef]
60. Long, Y.; He, P.; Shao, Z.; Li, Z.; Kim, H.; Yao, A.M.; Peng, Y.; Xu, R.; Ahn, C.H.; Lee, S.-W.; et al. Moisture-induced autonomous surface potential oscillations for energy harvesting. *Nat. Commun.* **2021**, *12*, 5287. [CrossRef]
61. Dangnan, F.; Espejo, C.; Liskiewicz, T.; Gester, M.; Neville, A. Water barrier performance of additively manufactured polymers coated with diamond-like carbon films. *Diam. Relat. Mater.* **2021**, *119*, 108541. [CrossRef]
62. Ighalo, J.O.; Yap, P.-S.; Iwuozor, K.O.; Aniagor, C.O.; Liu, T.; Dulta, K.; Iwuchukwu, F.U.; Rangabhashiyam, S. Adsorption of persistent organic pollutants (POPs) from the aqueous environment by nano-adsorbents: A review. *Environ. Res.* **2022**, *212*, 113123. [CrossRef]
63. Awad, A.M.; Jalab, R.; Benamor, A.; Nasser, M.S.; Ba-Abbad, M.M.; El-Naas, M.; Mohammad, A.W. Adsorption of organic pollutants by nanomaterial-based adsorbents: An overview. *J. Mol. Liq.* **2020**, *301*, 112335. [CrossRef]
64. Topolovec-Miklozic, K.; Lockwood, F.; Spikes, H. Behaviour of boundary lubricating additives on DLC coatings. *Wear* **2008**, *265*, 1893–1901. [CrossRef]
65. Erdemir, A.; Eryilmaz, O.L.; Nilufer, I.B.; Fenske, G.R. Effect of source gas chemistry on tribological performance of diamond-like carbon films. *Diam. Relat. Mater.* **2000**, *9*, 632–637. [CrossRef]
66. Erdemir, A.; Eryilmaz, O.L.; Fenske, G. Synthesis of diamondlike carbon films with superlow friction and wear properties. *J. Vac. Sci. Technol. A* **2000**, *18*, 1987–1992. [CrossRef]
67. Al-Azizi, A.A.; Eryilmaz, O.; Erdemir, A.; Kim, S.H. Surface structure of hydrogenated diamond-like carbon: Origin of run-in behavior prior to superlubricious interfacial shear. *Langmuir* **2015**, *31*, 1711–1721. [CrossRef]
68. Konicek, A.R.; Grierson, D.S.; Gilbert, P.U.; Sawyer, W.G.; Sumant, A.V.; Carpick, R.W. Origin of ultralow friction and wear in ultrananocrystalline diamond. *Phys. Rev. Lett.* **2008**, *100*, 235502. [CrossRef] [PubMed]
69. Zhang, X.; Schneider, R.; Müller, E.; Mee, M.; Meier, S.; Gumbsch, P.; Gerthsen, D. Electron microscopic evidence for a tribologically induced phase transformation as the origin of wear in diamond. *J. Appl. Phys.* **2014**, *115*, 063508. [CrossRef]
70. Prins, R. Eley–Rideal, the Other Mechanism. *Top. Catal.* **2018**, *61*, 714–721. [CrossRef]
71. Lin, N.; Lan, H.-Q.; Xu, Y.-G.; Cui, Y.; Barber, G. Coupled Effects between Solid Particles and Gas Velocities on Erosion of Elbows in Natural Gas Pipelines. *Procedia Eng.* **2015**, *102*, 893–903. [CrossRef]
72. Liu, Y.; Jiang, Y.; Sun, J.; Wang, L.; Liu, Y.; Chen, L.; Zhang, B.; Qian, L. Durable superlubricity of hydrogenated diamond-like carbon film against different friction pairs depending on their interfacial interaction. *Appl. Surf. Sci.* **2021**, *560*, 150023. [CrossRef]
73. Erdemir, A.; Ramirez, G.; Eryilmaz, O.L.; Narayanan, B.; Liao, Y.; Kamath, G.; Sankaranarayanan, S.K. Carbon-based tribofilms from lubricating oils. *Nature* **2016**, *536*, 67–71. [CrossRef] [PubMed]
74. De Barros Bouchet, M.I.; Martin, J.M.; Avila, J.; Kano, M.; Yoshida, K.; Tsuruda, T.; Bai, S.; Higuchi, Y.; Ozawa, N.; Kubo, M.; et al. Diamond-like carbon coating under oleic acid lubrication: Evidence for graphene oxide formation in superlow friction. *Sci. Rep.* **2017**, *7*, 46394. [CrossRef]
75. Kuwahara, T.; Romero, P.A.; Makowski, S.; Weihnacht, V.; Moras, G.; Moseler, M. Mechano-chemical decomposition of organic friction modifiers with multiple reactive centres induces superlubricity of ta-C. *Nat. Commun.* **2019**, *10*, 151. [CrossRef]
76. Liu, Y.; Chen, L.; Jiang, B.; Liu, Y.; Zhang, B.; Xiao, C.; Zhang, J.; Qian, L. Origin of low friction in hydrogenated diamond-like carbon films due to graphene nanoscroll formation depending on sliding mode: Unidirection and reciprocation. *Carbon* **2021**, *173*, 696–704. [CrossRef]
77. Luo, C.; Jiang, Y.; Liu, Y.; Wang, Y.; Sun, J.; Qian, L.; Chen, L. Role of Interfacial Bonding in Tribochemical Wear. *Front. Chem.* **2022**, *10*, 852371. [CrossRef] [PubMed]
78. Shi, P.; Sun, J.; Liu, Y.; Zhang, B.; Zhang, J.; Chen, L.; Qian, L. Running-in behavior of a H-DLC/Al2O3 pair at the nanoscale. *Friction* **2020**, *9*, 1464–1473. [CrossRef]
79. De Barros Bouchet, M.I.; Matta, C.; Vacher, B.; Le-Mogne, T.; Martin, J.M.; von Lautz, J.; Ma, T.; Pastewka, L.; Otschik, J.; Gumbsch, P.; et al. Energy filtering transmission electron microscopy and atomistic simulations of tribo-induced hybridization change of nanocrystalline diamond coating. *Carbon* **2015**, *87*, 317–329. [CrossRef]
80. Obeisun, O.A.; Finegan, D.P.; Engebretsen, E.; Robinson, J.B.; Taiwo, O.O.; Hinds, G.; Shearing, P.R.; Brett, D.J.L. Ex-situ characterisation of water droplet dynamics on the surface of a fuel cell gas diffusion layer through wettability analysis and thermal characterisation. *Int. J. Hydrog. Energy* **2017**, *42*, 4404–4414. [CrossRef]
81. Wijshoff, H. Drop dynamics in the inkjet printing process. *Curr. Opin. Colloid Interface Sci.* **2018**, *36*, 20–27. [CrossRef]

82. Launay, G.; Sadullah, M.S.; McHale, G.; Ledesma-Aguilar, R.; Kusumaatmaja, H.; Wells, G.G. Self-propelled droplet transport on shaped-liquid surfaces. *Sci. Rep.* **2020**, *10*, 14987. [CrossRef]
83. Villegas, M.; Zhang, Y.; Abu Jarad, N.; Soleymani, L.; Didar, T.F. Liquid-Infused Surfaces: A Review of Theory, Design, and Applications. *ACS Nano* **2019**, *13*, 8517–8536. [CrossRef] [PubMed]
84. Ersoy, N.E.; Eslamian, M. Phenomenological study and comparison of droplet impact dynamics on a dry surface, thin liquid film, liquid film and shallow pool. *Exp. Therm. Fluid Sci.* **2020**, *112*, 109977. [CrossRef]
85. Chen, Z.T.; Nguyen, T.H.; Rumrill, S.M.; Lau, K.K.S. One-Step Bottom-Up Growth of Highly Liquid Repellent Worm-Like Surfaces on Planar Substrates. *Adv. Mater. Interfaces* **2022**, *9*, 2101961. [CrossRef]
86. Kunze, T.; Gemming, S.; Posselt, M.; Seifert, G. Tribological Aspects of Carbon-Based Nanocoatings–Theory and Simulation. *Z. Für Phys. Chem.* **2011**, *225*, 379–387. [CrossRef]
87. Ma, T.-B.; Hu, Y.-Z.; Xu, L.; Wang, L.-F.; Wang, H. Shear-induced lamellar ordering and interfacial sliding in amorphous carbon films: A superlow friction regime. *Chem. Phys. Lett.* **2011**, *514*, 325–329. [CrossRef]
88. Kunze, T.; Posselt, M.; Gemming, S.; Seifert, G.; Konicek, A.R.; Carpick, R.W.; Pastewka, L.; Moseler, M. Wear, Plasticity, and Rehybridization in Tetrahedral Amorphous Carbon. *Tribol. Lett.* **2013**, *53*, 119–126. [CrossRef]
89. Argibay, N.; Babuska, T.F.; Curry, J.F.; Dugger, M.T.; Lu, P.; Adams, D.P.; Nation, B.L.; Doyle, B.L.; Pham, M.; Pimentel, A.; et al. In-situ tribochemical formation of self-lubricating diamond-like carbon films. *Carbon* **2018**, *138*, 61–68. [CrossRef]
90. Hayashi, K.; Tezuka, K.; Ozawa, N.; Shimazaki, T.; Adachi, K.; Kubo, M. Tribochemical reaction dynamics simulation of hydrogen on a diamond-like carbon surface based on tight-binding quantum chemical molecular dynamics. *J. Phys. Chem. C* **2011**, *115*, 22981–22986. [CrossRef]
91. Schall, J.D.; Gao, G.; Harrison, J.A. Effects of Adhesion and Transfer Film Formation on the Tribology of Self-Mated DLC Contacts. *J. Phys. Chem. C* **2010**, *114*, 5321–5330. [CrossRef]
92. Li, L.; Song, W.; Ovcharenko, A.; Xu, M.; Zhang, G. Effects of atomic structure on the frictional properties of amorphous carbon coatings. *Surf. Coat. Technol.* **2015**, *263*, 8–14. [CrossRef]
93. Bai, S.; Onodera, T.; Nagumo, R.; Miura, R.; Suzuki, A.; Tsuboi, H.; Hatakeyama, N.; Takaba, H.; Kubo, M.; Miyamoto, A. Friction Reduction Mechanism of Hydrogen- and Fluorine-Terminated Diamond-Like Carbon Films Investigated by Molecular Dynamics and Quantum Chemical Calculation. *J. Phys. Chem. C* **2012**, *116*, 12559–12565. [CrossRef]
94. Wolloch, M.; Levita, G.; Restuccia, P.; Righi, M.C. Interfacial Charge Density and Its Connection to Adhesion and Frictional Forces. *Phys. Rev. Lett.* **2018**, *121*, 026804. [CrossRef] [PubMed]
95. Cui, L.; Lu, Z.; Wang, L. Toward Low Friction in High Vacuum for Hydrogenated Diamondlike Carbon by Tailoring Sliding Interface. *ACS Appl. Mater. Interfaces* **2013**, *5*, 5889–5893. [CrossRef] [PubMed]
96. Wu, D.; Ren, S.; Pu, J.; Lu, Z.; Zhang, G.; Wang, L. A comparative study of tribological characteristics of hydrogenated DLC film sliding against ceramic mating materials for helium applications. *Appl. Surf. Sci.* **2018**, *441*, 884–894. [CrossRef]
97. Liu, Y.; Yu, B.; Cao, Z.; Shi, P.; Zhou, N.; Zhang, B.; Zhang, J.; Qian, L. Probing superlubricity stability of hydrogenated diamond-like carbon film by varying sliding velocity. *Appl. Surf. Sci.* **2018**, *439*, 976–982. [CrossRef]
98. Wang, L.; Cui, L.; Lu, Z.; Zhou, H. Understanding the unusual friction behavior of hydrogen-free diamond-like carbon films in oxygen atmosphere by first-principles calculations. *Carbon* **2016**, *100*, 556–563. [CrossRef]
99. Huu, T.L.; Zaïdi, H.; Paulmier, D. Lubricating properties of diamond-like coating. *Wear* **1995**, *181–183*, 766–770. [CrossRef]
100. Ma, T.B.; Wang, L.F.; Hu, Y.Z.; Li, X.; Wang, H. A shear localization mechanism for lubricity of amorphous carbon materials. *Sci. Rep.* **2014**, *4*, 3662. [CrossRef]
101. Hayashi, K.; Sato, S.; Bai, S.; Higuchi, Y.; Ozawa, N.; Shimazaki, T.; Adachi, K.; Martin, J.-M.; Kubo, M. Fate of methanol molecule sandwiched between hydrogen-terminated diamond-like carbon films by tribochemical reactions: Tight-binding quantum chemical molecular dynamics study. *Faraday Discuss.* **2012**, *156*, 137–146. [CrossRef] [PubMed]
102. Hirayama, T.; Eguchi, Y.; Saeki, K.; Matsuoka, T.; Kikegawa, T. Structural analysis of a-C:H and a-C:H:Si films under high-pressure and high-temperature by synchrotron X-ray diffraction. *Diam. Relat. Mater.* **2016**, *70*, 83–90. [CrossRef]
103. Liu, X.; Zhang, H.; Liu, C.; Zhang, L.; Wang, Q.; Hu, H.; Zheng, J. Influence of bias patterns on the tribological properties of highly hydrogenated PVD a-C:H films. *Surf. Coat. Technol.* **2022**, *442*, 128234. [CrossRef]
104. Boeira, C.D.; Cemin, F.; Leidens, L.M.; Weber, J.S.; Michels, A.F.; Aguzzoli, C.; Serra, R.; Evaristo, M.; Fernandes, F.; Alvarez, F.; et al. Adhesion of Amorphous Carbon Nanofilms on Ferrous Alloy Substrates Using a Nanoscale Silicon Interlayer: Implications for Solid-State Lubrication. *ACS Appl. Nano Mater.* **2022**, *5*, 3763–3772. [CrossRef]
105. Mano, H.; Ohana, T. Evaluation of Anti-Adhesion Characteristics of Diamond-Like Carbon Film by Combining Friction and Wear Test with Step Loading and Weibull Analysis. *Materials* **2021**, *14*, 2746. [CrossRef]
106. Tillmann, W.; Lopes Dias, N.F.; Stangier, D.; Maus-Friedrichs, W.; Gustus, R.; Thomann, C.A.; Moldenhauer, H.; Debus, J. Improved adhesion of a-C and a-C:H films with a CrC interlayer on 16MnCr5 by HiPIMS-pretreatment. *Surf. Coat. Technol.* **2019**, *375*, 877–887. [CrossRef]
107. Xu, M.; Cai, X.; Chen, Q.; Kwok, S.C.H.; Chu, P.K. Comparative study of mechanical properties of a-C:H films produced on tungsten pre-implanted stainless steel substrate by plasma immersion ion implantation and deposition. *Diam. Relat. Mater.* **2007**, *16*, 1304–1311. [CrossRef]

108. Zaharia, T.; Kudlacek, P.; Creatore, M.; Groenen, R.; Persoone, P.; van de Sanden, M.C.M. Improved adhesion and tribological properties of fast-deposited hard graphite-like hydrogenated amorphous carbon films. *Diam. Relat. Mater.* **2011**, *20*, 1266–1272. [CrossRef]
109. Xie, J.; Komvopoulos, K. Friction, nanostructure, and residual stress of single-layer and multi-layer amorphous carbon films deposited by radio-frequency sputtering. *J. Mater. Res.* **2016**, *31*, 1857–1864. [CrossRef]
110. Liu, F.-X.; Wang, Z.-L. Thickness dependence of the structure of diamond-like carbon films by Raman spectroscopy. *Surf. Coat. Technol.* **2009**, *203*, 1829–1832. [CrossRef]
111. Zheng, X.-H.; Wang, T.; Wang, G.-Q.; Zhang, W.-K.; Yang, F.-E. Effects of Mo Single-Layer Thickness on Microstructure and Tribological Behavior of WSx/Mo/a-C/Mo Multilayer Films. *Adv. Eng. Mater.* **2021**, *23*, 2001413. [CrossRef]
112. Hong, C.; Tu, J.; Gu, C.; Zheng, X.; Liu, D.; Li, R.; Mao, S.X. The Effect of Stress Relaxation on the Microstructure and Hardness Evolution of Pure Amorphous-Carbon and C/Ti Multilayer Films. *Adv. Eng. Mater.* **2010**, *12*, 920–925. [CrossRef]
113. Rashid, N.M.A.; Ritikos, R.; Othman, M.; Khanis, N.H.; Gani, S.M.A.; Muhamad, M.R.; Rahman, S.A. Amorphous silicon carbon films prepared by hybrid plasma enhanced chemical vapor/sputtering deposition system: Effects of r.f. power. *Thin Solid Film.* **2013**, *529*, 459–463. [CrossRef]
114. Cheng, H.-Y.; Wu, W.-Y.; Ting, J.-M. Microstructure and optical properties of chromium containing amorphous hydrogenated carbon thin films (a-C:H/Cr). *Thin Solid Film.* **2009**, *517*, 4724–4727. [CrossRef]
115. Awang, R.; Tong, G.B.; Ab. Gani, S.M.; Ritikos, R.; Rahman, S.A. The Effects of Deposition Pressure on the Optical and Structural Properties of d.c. PECVD Hydrogenated Amorphous Carbon Films. *Mater. Sci. Forum* **2006**, *517*, 81–84. [CrossRef]
116. Anishchik, V.M.; Uglov, V.V.; Kuleshov, A.K.; Filipp, A.R.; Rusalsky, D.P.; Astashynskaya, M.V.; Samtsov, M.P.; Kuznetsova, T.A.; Thiery, F.; Pauleau, Y. Electron field emission and surface morphology of a-C and a-C:H thin films. *Thin Solid Film.* **2005**, *482*, 248–252. [CrossRef]
117. Sheeja, D.; Tay, B.K.; Sze, J.Y.; Yu, L.J.; Lau, S.P. A comparative study between pure and Al-containing amorphous carbon films prepared by FCVA technique together with high substrate pulse biasing. *Diam. Relat. Mater.* **2003**, *12*, 2032–2036. [CrossRef]
118. de Moraes, M.A.B.; Durrant, S.F.; Rouxinol, F.P. Electron emission enhanced chemical vapor deposition (EEECVD) for the fabrication of diverse silicon-containing films. *Thin Solid Film.* **2001**, *398–399*, 591–596.
119. Xiao, Y.; Tan, X.; Jiang, L.; Xiao, T.; Xiang, P.; Yan, W. The effect of radio frequency power on the structural and optical properties of a-C:H films prepared by PECVD. *J. Mater. Res.* **2017**, *32*, 1231–1238. [CrossRef]
120. Turri, R.G.; Santos, R.M.; Rangel, E.C.; da Cruz, N.C.; Bortoleto, J.R.R.; Dias da Silva, J.H.; Antonio, C.A.; Durrant, S.F. Optical, mechanical and surface properties of amorphous carbonaceous thin films obtained by plasma enhanced chemical vapor deposition and plasma immersion ion implantation and deposition. *Appl. Surf. Sci.* **2013**, *280*, 474–481. [CrossRef]
121. Du, N.; Wei, X.; Li, X.; Chen, Z.; Lu, S.; Ding, J.; Feng, C.; Chen, K.; Qiao, J.; Zhang, D.; et al. Friction reactions induced by selective hydrogenation of textured surface under lubricant conditions. *Friction* **2023**, *12*, 174–184. [CrossRef]
122. Cui, L.; Bi, C.; Peng, X.; Fan, Y. Probing the role of sp2-C in high-temperature tribology of a-C films by a comparative study of ta-C film and a-C film. *Diam. Relat. Mater.* **2023**, *137*, 110127. [CrossRef]
123. Xiong, W.; Feng, X.; Xiao, Y.; Huang, T.; Li, X.; Huang, Z.; Ye, S.; Li, Y.; Ren, X.; Wang, X.; et al. Fluorine-free prepared two-dimensional molybdenum boride (MBene) as a promising anode for lithium-ion batteries with superior electrochemical performance. *Chem. Eng. J.* **2022**, *446*, 137466. [CrossRef]
124. Ishikawa, T.; Choi, J. The effect of microstructure on the tribological properties of a-C:H films. *Diam. Relat. Mater.* **2018**, *89*, 94–100. [CrossRef]
125. Du, N.; Feng, C.; Chen, K.; Qiao, J.; Zhang, D.; Li, X. Friction dependence on the textured structure of an amorphous carbon surface: A reactive molecular dynamics study. *Appl. Surf. Sci.* **2023**, *610*, 155584. [CrossRef]
126. Chen, X.; Kato, T.; Kawaguchi, M.; Nosaka, M.; Choi, J. Structural and environmental dependence of superlow friction in ion vapour-deposited a-C : H : Si films for solid lubrication application. *J. Phys. D Appl. Phys.* **2013**, *46*, 255304. [CrossRef]

lubricants

MDPI

Article

Mechanical and Tribological Behavior of Nitrided AISI/SAE 4340 Steel Coated with NiP and AlCrN

Marcos E. Soares [1], Qianxi He [2,*], Jose M. DePaiva [2,*], Bruna M. de Freitas [3], Paulo Soares [3], Stephen C. Veldhuis [2], Fred L. Amorim [3] and Ricardo D. Torres [3]

1 Mechanical Engineering Department, Universidade Tecnológica Federal do Paraná (UTFPR), Ponta Grossa 84017-220, Brazil
2 McMaster Manufacturing Research Institute (MMRI), McMaster University, Hamilton, ON L8S 4L8, Canada
3 Mechanical Engineering Department, Pontifícia Universidade Católica do Paraná (PUCPR), Curitiba 80215-901, Brazil
* Correspondence: heq19@mcmaster.ca (Q.H.); paivajj@mcmaster.ca (J.M.D.)

Abstract: In this study, novel surface engineering strategies to improve the wear performance of AISI 4340 were investigated. The strategies were as follows: (i) NiP deposition on a previously nitrided steel substrate, followed by NiP interdiffusion heat treatment at either 400 °C or 610 °C (referred to as duplex treatment); (ii) the deposition of AlCrN PVD coating on NiP layers on a previously nitrided steel substrate (referred to as triplex treatment). Prior to the deposition of AlCrN, the NiP was subjected to the interdiffusion heat treatment at either 400 °C or 610 °C. These strategies were compared with the performance of the AlCrN coating directly applied on nitrided steel. To characterize the microstructural features of each layer, X-ray diffraction (XRD) and scanning electron microscopy (SEM) coupled with energy-dispersive X-ray spectroscopy (EDS) analysis were conducted. We also carried out mechanical and tribological behavior assessments. The tribological tests were carried out using a ball-on-disc tribometer under a constant load of 20 N and a tangential speed of 25 cm/s; cemented carbide spheres with a diameter of 6 mm were the counterpart body. The friction coefficient was continuously monitored throughout the tests. The results reveal that the wear mechanism for the AlCrN coating is predominantly oxidative. The most wear-resistant surface architecture was the one comprising AlCrN over the NiP layer subjected to interdiffusion heat treatment at either 400 °C or 610 °C.

Keywords: AlCrN/NiP tribological behavior; AlCrN/NiP mechanical properties; duplex treatment; triplex treatment; electroless nickel

1. Introduction

Offshore oil exploration in the pre-salt region of the Brazilian continental platform, which can reach depths of 2000 to 5000 m, requires equipment with exceptional mechanical and corrosion properties. To withstand the demanding conditions of pre-salt oil exploration, various strategies are employed, including the use of corrosion-resistant alloys (CRA) such as Inconel 625 and super-duplex stainless steel, as well as the application of protective coatings, for instance, the application of carbon steel cladded with Inconel 625 [1,2].

Ni-based coatings, particularly those containing phosphorous, are widely used in oil exploration equipment. The NiP coating, obtained through an acid–aqueous solution containing nickel sulfate and sodium hypophosphite, grows on the metal surface via an autocatalytic reaction, resulting in a uniform thickness on the steel surface [3,4]. In the oil exploration industry, a minimum P content of around 8% and a Ni-P thickness exceeding 75 μm are required to provide effective cathodic protection to the steel surface. However, as-deposited coatings with a high P content exhibit an amorphous structure with low mechanical properties [3,5].

To improve the mechanical and tribological performance of the NiP coating, an interdiffusion post-heat treatment (IPHT) is commonly employed. The IPHT creates a metallurgical bond between the NiP deposit and the steel substrate, leading to the crystallization of the amorphous structure and the precipitation of the Ni_3P phase, thereby enhancing the coating's properties [5–11]. However, when the IPHT is conducted at the recommended temperature of 610 °C for an extended period (up to 10 h), a softening of the low-alloy steel substrate occurs due to the tempering temperature being lower (around 370 °C) than the IPHT temperature [12]. Consequently, the steel substrate loses the ability to sustain the load, which could cause the failure of the NiP coating.

To address this issue, previous studies have proposed matching the IPHT and tempering temperature at 600 °C, along with a shorter duration of 2 to 4 h [13]. This approach has been shown to enable the NiP-coated AISI/SAE 4140 steel to meet the mechanical requirements of ASTM 320 L7, which specifies alloy steel and stainless-steel bolting for low-temperature service. Additionally, plasma nitriding of the quenched/tempered substrate before NiP deposition, followed by IPHT at 400 °C or 610 °C, has been investigated [12]. This approach has been shown to enable the NiP-coated AISI/SAE 4140 steel to meet the mechanical requirements of ASTM 320 L7, which specifies alloy steel and stainless-steel bolting for low-temperature service. The results demonstrated that nitriding effectively prevents substrate softening, regardless of the IPHT temperature, with the nitriding effect being more prominent at 400 °C.

In contrast, physical vapor deposition (PVD) coatings, such as those deposited by cathodic arc evaporation (CAE) [14–16], sputtering, balanced magnetron sputtering (BMS), unbalanced magnetron sputtering (UBMS), and high-power impulse magnetron sputtering (HiPIMS) [17,18], among others, are rarely used in highly corrosion-aggressive environments such as the conditions found in offshore oil exploration. Direct application of PVD coatings on carbon or low-alloy steels presents limitations, including a low thickness (often less than 5 μm) that increases the risk of a coating system failure and epitaxial growth, which allows corrosion media to permeate into the coating and reach the steel substrate [17,19]. Droplets, especially in CAE-deposited coatings, often result in a localized absence of the coating material, promoting severe corrosion when exposed to corrosive media [20,21]. However, certain PVD coatings, such as TiAlN, AlTiN, AlCrN, and TiSiN/AlCrN, exhibit high hardness (up to 35 GPa), offering excellent wear and fatigue resistance. These coatings demonstrate enhanced performance when deposited on a nitrided substrate [22–25].

This study aims to investigate the tribological behavior of an AlCrN PVD coating on the NiP deposit on plasma-nitrided AISI/SAE 4340 steel specimens (Triplex). AlCrN is a PVD coating known for its high hardness and favorable tribological properties. The mechanical and tribological performance of NiP coatings, subjected to two different IPHT temperatures (400 °C and 610 °C) in the duplex architecture, are initially explored. Subsequently, in the triplex architecture, the impact of IPHT on the ability of the NiP deposit to provide mechanical support for the AlCrN coating and enhance its wear resistance is investigated.

2. Experimental Procedures

This study aims to investigate the tribological performance of the AlCrN on the NiP coating system with 5 different layer architectures. The experimental setup involved the following steps:

- Preparation of specimens: Several 5 mm disc specimens were obtained by cutting a 25 mm diameter AISI/SAE 4340 bar. These specimens served as the base material for the coating system investigation.
- Heat treatment: The steel specimens underwent heat treatment processes, including quenching and tempering (Q/T). This treatment optimized the mechanical properties of the specimens to ensure consistent and reliable results.

- Plasma nitriding: After the Q/T process, the quenched and tempered samples were subjected to plasma nitriding.

Figure 1 illustrates the 5 different layer architectures investigated in this project. Each architecture represents a specific combination of coatings and heat treatments:

- Duplex architecture 1 (AlCrN on nitrided steel): In this architecture, an approximately 3-micrometer-thick AlCrN coating was directly deposited onto the nitrided steel specimens. This architecture aimed to evaluate the tribological performance of the AlCrN coating on a nitrided substrate.
- Duplex architecture 2 and 3 (NiP on nitrided steel): In the second and third architectures, a 30-micrometer-thick NiP coating was deposited onto the nitrided substrates. Subsequently, interdiffusion post-heat treatment (IPHT) was performed at 400 °C or 610 °C. These architectures aimed to assess the effect of the IPHT temperature on the mechanical and tribological properties of the NiP coating.
- Triplex architecture 4 and 5 (AlCrN on NiP on nitrided steel): In the fourth and fifth architectures, an AlCrN coating was deposited using the cathodic arc evaporation (CAE) technique onto the NiP-coated substrates, which had undergone IPHT at 400 °C or 610 °C. These architectures aimed to investigate the influence of the IPHT on the ability of the NiP deposit to provide mechanical support for the AlCrN coating and enhance its wear resistance.

2.1. *Steel Heat Treatment and Nitriding*

The AISI/SAE 4340 steel disc specimens were subjected to a controlled heat treatment. The specimens were heated to 870 °C and held at this temperature for 1 h, followed by quenching in oil. Subsequently, the quenched specimens underwent tempering at 370 °C for 2 h. The resulting average hardness was approximately 45 HRC. Following this, the samples were cleaned and prepared metallographically to obtain a polished surface.

After the heat treatment, the steel specimens were submitted to plasma nitriding. The nitriding process was conducted in an in-house chamber. The nitriding environment consisted of 75% N_2 and 25% H_2, working pressure of 2 Torr, at 470 °C for 6 h. The average hardness of the hardened layer was 8.6 GPa, with a thickness around 300 μm. The surface of the nitrided specimens was prepared using 1200-grit sandpaper.

2.2. *Electroless Nickel Deposition Process and AlCrN Coating*

The electroless nickel (NiP) deposition process was carried out by immersing the nitrided steel specimens in an aqueous solution, with the following composition: 34 g/L of nickel sulfate, 35 g/L of sodium hypophosphite, 35 g/L malic acid, 10 g/L acid succinic, 5% *v*/*v* of ammonia hydroxide, and 1 ppm of thiourea. The nitrided specimens were kept for 2 h in the solution, with the solution temperature maintained at 90 °C and the pH controlled within a range of 4.5 to 5.5. The surface-to-solution volume ratio for the specimens was 0.67 dm^2/L. This process resulted in an average NiP layer thickness of 20 μm, with a phosphorous content exceeding 10%. Subsequently, the NiP-coated samples were prepared using 1200-grit sandpaper.

The PVD coating applied on both nitrided and NiP-coated specimens was an AlCrN system deposited by the CAE/PVD method, according to the Alcrona-PRO® process at the Oerlikon Balzers facility (SJ Pinhais, Brazil). The AlCrN coating contains approximately 26 wt.% nitrogen, 38 wt.% aluminum, and 38 wt.% chromium.

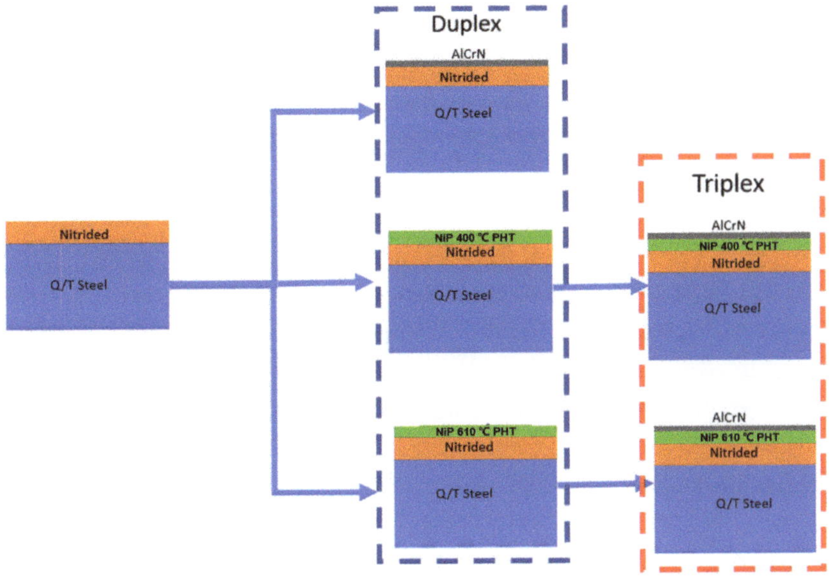

Figure 1. The duplex architectures: The AlCrN/nitrided steel, NiP IPHT at 400 °C/nitrided steel, NiP IPHT at 610 °C/nitrided steel. The triplex architectures: AlCrN/NiP IPHT at 400 °C/nitrided steel, AlCrN/NiP IPHT at 610 °C/nitrided steel architectures.

2.3. *Characterization Techniques*

2.3.1. Microstructural Characterization

The phase identification of the phases in the coating layers was obtained using X-ray diffraction (XRD) with a Shimadzu XRD-7000 instrument (Shimadzu—Kyoto, Japan), employing CuKα radiation (λ = 1.54 nm), with Grazing Incidence Angle geometry (GIXRD). We used a thin-film setting at either 1° or 2° incidence angle, scan speed of 0.5°/min and 0.02° step, with a 2θ scan range from 20° to 80°. The diffracted peaks were indexed using the match function in the software and the ICDD PDF2+ database.

Microstructural analysis (morphology and thickness) of the coatings was carried out by inspecting the specimens' cross-section in a scanning electron microscope (Tescan—Vega3 SEM—Brno, Czech Republic).

2.3.2. Hardness and Elastic Modulus Determination

The determination of hardness (H) and reduced elastic modulus (E) was carried out by performing instrumented indentation tests normal to the outer layer of each architecture with a Berkovich tip, using a Dynamic Ultra Micro Hardness Tester Shimadzu DUH-211S (Shimadzu—Kyoto, Japan). During the indentation tests, a load of 50 mN was applied and held for 15 s at the maximum load. At least 10 indentations were made for each surface architecture. Hardness and reduced Young's modulus (E) values were extracted from load–displacement curves, using the Oliver and Pharr method.

2.3.3. Adhesion Tests

Adhesion was measured through scratch tests conducted on the surface of the coated specimens. These tests were performed under a scratch tester (Revestest, Anton-Paar, Corcelles, Switzerland), with a gradual application of force ranging from 1 N to 75 N with an HRC indenter, for a distance of 3 mm and at a speed of 6 mm/min. Subsequently, the failure modes were identified, and critical loads were assigned in accordance with the ASTM C 1624-5 standard [26].

2.3.4. Tribological Tests

The tribological tests were carried out using a ball-on-disc configuration in a universal tribometer (Revestest, Anton-Paar, Corcelles, Switzerland), according to the ASTM G99-17 Standard [27]. The tribological counterpart was a 6 mm cemented carbide ball. Sliding distances of 250, 500, and 1000 m, a speed of 25 cm/s, and a load of 20 N were employed for these tests. The coefficient of friction (CoF) was continuously recorded during the tests. Following the tests, the cross-sectional area of the wear track was measured using an optical profiler (Taylor-Robson CCI Lite—Leiscester, England). The volume loss was determined by multiplying the cross-section area by the wear track's perimeter. Additionally, the wear mechanisms on the tracks were examined by SEM/EDS (Tescan, Vega3). The wear rate (*WR*) was calculated using the following equation:

$$WR = \frac{V}{F * L} \tag{1}$$

where V (mm^3) is the total volumetric wear, F (N) is the normal load, and L (m) is the sliding distance.

3. Results and Discussion

The Ra roughness values for all five surface architectures are presented in Table 1. The lowest roughness is for the duplex reference specimen, where the AlCrN coating was deposited onto a nitrided steel substrate. For polished plasma-nitrided steel surfaces, the roughness can be as low as 0.05 Ra. However, the CAE/PVD deposition process increases the roughness of the specimens due to the presence of droplets, which are large amounts of the deposited material that evaporate from the target and subsequently condense on the specimen's surface [20,21].

Table 1. Average roughness (Ra) for the 5 surface conditions.

Architecture of Coatings	Roughness, Ra (μm)
AlCrN-NiP400-Nitrided Steel	0.19 ± 0.01
AlCrN-NiP610-Nitrided Steel	0.21 ± 0.04
NiP400-Nitrided Steel	0.23 ± 0.07
NiP610-Nitrided Steel	0.31 ± 0.03
AlCrN-Nitrided Steel	0.15 ± 0.02

In cases where NiP was deposited on the nitrided surface, the roughness measured around 0.23 Ra and 0.31 Ra for the IPHT performed at 400 °C and 610 °C, respectively. The higher roughness of the NiP coatings is attributed to the surface topography that results from the NiP deposition process, which exhibits a cauliflower-like texture [28–30]. Furthermore, when the IPHT is performed at 610 °C, the Ni_3P coalesces, giving lower hardness than the NIP IPHT at 400 °C. Consequently, the IPHT specimens at 610 °C plastically deform, resulting in higher surface hardness, e.g., 0.31 Ra, than the specimens heat-treated at 400 °C, e.g., 0.23 Ra. However, when the AlCrN is deposited over the IPHT substrates at 400 °C or 610 °C, the surface roughness decreases and, practically, evens out.

The XRD patterns in Figure 2 depict the phases present in the five different architectural configurations. In the nitrided case, the ε-$Fe_{2,3}N$ phases were observed. The NiP layer on the nitrided substrates exhibits similar phases, regardless of the post-heat treatment temperatures: Ni_3P and Ni phases. However, the NiO phase is only present in the NiP deposit subjected to IPHT at 610 °C. Notably, in Figure 3b, it can be observed that Ni_3P is an incoherent precipitate when the NiP deposit is IPHT at 610 °C. Conversely, the Ni_3P is not visible in Figure 3a, but its presence is confirmed by the XRD pattern in Figure 2, suggesting coherent precipitation.

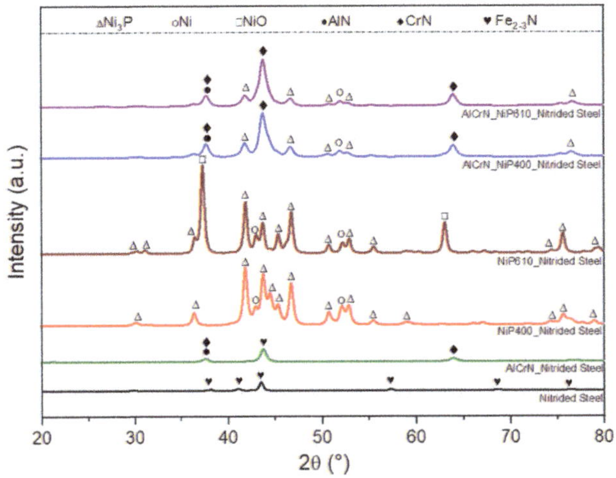

Figure 2. XRD of the nitrided steel plus XRD of the five architectures.

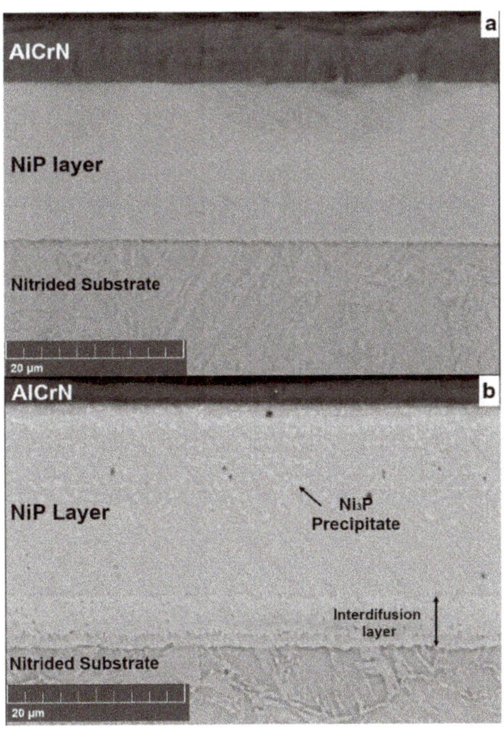

Figure 3. Cross-section of the (**a**) AlCrN/NiP IPHT at 400 °C/nitrided steel and (**b**) AlCrN/NiP IPHT at 610 °C/nitrided architectures.

The XRD pattern in Figure 2 shows AlCrN peaks as AlN and CrN, both of which have an FCC crystal structure. Al and Cr interchange within this structure. The triplex architectures exhibit XRD diffraction patterns for both the AlCrN and NiP layers. Figure 3a,b illustrate the cross-sections of the two triplex architectures. The thickness of AlCrN is 5 µm and 3 µm for NiP IPHT at 400 °C and 610 °C, respectively. In contrast, the NiP thickness is 18 μm and 20 μm for NiP IPHT at 400 °C and 610 °C, respectively. Furthermore, the interdiffusion layer, formed during IPHT at 610 °C, is approximately 5 μm thick and is not detectable in the coated system when the IPHT is performed at 400 °C, as shown in Figure 3a.

Table 2 displays the hardness (H), reduced elastic modulus (E), and H/E ratio for the five architectures, along with the hardness of the nitrided case. In the nitrided case, the values for H and E are 6.62 and 262 GPa, respectively. These values align with recent findings for nitrided cases where the outer layer is ε-$Fe_{2,3}N$, as indicated by the XRD diffraction pattern in Figure 2 [2,31]. The NiP IPHT at 610 °C shows H and E values of 11 and 206 GPa, respectively. The implementation of IPHT at this temperature results in four main effects: (i) the formation of an interdiffusion layer with a thickness of approximately 5 μm (Figure 3b); (ii) crystallization of the NiP coating; (iii) incoherent precipitation of Ni_3P (Figure 3b); (iv) formation of NiO due to the exposition of the coating to a rich air environment [6,32,33] (see XRD diffraction pattern in Figure 2). The first two effects contribute to an increase in the hardness of the NiP coating compared to the amorphous arrangements of P and Ni species. On the other hand, when the IPHT is conducted at 400 °C, the interdiffusion layer has a submicron thickness. The Ni_3P precipitates coherently, resulting in higher mechanical properties, i.e., 14.67 and 227 GPa for H and E, respectively, compared to the NiP IPHT at 610 °C [32]. For the AlCrN on the nitrided steel substrate, the H and E values are 35.4 and 353 GPa, respectively. Typical H and E values for the AlCrN system range between 28 and 35 GPa and from 250 to 300 GPa, respectively, depending on the coating stoichiometry, deposition parameters, and other factors [19,34]. The H and E of the AlCrN coating on NiP IPHT at 400 °C are 34.88 and 348 GPa, respectively, while for AlCrN on the NiP IPHT at 610 °C are 30.79 and 338 GPa, respectively. The hardness and reduced elastic modulus are higher for the NiP IPHT at 400 °C due to the coherent precipitation of Ni_3P, consequently providing a higher load-bearing capacity for the AlCrN coating.

Table 2. Average values of H, E, and H/E ratio for the five architectures.

Layer Architecture	H (GPa)	E (GPa)	H/E
Nitrided Steel	6.62 ± 0.31	262 ± 5	0.025 ± 0.0012
NiP 400/Nitrided Steel	14.67 ± 1.26	227 ± 9	0.064 ± 0.003
NiP 610/Nitrided Steel	11.00 ± 3.74	206 ± 27	0.053 ± 0.016
AlCrN/Nitrided Steel	35.50 ± 4.80	353 ± 38	0.100 ± 0.013
AlCrN/NiP 400/Nitrided Steel	34.88 ± 4.13	348 ± 29	0.099 ± 0.0038
AlCrN/NiP 610/Nitrided Steel	30.79 ± 3.79	338 ± 41	0.091 ± 0.0074

Figure 4 illustrates the scratch grooves for the five different architectures, with Table 3 detailing the failure modes and corresponding critical loads of the scratch tests. Specifically, Figure 4a,b display the scratch groove characteristics for NiP interdiffusion post-heat treated (IPHT) at 400 °C and 610 °C, respectively. Notably, both the NiP IPHT at 400 °C and 610 °C exhibit no delamination of the coating in the scratch grooves. This absence of delamination is attributed to two factors: (i) the NiP coating thickness is approximately 20 μm, and (ii) the NiP coating undergoes plastic deformation, rendering the standard adhesion principles applied to ceramic coatings on steel, such as nitrides, inapplicable to the NiP deposit. The scratch tests conducted in the NiP system evaluate the coating's ability to withstand stress generated by the HRC diamond stylus moving across the specimen surface. For instance, in the case of NiP IPHT at 400 °C, stress accommodation at the test's onset results in the formation of arc tensile cracks ($Lc1 = 3$ N), occurring just after the HRC stylus has passed due to strain release. Subsequently, chevron tensile cracks form ($Lc2 = 14$ N), followed by wedging spallation ($Lc3 = 33.5$ N).

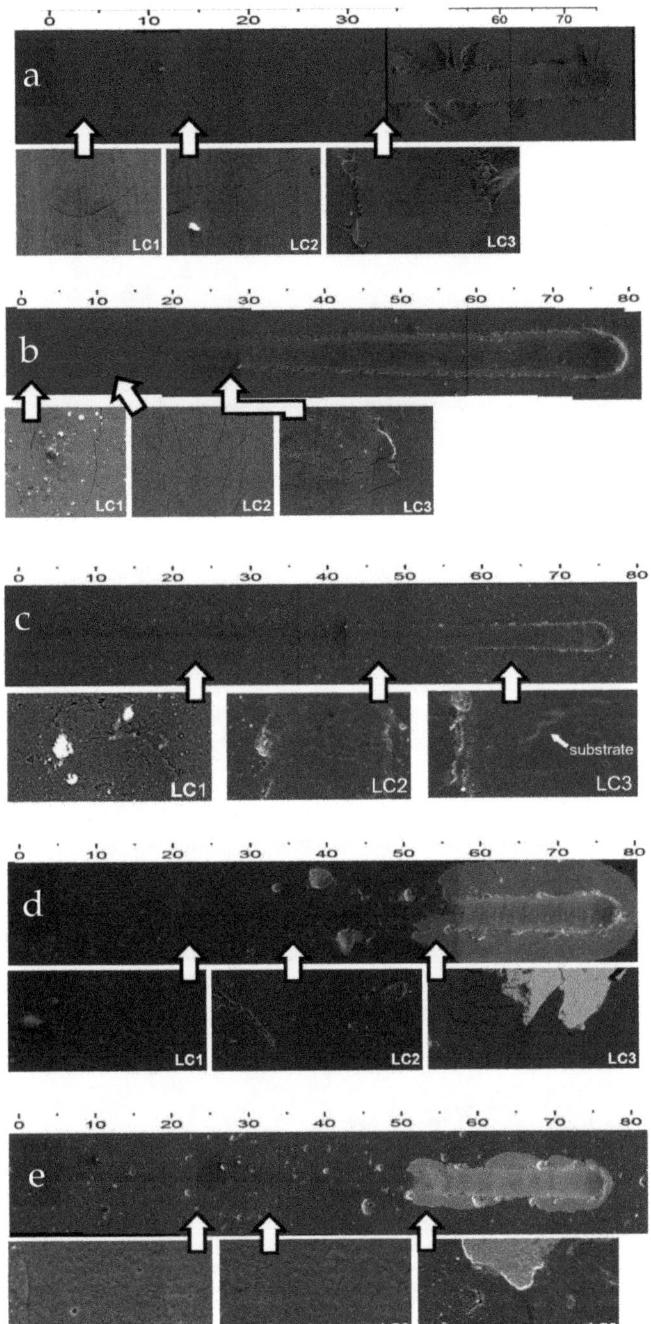

Figure 4. Aspect of the scratch groove: (**a**) NiP IPHT at 400 °C/nitrided Steel (**b**) NiP IPHT at 610 °C/nitrided steel, (**c**) AlCrN/nitrided steel (**d**) AlCrN/NiP IPHT at 400 °C/nitrided steel and (**e**) AlCrN/NiP IPHT at 610 °C/nitrided steel.

Table 3. Scratch test critical loads and failure modes of the five architectures.

Architecture	Lc1 (N)	Failure Mode	Lc2 (N)	Failure Mode	Lc3 (N)	Failure Mode
NiP IPHT 400 °C	3.0	arc tensile cracks	14.0	chevron tensile cracks	33.5	wedging spallation
NiP IPHT 610 °C	1.0	lateral cracks	10.0	chevron tensile cracks	28.0	wedging spallation
AlCrN/Nitrided Steel	22.5	conformal cracks	46.0	recovery spallation	63.0	buckling spallation
AlCrN/NiP 400 °C Nitrided Steel	22.0	lateral cracks	32.5	chevron tensile cracks	51.0	gross spallation
AlCrN/NiP 610 °C Nitrided Steel	25.0	lateral cracks	35.0	chevron tensile cracks	51.0	gross spallation

Conversely, NiP IPHT at 610 °C displays lateral cracks at the scratch groove's beginning (Lc1 = 1 N), succeeded by more substantial chevron tensile cracks (Lc2 = 10 N). This could be attributed to the interaction of the HRC stylus with the microstructure consisting of a Ni matrix with incoherent Ni_3P particles, providing an easy path for crack propagation. Finally, wedging spallation is also observed (Lc3 = 28 N). In Figure 4c, the scratch groove of AlCrN on the nitrided substrate reveals common failure modes for hard coatings over nitrided substrates. Conformal cracks emerge at 22.5 N (Lc1), followed by recovery spallation at 46 N (Lc2) and buckling spallation at 63 N (Lc3). The nitrided substrate enhances the adhesion of PVD coatings [35]. Finally, Figure 4d,e depict the scratch grooves for the two triplex architectures. These architectures exhibit failure modes, including lateral cracks, chevron tensile cracks, and gross spallation (Lc3) at a critical load of 51 N, indicating poor adhesion of AlCrN on the NiP layer, irrespective of the IPHT temperature. The root cause of this poor adhesion lies in the plastic deformation of the NiP deposit due to the presence of Ni in the microstructure, as depicted in Figure 2.

The coefficient of friction behavior, as a function of sliding distance for all architectures, is depicted in Figure 5. The friction coefficient behavior of the duplex structure and AlCrN on nitrided steel exhibits a notable increase, reaching its maximum between 100 and 130 m, indicative of the running-in stage associated with the removal of AlCrN droplets. This phenomenon is common in coatings deposited by cathodic arc evaporation (CAE/PVD). Subsequently, the coefficient of friction stabilizes at a steady-state condition of around 0.53, observed at a sliding distance of approximately 300 m. The primary wear mechanism is oxidative because chromium, aluminum, and oxygen signals overlap, according to energy-dispersive spectroscopy (EDS), not shown here, suggesting the formation of chromium–aluminum oxides (refer to Figure 6c). The wear rate (Figure 7) remains constant throughout the sliding distance.

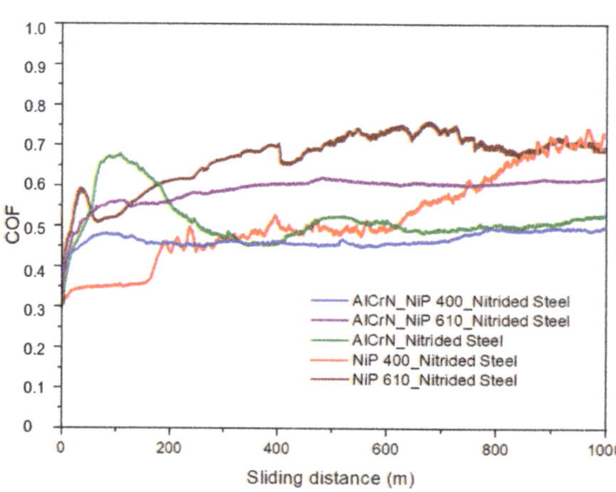

Figure 5. Friction coefficient behavior of the five architectures.

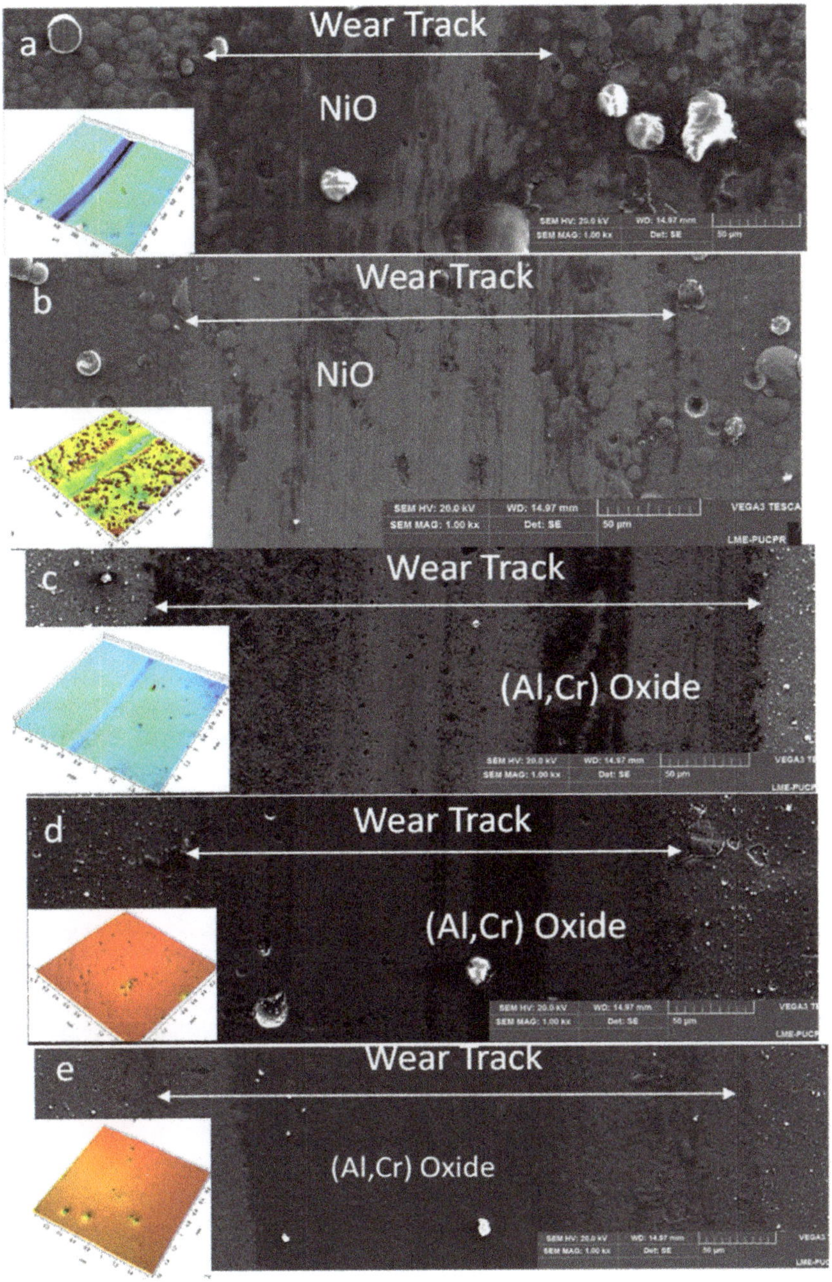

Figure 6. Aspect of the wear tracks: (**a**) NiP IPHT at 400 °C/nitrided steel; (**b**) NiP IPHT at 610 °C/nitrided steel; (**c**) AlCrN/nitrided steel; (**d**) AlCrN/NiP IPHT at 400 °C/nitrided steel; and (**e**) AlCrN/NiP IPHT at 610 °C/nitrided steel architectures.

For the NiP IPHT at 400 °C, the coefficient of friction behavior reveals a plateau at 0.34, associated with the removal of the cauliflower layer. The outer cauliflower layer exhibits poor adhesion to the NiP deposit, easily sheared from the specimen surface. Following this initial phase, an increase in the coefficient of friction is observed, likely attributed to the work hardening of the NiP deposit and the formation of NiO (according to EDS results) due to the sliding contact, as depicted in Figure 6a. The wear rate (Figure 7) is not constant; it increases with sliding distance, reaching its maximum value at 500 m, coinciding with the removal of the cauliflower layer. Subsequently, the formation of NiO and the work hardening contribute to a decrease in the wear rate, observed at 1000 m of sliding.

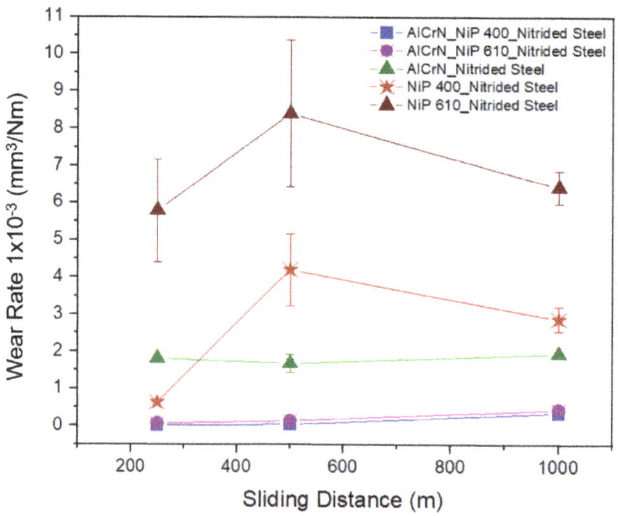

Figure 7. Wear rate with the sliding distance of the five architectures.

On the contrary, when the NiP deposit undergoes interdiffusion post-heat treatment (IPHT) at 610 °C, a shift in the friction coefficient behavior is observed compared to NiP IPHT at 400 °C (Figure 5). The friction coefficient exhibits an unstable pattern, potentially attributed to the presence of the NiO layer (according to EDS results) formed due to exposure to the elevated temperature of 610 °C and the concurrent removal and redeposition of NiP. The reported thickness of the NiO layer formed during the interdiffusion heat treatment ranges between 0.5 and 1 µm [13]. Plow lines, indicative of a two-body wear mechanism, are observed in Figure 6b. The wear rate in Figure 7 follows a similar trend observed for NiP IPHT at 400 °C.

Comparing the friction coefficient of AlCrN over NiP IPHT at 400 °C in Figure 5 with AlCrN directly over the nitrided substrate in Figure 6a reveals that AlCrN over NiP IPHT at 400 °C exhibits a lower friction coefficient. The steady-state coefficient of friction (COF) is 0.47 when AlCrN is over NiP IPHT at 400 °C, while the COF is 0.53 when AlCrN is over a nitrided steel substrate. Another distinction between the two substrates is the absence of an increase in COF during the initial sliding meters when AlCrN is over NiP substrates.

A clear difference in the friction coefficient is evident between AlCrN over NiP IPHT at 400 °C and 610 °C (Figure 5). Moreover, the steady-state friction coefficient for AlCrN over NiP IPHT at 610 °C is approximately 0.6 higher than AlCrN over NiP at 400 °C (i.e., 0.47). The primary wear mechanism is oxidative because chromium, aluminum, and oxygen signals overlap, according to EDS, consistent with the behavior observed when AlCrN is over nitrided steel, as indicated in Figure 6d,e, respectively. The wear rate for AlCrN over NiP IPHT at 400 °C or 610 °C (Figure 7) is practically identical, and the wear rate remains constant throughout the sliding distance.

The H/E ratio, a widely used predictor of wear performance, is presented in Table 2 for all five architectures, along with the H/E for the nitrided case [36]. One can realize a sound correlation between the wear rate of the duplex architectures and their H/E ratio. For instance, the H/E ratios of the NiP IPHT at 400 °C, NiP IPHT at 610 °C, and AlCrN over nitrided are 0.064, 0.053, and 0.1, respectively. Examining the wear rates (Figure 7), it becomes evident that the AlCrN over the nitrided substrate exhibits the lowest wear rate, while the wear rate is highest for the NiP IPHT at 610 °C. On the other hand, the triplex architecture shows an H/E ratio of around 0.09, regardless of the NiP microstructure, and, interestingly, the wear rates for both architectures are comparable.

4. Conclusions

The tribological behavior of duplex and triplex architectures was meticulously investigated. The duplex configuration, whether it was AlCrN or NiP coatings on a nitrided steel substrate, and the triplex architecture, where AlCrN was deposited on the NiP-coated surface in the previously nitrided steel substrate, were thoroughly examined. In both duplex and triplex architectures, the NiP underwent interdiffusion post-heat treatment at either 400 °C or 610 °C. The following conclusions can be drawn:

The interdiffusion post-heat treatment (IPHT) played a pivotal role in the mechanical behavior of the duplex and triplex architectures, particularly in the NiP deposits. The microstructure resulting from the IPHT at 400 °C comprised Ni and coherent Ni_3P, while the IPHT at 610 °C led to a microstructure consisting of Ni, NiO, and incoherent Ni_3P. This significant difference in microstructure resulted in higher H and E for the NiP IPHT at 400 °C. Comparatively, AlCrN over a nitrided substate showed H and E higher than AlCrN over the NiP deposit.

One of the key findings of this research was the poor adhesion of the AlCrN over the NiP deposit, a significant observation, regardless of the IPHT temperature. The adhesion of AlCrN on the NiP deposit showed gross spallation, indicating poor adhesion.

The coefficient of friction of the AlCrN over a nitrided steel surface showed an initial increase during the early stages of sliding. This was attributed to the presence of droplets on the coated surface, which act as frictional hotspots. Once these droplets were removed, the COF decreased and reached a steady state, stabilizing at approximately 0.53. The lowest COF was observed when AlCrN was over NiP IPHT at 400 °C, indicating the superior tribological performance of this architecture. The wear mechanism remained oxidative, regardless of the NiP IPHT temperature. The wear rate of AlCrN over NiP IPHT at 400 °C was the lowest, while NiP over the nitrided steel substrate exhibited the highest wear rate. The formation of NiO due to sliding contact contributed to increased wear resistance in the NiP deposit. Notably, a strong correlation was observed between the H/E ratio and the wear rates in both the duplex and triplex architectures, highlighting the importance of this ratio in predicting the wear behavior of the coatings.

Author Contributions: Conceptualization, M.E.S.; Methodology, Q.H. and B.M.d.F.; Formal analysis, B.M.d.F. and P.S.; Resources, F.L.A.; Data curation, F.L.A.; Writing—original draft, P.S.; Writing—review & editing, J.M.D. and S.C.V.; Supervision, R.D.T. All authors have read and agreed to the published version of the manuscript.

Funding: This research received no external funding. The APC was funded by McMaster and MDPI.

Data Availability Statement: Data are contained within the article.

Conflicts of Interest: The authors declare no conflict of interest.

References

1. Henriques, C.C.D.; Joia, C.-J.B.; Guedes, F.; Baptista, I.P. Material Selection for Brazilian Presalt Fields. In Proceedings of the Offshore Technology Conference, Houston, Texas, USA, 30 April–3 May 2012; p. OTC-23320-MS. [CrossRef]
2. Iannuzzi, M.; Barnoush, A.; Johnsen, R. Materials and corrosion trends in offshore and subsea oil and gas production. *Npj Mater. Degrad.* **2017**, *1*, 2. [CrossRef]
3. Sudagar, J.; Lian, J.; Sha, W. Electroless nickel, alloy, composite and nano coatings—A critical review. *J. Alloys Compd.* **2013**, *571*, 183–204. [CrossRef]
4. Mallory, G. The Fundamental Aspects of Electroless Nickel Plating. *Electroless Plat. Fundam. Appl.* **1991**, 1–56.
5. Fayyad, E.M.; Abdullah, A.M.; Hassan, M.K.; Mohamed, A.M.; Jarjoura, G.; Farhat, Z. Recent advances in electroless-plated Ni-P and its composites for erosion and corrosion applications: A review. *Emergent Mater.* **2018**, *1*, 3–24. [CrossRef]

6. Rabizadeh, T.; Allahkaram, S.R.; Zarebidaki, A. An investigation on effects of heat treatment on corrosion properties of Ni-P electroless nano-coatings. *Mater. Des.* **2010**, *31*, 3174–3179. [CrossRef]
7. Taheri, R.; Oguocha, I.N.A.; Yannacopoulos, S. The tribological characteristics of electroless NiP coatings. *Wear* **2001**, *249*, 389–396. [CrossRef]
8. Keong, K.G.; Sha, W.; Malinov, S. Hardness evolution of electroless nickel-phosphorus deposits with thermal processing. *Surf. Coatings Technol.* **2003**, *168*, 263–274. [CrossRef]
9. Keong, K.G.; Sha, W. Crystallisation and Phase Transformation Behaviour of Electroless Nickel-Phosphorus Deposits and Their Engineering Properties. *Surf. Eng.* **2002**, *18*, 329–343. [CrossRef]
10. Ramalho, A.; Miranda, J.C. Friction and wear of electroless NiP and NiP + PTFE coatings. *Wear* **2005**, *259*, 828–834. [CrossRef]
11. Zangeneh-Madar, K.; Vaghefi, S.M.M. The effect of thermochemical treatment on the structure and hardness of electroless Ni-P coated low alloy steel. *Surf. Coatings Technol.* **2004**, *182*, 65–74. [CrossRef]
12. Soares, M.E.; Soares, P.; Souza, P.R.; Souza, R.M.; Torres, R.D. The effect of nitriding on adhesion and mechanical properties of electroless Ni–P coating on AISI 4140 steel. *Surf. Eng.* **2016**, *33*, 1–6. [CrossRef]
13. Correia, K.S.; Greca, L.G.; Sopchenski, L.; Soares, P.; Amorim, F.L.; Torres, R.D. Strength and Deformation Properties of Low-Alloy Steel Bolts with Electroless Ni-P Coating: An Investigation of Two Thermal Routes. *J. Mater. Eng. Perform.* **2020**, *29*, 6025–6032. [CrossRef]
14. Hsu, C.; Chiu, S.; Shih, Y. Effects of thickness of electroless Ni-P deposit on corrosion fatigue damage of 7075-T6 under salt spray atmosphere. *Mater. Trans.* **2004**, *45*, 3201–3208. [CrossRef]
15. Hsu, C.H.; Lee, C.Y.; Chen, K.L.; Lu, J.H. Effects of CrN/EN and Cr2O3/EN duplex coatings on corrosion resistance of ADI. *Thin Solid Films* **2009**, *517*, 5248–5252. [CrossRef]
16. Hudak, O.E.; Kutrowatz, P.; Wojcik, T.; Ntemou, E.; Primetzhofer, D.; Shang, L.; Ramm, J.; Hunold, O.; Kolozsv, S.; Polcik, P.; et al. Improved corrosion resistance of cathodic arc evaporated $Al_{0.7}Cr_{0.3-x}V_xN$ coatings in NaCl-rich media. *Corros. Sci.* **2023**, *221*, 11376. [CrossRef]
17. Li, Y.Y.; Wu, F.B. Microstructure and corrosion characteristics of CrN/NiP sputtering thin films. *Thin Solid Films* **2010**, *518*, 7527–7531. [CrossRef]
18. Biswas, B.; Purandare, Y.; Khan, I.; Hovsepian, P.E. Effect of substrate bias voltage on defect generation and their influence on corrosion and tribological properties of HIPIMS deposited CrN/NbN coatings. *Surf. Coatings Technol.* **2018**, *344*, 383–393. [CrossRef]
19. Sanchéz, J.E.; Sanchéz, O.M.; Ipaz, L.; Aperador, W.; Caicedo, J.C.; Amaya, C.; Landaverde, M.A.H.; Beltran, F.E.; Muñoz-Saldaña, J.; Zambrano, G. Mechanical, tribological, and electrochemical behavior of $Cr_{1-x}Al_xN$ coatings deposited by r.f. reactive magnetron co-sputtering method. *Appl. Surf. Sci.* **2010**, *256*, 2380–2387. [CrossRef]
20. Petrogalli, C.; Montesano, L.; Gelfi, M.; La Vecchia, G.M.; Solazzi, L. Tribological and corrosion behavior of CrN coatings: Roles of substrate and deposition defects. *Surf. Coatings Technol.* **2014**, *258*, 878–885. [CrossRef]
21. Panjan, P.; Čekada, M.; Panjan, M.; Kek-Merl, D.; Zupanič, F.; Čurković, L.; Paskvale, S. The Surface density of growth defects in different PVD hard coatings prepared by sputtering. *Vacuum* **2012**, *86*, 794–798. [CrossRef]
22. Mo, J.L.; Zhu, M.H.; Leyland, A.; Matthews, A. Impact wear and abrasion resistance of CrN, AlCrN and AlTiN PVD coatings. *Surf. Coatings Technol.* **2013**, *215*, 170–177. [CrossRef]
23. Liew, W.Y.H.; Jie, J.L.L.; Yan, L.Y.; Dayou, J.; Sipaut, C.S.; Madlan, M.F.B. Frictional and wear behaviour of AlCrN, TiN, TiAlN single-layer coatings, and TiAlN/AlCrN, AlN/TiN nano-multilayer coatings in dry sliding. *Procedia Eng.* **2013**, *68*, 512–517. [CrossRef]
24. Ballesteros-Arguello, A.; Ramírez-Reyna, F.O.; Rodríguez-Castro, G.A.; Meneses-Amador, A.; Fernández-Valdés, D.; Reyes-Carcaño, O. Experimental and numerical evaluation of the contact fatigue resistance of AlCrN, FexN and AlCrN/FexN coatings on AISI 4140 steel. *Surf. Coatings Technol.* **2021**, *423*, 127620. [CrossRef]
25. Mo, J.L.; Zhu, M.H.; Lei, B.; Leng, Y.X.; Huang, N. Comparison of tribological behaviours of AlCrN and TiAlN coatings-Deposited by physical vapor deposition. *Wear* **2007**, *263*, 1423–1429. [CrossRef]
26. *ASTM C 1624-22*; Standard Test Method for Adhesion Strength and Mechanical Failure Modes of Ceramic Coatings by Quantitative Single Point Scratch Testing. ASTM International: West Conshohocken, PA, USA, 2022. [CrossRef]
27. *ASTM G99-17;* Standard Test Method for Wear Testing with a Pin-on-Disk Apparatus. ASTM International: West Conshohocken, PA, USA, 2020. [CrossRef]
28. Islam, M.; Azhar, M.R.; Fredj, N.; Burleigh, T.D. Electrochemical impedance spectroscopy and indentation studies of pure and composite electroless Ni-P coatings. *Surf. Coatings Technol.* **2013**, *236*, 262–268. [CrossRef]
29. Sahoo, P.; Das, S.K. Tribology of electroless nickel coatings—A review. *Mater. Des.* **2011**, *32*, 1760–1775. [CrossRef]
30. Rajendran, R.; Sha, W.; Elansezhian, R. Abrasive wear resistance of electroless Ni-P coated aluminium after post treatment. *Surf. Coatings Technol.* **2010**, *205*, 766–772. [CrossRef]
31. Chen, J.-S.; Yu, C.; Lu, H. Phase stability, magnetism, elastic properties and hardness of binary iron nitrides from first principles. *J. Alloys Compd.* **2015**, *625*, 224–230. [CrossRef]
32. Apachitei, I.; Tichelaar, F.D.; Duszczyk, J.; Katgerman, L. The effect of heat treatment on the structure and abrasive wear resistance of autocatalytic NiP and NiP–SiC coatings. *Surf. Coatings Technol.* **2002**, *149*, 263–278. [CrossRef]

33. Malfatti, C.F.; Veit, H.M.; Santos, C.B.; Metzner, M.; Hololeczek, H.; Bonino, J.P. Heat treated NiP-SiC composite coatings: Elaboration and tribocorrosion behaviour in NaCl solution. *Tribol. Lett.* **2009**, *36*, 165–173. [CrossRef]
34. Reiter, A.E.; Derflinger, V.H.; Hanselmann, B.; Bachmann, T.; Sartory, B. Investigation of the properties of $Al_{1-x}Cr_xN$ coatings prepared by cathodic arc evaporation. *Surf. Coatings Technol.* **2005**, *200*, 2114–2122. [CrossRef]
35. Ramírez-Reyna, F.O.; Rodríguez-Castro, G.A.; Figueroa-López, U.; Morón, R.C.; Arzate-Vázquez, I.; Meneses-Amador, A. Effect of nitriding pretreatment on adhesion and tribological properties of AlCrN coating. *Mater. Lett.* **2021**, *284*, 128931. [CrossRef]
36. Kumar, A.; Li, D.Y. Can the H/E ratio be generalized as an index for the wear resistance of materials? *Mater. Chem. Phys.* **2022**, *275*, 125245. [CrossRef]

lubricants

MDPI

Article

Wear Characterization of Cold-Sprayed HEA Coatings by Means of Active–Passive Thermography and Tribometer

Raffaella Sesana [1], Luca Corsaro [1,*], Nazanin Sheibanian [1,2], Sedat Özbilen [3] and Rocco Lupoi [3]

[1] Department of Mechanical and Aerospace Engineering, Politecnico di Torino, Corso Duca degli Abruzzi 24, 10129 Torino, Italy; raffaella.sesana@polito.it (R.S.); nazanin.sheibanian@polito.it (N.S.)
[2] TSUBAKI NAKASHIMA Central Laboratory (TN ITALY), Corso Torino 378, Pinerolo, 10064 Torino, Italy
[3] Department of Mechanical, Manufacturing & Biomedical Engineering, Trinity College Dublin, The University of Dublin, Parsons Building, 2 Dublin, Ireland; ozbilen.sedat@gmail.com (S.Ö.); lupoir@tcd.ie (R.L.)
* Correspondence: luca.corsaro@polito.it

Abstract: The aim of this work is to verify the applicability of thermography as a non-destructive technique to quantify the wear performance of several high-entropy alloy coatings. Thermal profiles obtained from passive and active thermography were analyzed and the results were correlated with the classical tribological approaches defined in standards. HEA coatings made of several chemical compositions ($Al_xCoCrCuFeNi$ and MnCoCrCuFeNi) and realized by using different cold spray temperatures (650 °C, 750 °C, and 850 °C) were tested in a pin-on-disk configuration, with a dedicated pin developed for the wear tests. Then, the wear performances of each sample were analyzed with the hardness and wear parameter results. The thermal profiles of passive and active thermography allowed a complete characterization of the wear resistance and performance analysis of the coatings analyzed. The results are also compared with those presented in the literature.

Keywords: cold spray (CS); high-entropy alloy (HEA); non-destructive technique (NDT); active thermography (AT); passive thermography (TP); pulsed technique; tribometer; wear characterization

1. Introduction

The remarkable mechanical properties, fatigue resistance, and ability to withstand high temperatures of traditional materials such as nickel, cobalt, and titanium-based alloys are widely applied in high-temperature scenarios, such as aeroengines, gas turbines, and the nuclear industry.

However, these materials may fail prematurely under extreme environmental conditions due to their insufficient resistance to friction and wear [1–6]. It is estimated that maintenance and replacement expenses as a result of this wear and tear can amount to approximately 1–2% of the GDP of industrialized nations.

In order to improve engineering components' performance, dependability, and longevity, it is imperative to develop materials which are mechanically strong and exhibit superior wear resistance.

Previous research has suggested that materials possessing favorable mechanical qualities, such as high hardness, elastic modulus, and fracture toughness, can lead to improved tribological performance in various applications [7,8]. High-entropy alloys (HEAs), a recent advancement in alloy development, have garnered considerable interest due to their potential properties [9–13].

HEAs are alloys comprising five or more primary elements in equimolar or near-equimolar ratios [9,11,14]. These alloys can exhibit outstanding characteristics which surpass those of traditional alloys, including superior strength and hardness, notable resistance to wear, exceptional resilience under high-temperature conditions, robust structural stability, and excellent resistance to corrosion and oxidation [15].

The tribological characteristics of HEAs are of large significance in industrial applications, leading to extensive research into their friction and wear performances under various conditions. For example, studies have investigated the adhesive wear behavior of $Al_xCoCrCuFeNi$ HEAs under atmospheric conditions [16]. These findings have revealed that higher aluminum concentrations are associated with increased hardness, resulting in significantly enhanced wear resistance [17]. Wu and colleagues [16] examined the adhesive wear characteristics of $Al_xCoCrCuFeNi$ high-entropy alloys and found that at aluminum fractions of 0.05 and 1.0 molar ratios, the predominant wear mechanism was delamination wear. However, increasing the aluminum content to a 2.0 molar ratio shifted the wear mechanism to predominantly oxidative wear. Similarly, Hsu et al. [18] analyzed the effect on sliding wear resistance of adding iron to $AlCoCrFe_XMo_{0.5}Ni$. They noted a reduction in wear rate with increasing iron content, attributed to heightened oxidation and sigma phase formation [19].

The coating quality plays a significant role in determining the mechanical properties and wear resistance of coated contacts [20,21]. It is possible to investigate the characterization of coatings using a variety of methodologies depending on the type, composition, and thickness of the coating and its substrate. A destructive technique can be used to directly characterize materials. To estimate mechanical properties and superficial qualities, measurements of hardness and roughness are widely used. In recent decades, active thermography (AT) techniques have become increasingly popular due to their non-contact nature and the ability to cover a wide area. Thermography has been used in many research fields related to coatings over the years, both in passive and active configurations, with the aim of investigating mechanical properties or coating quality. Using passive thermography (PT), it has been possible to assess the quality and performance of several plasma sprayed coatings on the contact fatigue life [21,22]. In these papers, the temperature of the contact point was chosen as the optimal infrared signal since it is only the preheating that is associated with the failure of the coating. The authors of [21,23] and [21,24] discuss interesting topics related to AT, such as microscopic cracks and coating adhesions. Accordingly, [23] highlighted how thermal maps generated by vibrothermography stimulated by ultrasound can be used to detect microcracks.

The purpose of this research is to investigate the feasibility of using AT techniques to characterize the wear performance of HEA coatings as an innovative process. A variety of HEA coatings ($Al_xCoCrCuFeNi$ and MnCoCrCuFeNi) deposited at different cold spray temperatures (650 °C, 750 °C, and 850 °C) were analyzed. The wear performance of HEA coatings using a pin-on-disk configuration and using PT and AT was analyzed. Thermal profiles were analyzed and correlated with classical tribological approaches as defined by standards.

2. Materials and Methods

Samples adopted in this study are made of a substrate and an HEA coating deposited with a cold spray technique. In particular, the substrate utilized for each sample is magnesium, while the HEA coating is generated with several HEA chemical composition and cold spray deposition temperatures. Figure 1 shows the tested samples and also a scheme of the coating geometry over the substrate, while in Table 1 the geometrical, chemical, and process parameters are summarized.

A detailed characterization of the hardness and roughness of the same coatings tested in this work was performed in [15] by the same authors.

Wear tests were performed by using a tribometer with a setup close to a classical pin-on-disk configuration. A dedicated device similar to a pin was designed with the aim to ensure the clamping of tested samples and to guarantee the visibility of the contact surfaces during PT acquisitions. Concerning the counterpart utilized during the wear tests, disks were made of quenched 100Cr6 steel with a grinding surface. A hardness of 850 HV with a roughness of 0.05 (Ra) was obtained for all counterparts.

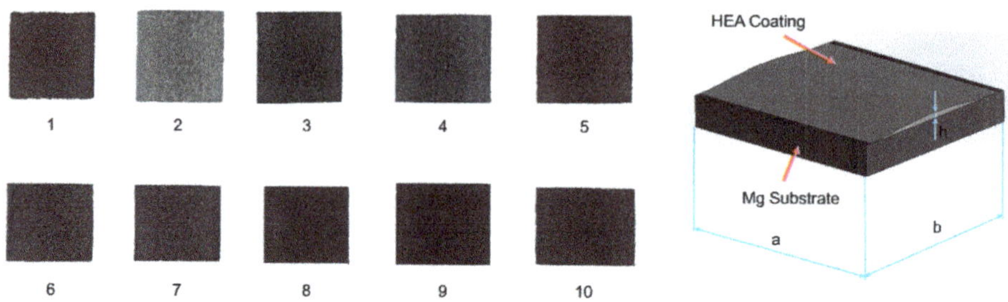

Figure 1. Tested samples (**left** side) and geometrical shape (**right** side).

Table 1. Process parameters and geometrical dimensions.

Sample	Chemical Composition	Nominal Chemical Composition in Atomic Percentage (at%)						Cold Spray Temp. [°C]	Geometrical Dimensions [mm]		
		Al	Co	Cr	Cu	Fe	Ni		a	b	h
1	$Al_{0.1}CoCrCuFeNi$	1.96	19.6	19.6	19.6	19.6	19.6	650	10.09	9.39	0.313
2		1.96	19.6	19.6	19.6	19.6	19.6	750	10.01	9.43	0.776
3	$Al_{0.2}CoCrCuFeNi$	3.85	19.2	19.2	19.2	19.2	19.2	650	10.11	9.42	0.507
4		3.85	19.2	19.2	19.2	19.2	19.2	750	10.73	10.01	0.613
5		3.85	19.2	19.2	19.2	19.2	19.2	850	10.72	10.01	0.627
6	$Al_{0.5}CoCrCuFeNi$	9.10	18.2	18.2	18.2	18.2	18.2	650	11.27	10.01	0.497
7		9.10	18.2	18.2	18.2	18.2	18.2	750	11.19	9.97	0.553
8	MnCoCrCuFeNi	16.7	16.7	16.7	16.7	16.7	16.7	650	10.09	9.17	0.386
9		16.7	16.7	16.7	16.7	16.7	16.7	750	10.59	9.21	0.604
10		16.7	16.7	16.7	16.7	16.7	16.7	850	11.73	9.23	0.632

Figure 2 illustrates the designed pin with the corresponding 3D CAD geometry and also the experimental setup utilized during the wear tests. Wear testing parameters are illustrated in Table 2 and the tests were performed with a unidirectional rotation in ambient conditions.

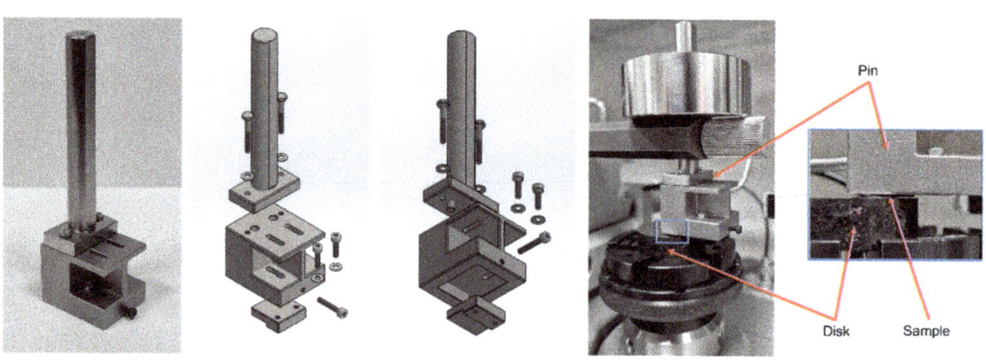

Figure 2. Designed pin and experimental setup during wear tests.

Table 2. Wear testing parameters.

Wear Testing Parameters	
Rotational speed [rpm]	250
Normal load [N]	10
Sliding length [m]	200
Linear speed [mm/s]	235.6

The wear volume computation for the tested samples was realized by using a digital microscope with a resolution of 0.01 µm. The mean value of the surface profile was generated by considering 60 lines equally spaced over the 3D surface of the sample. This procedure was carried out before and after the wear test acquiring the surface of the sample by using the 3D Stitching function available in the digital microscope. This way, the wear area ($\Delta Area$) is computed as the difference between surface profiles before and after the wear test conditions. The corresponding wear volume ($\Delta Volume$) was obtained by multiplying the $\Delta Area$ with the length of the sample. Figure 3 shows, as an example for Sample 5, the $\Delta Area$ comparison (left side) and the coating surfaces (right side) before and after the wear test. The same results for each sample are shown in Supplementary Materials.

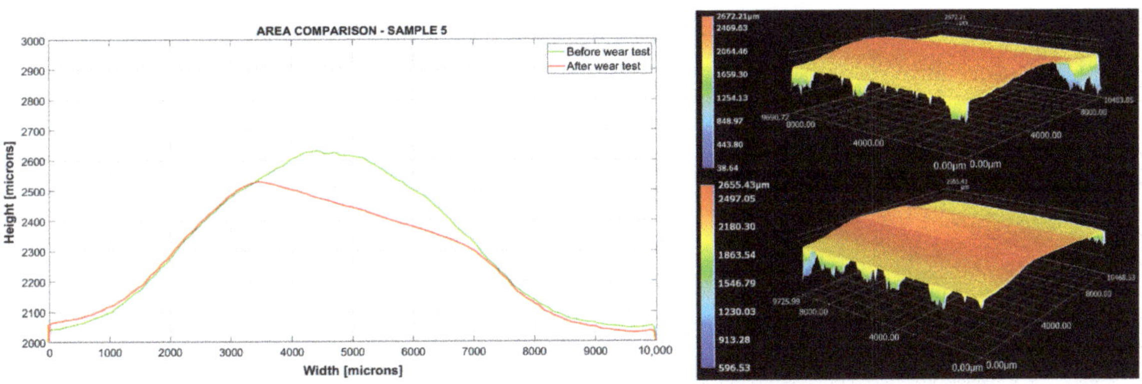

Figure 3. Area (**left side**) and 3D optical microscope comparison (**right side**) for Sample 5.

The mass reduction in the HEA samples, generated by the wear process, was also evaluated by means of a balance with a resolution of 0.0001 g. The coating was properly cleaned after the wear test to avoid debris during the mass measurement.

The wear characterization was obtained by using the well-known parameters such as the wear rate, the Archard's coefficient, and the wear resistance [25,26]. The specific wear rate parameter is very useful for comparing experimental results obtained in case of different wear testing conditions [27]. More in detail, the formula is illustrated in Equation (1):

$$W_s = \frac{V}{F_N\,L} \tag{1}$$

where W_s is the specific wear rate in [mm^3/Nm], V is the wear volume in [mm^3], F_N is the normal load in [N], and L is the sliding distance in [mm]. The wear resistance W_R is its reciprocal. In the literature [28,29], the wear resistance is related to the hardness of the sample. This way, the comparison between the wear resistance and the structure of the coating, which is related to the production process, can be investigated. The Archard's coefficient allows evaluation of the wear resistance of coating during tests. This parameter considers both tested materials and wear testing conditions, and can be computed as in Equation (2):

$$k = \frac{V \cdot HV}{F_N \cdot L} \quad (2)$$

where k is the dimensionless Archard's coefficient and HV is the less hard material in [MPa].

The wear characterization with AT and PT was realized with thermal profiles extracted from thermograms. The PT approach was used to compare the temperature increment generated during the wear test with the corresponding coefficient of friction (COF) measured with the tribometer. For what concerns the AT approach, it was investigated to evaluate its feasibility as non-destructive technique for the wear volume estimation by comparing the thermal responses of the same ROI from the thermograms obtained before and after the wear damage. Thermography analyses were performed by means of two different thermal cameras.

Figures 4 and 5 show experimental setups during PT and AT. An example of the thermogram during PT is illustrated on the right side of Figure 4. In particular, an IR Thermal camera (NETD 0.08 °C) was utilized to acquire thermal profiles and the reference specimen was allowed to control the environmental temperature during the test. Thermal profiles were extracted with a cursor located in the contact point between the disk and the tested sample. The absolute temperature was directly used for the analysis since no relevant environmental temperature variations were observed during the wear tests.

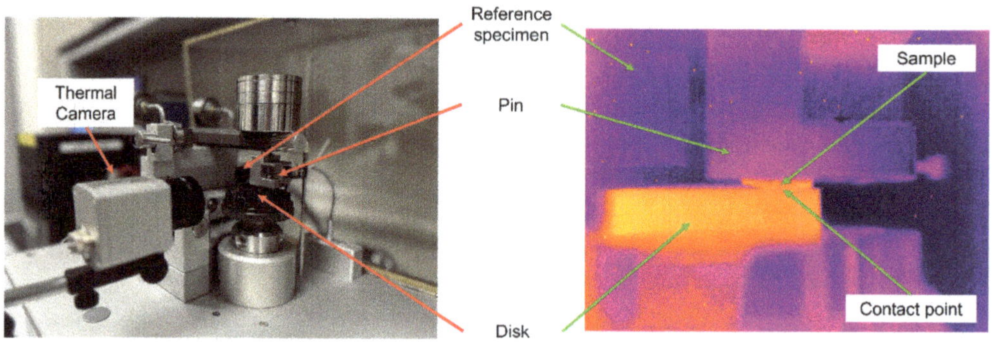

Figure 4. Passive thermography experimental setup.

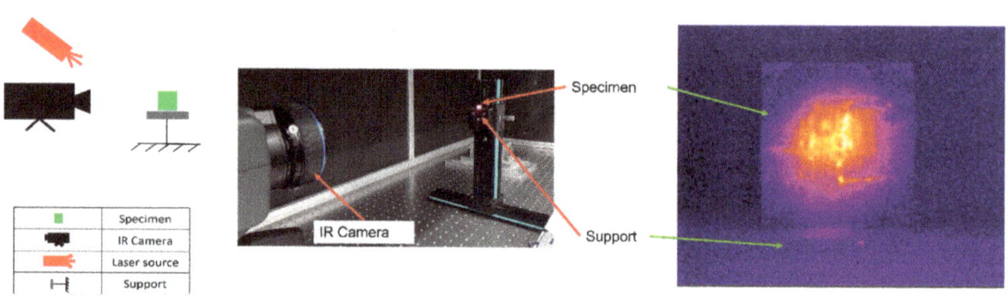

Figure 5. Active thermography equipment and experimental setup.

AT response was acquired by means of another IR thermal camera with a sensitivity lower than 20 mK and a 3–5 μm spectral range. The heating source was a laser beam with a maximum power of 50 W. The reflection mode configuration was adopted, and samples were located 440 mm distance from the thermal camera. The maximum frame rate acquisition was selected, corresponding to 785.67 Hz. Environmental conditions (room temperature and humidity) were monitored during each test, while the corresponding reflected temperature was estimated according to the standard [30]. The pulsed technique was utilized and an energy contribution made of a pulse period of 250 ms and a laser power of 30 Watt were utilized for each sample. The diameter of the laser spot was 6 mm and the laser power distribution over the spot had a Gaussian distribution. Then, a region of interest (ROI) with 4288 pixels was chosen to extract thermal profiles in radiance unit measurements. Thanks to the same superficial aspect of each sample, radiance was adopted instead of temperature to avoid emissivity problems related to its computation. Four replications were performed and then the corresponding mean value was computed.

The substrate zone was chosen as the heated zone since it remained undamaged before and after wear tests. As a matter of fact, the samples were tested by AT before and after the wear tests with the aim to characterize wear behavior by means of a non-destructive technique. The same testing conditions were adopted for both parts of the analysis (before and after wear tests).

This way, PT was used for online thermal characterization in conjunction with COF measurements, while AT was adopted for the wear damage characterization. In the specific case of PT, thermal responses obtained from the thermogram were adopted directly for the analysis. On the other hand, a dedicated data procedure was developed for the AT technique. More in detail, the samples were at first tested with the pulsed technique (before-wear condition), then the wear damage at the tribometer was realized and, subsequently, the samples were tested again with pulsed AT (after-wear condition). Four replications were conducted of each measurement and absolute radiance profiles were extracted from the target laser ROI.

Then, the relative radiance profile was computed for each replication as the difference between the absolute radiance profile and the environmental radiance. The environmental radiance was directly estimated from thermal responses, and it corresponds to the radiance value before the heating profile. Then, the corresponding mean value of the four replications was computed, obtaining the averaged relative radiance profile. The wear characterization was analyzed by following two different approaches.

The first considers all tested samples while, in the second case, the wear characterization is classified on the basis of the chemical composition of the coating.

For what concerns the first approach, once the averaged relative radiance profiles for both the before-wear and after-wear conditions are evaluated, the normalization with respect to each maximum value is generated. Figure 6 shows, as an example for samples 2 (left side) and 7 (right side), the normalized relative radiance profiles for the before-wear and after-wear conditions. This way, the wear damage in terms of thickness decrease, which means a volume reduction, is emphasized on both the heating and cooling phases of the thermal profiles.

The difference between the thermal area ($\Delta A_{Thermal}$) parameters was chosen to quantify this difference since it is easy to calculate and to correlate with respect to a wear volume reduction. More in detail, the $\Delta A_{\text{Thermal}}$ computation with respect to a certain time is illustrated in Equation (3).

$$\Delta A_{Thermal} = \int_{t_0}^{t_f} R_{aw} dt - \int_{t_0}^{t_f} R_{bw} dt \tag{3}$$

In particular, t_0 corresponds to the initial time, t_f corresponds to the final time, R_{aw} is the normalized relative radiance for the after-wear test condition, and R_{bw} is the normalized relative radiance for the before-wear test condition. The final time, t_f, considered in this study is 10 s and it was properly chosen with the aim of measuring for a long time the differences between the before-wear and after-wear conditions during the cooling phase.

A second approach was also proposed, and it is based on the chemical composition of the coating. In this case, the wear characterization is performed by using the bare cooling phase obtained from the thermal response. Also in this case, the difference between the area of the cooling phases ($\Delta A_{Cooling}$) was utilized as the thermal parameter.

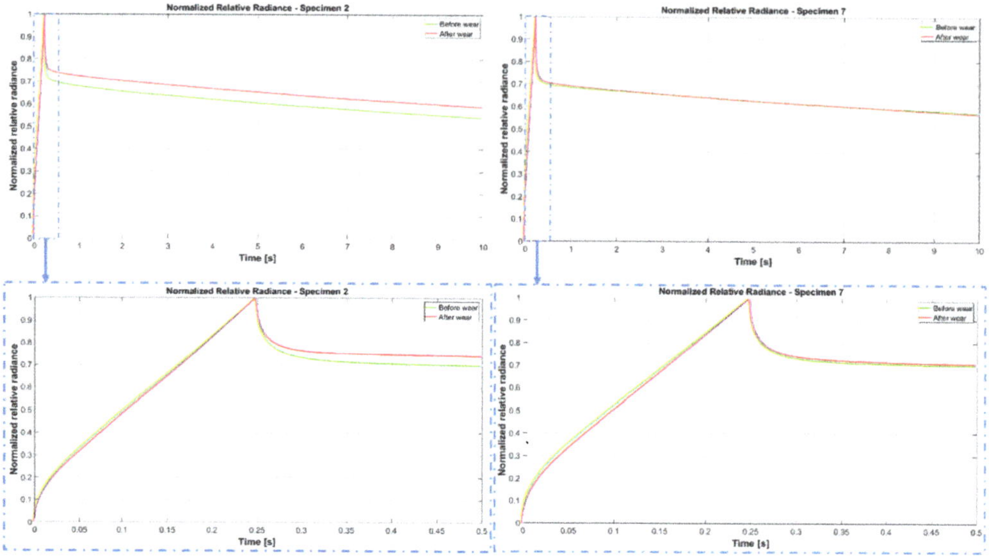

Figure 6. Normalized relative radiance profiles for samples 2 and 7.

3. Results and Discussion

In this section, results regarding the wear tests and the thermography characterization are illustrated and commented on.

The wear tests results are shown in Figure 7 and in Table 3. In particular, variations in terms of volume (Δ*Volume*) and area (Δ*Area*) were calculated as the difference between the corresponding before-wear and after-wear conditions. Volume and area were calculated as illustrated in the Section 2 (see Figure 3). In particular, Figure 7 illustrates the Δ*Volume* and Δ*Area* comparison on the left side, while the corresponding abraded mass is shown on the right side. From the analysis in Figure 7, a similar trend can be observed for both wear test results, except in the case of samples 6 and 7. As illustrated in Supplementary Materials, samples 6 and 7 show a similar Δ*Volume* and Δ*Area*, but in sample 6, an imprecise deposition of the coating can be appreciated. Probably, this non-uniform coating deposition affects the results of the removed mass.

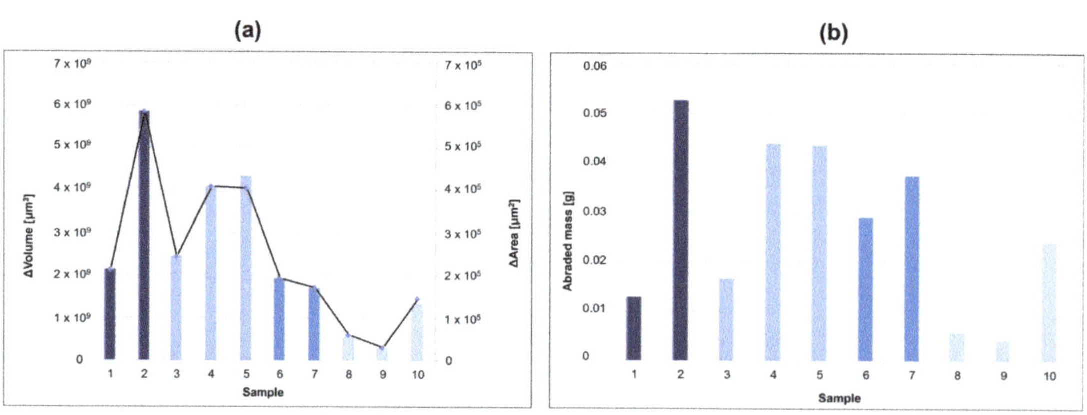

Figure 7. Measured Δ*Volume* (column) and Δ*Area* (line) (**a**) and abraded mass (**b**) for different samples.

Table 3. Wear tests result.

Sample	Area [μm²]			Volume [μm³]			Mass Variation [g]
	Before Wear Test (×10⁶)	After Wear Test (×10⁶)	Δ*Area* (×10⁶)	Before Wear Test (×10¹⁰)	After Wear Test (×10¹⁰)	Δ*Volume* (×10¹⁰)	
1	1.85	1.63	0.22	1.86	1.65	0.21	0.0130
2	4.06	3.48	0.58	4.07	3.48	0.59	0.0527
3	3.41	3.17	0.24	3.45	3.21	0.24	0.0167
4	3.55	3.14	0.41	3.55	3.14	0.41	0.0440
5	3.15	2.74	0.41	3.37	2.94	0.43	0.0436
6	3.58	3.39	0.19	3.58	3.39	0.19	0.0290
7	3.77	3.59	0.18	3.75	3.58	0.17	0.0375
8	2.46	2.40	0.06	2.25	2.20	0.05	0.0057
9	3.32	3.29	0.03	3.06	3.03	0.03	0.0042
10	3.89	3.74	0.15	3.59	3.45	0.14	0.0242

Wear parameters (the wear rate, the Archard's coefficient, and the wear resistance) were calculated according to Equations (1) and (2); the results are illustrated in Table 4. More in detail, the left side of Figure 8 compares both the wear rate and the Archard's coefficient of each sample. Moreover, the wear resistances were correlated with the corresponding hardness and the results are illustrated on the right side of Figure 8. In Figure 8a it can be observed that with increasing process temperature the wear rate and the Archard's coefficient increase for Al samples, while for Mn samples an oscillating trend is obtained. The only difference is observed in samples 6 and 7 ($Al_{0.5}$ samples) which show a behavior similar to the Mn samples.

In Figure 8b the HV and the wear resistance show a coherent behavior for all samples and, in particular, for Mn samples it can be observed that there is a process temperature which optimizes the parameters, corresponding to 750 °C. This optimal temperature is not obtained for the Al samples where the investigated parameters increase with increasing process temperature. In particular, for the $Al_{0.5}$ samples the increment of wear rate with process temperature is not so evident. According to [31], Al and Mn are usually added to improve strength, hardness, high temperature wear, fatigue resistance, corrosion resistance, etc. Al is especially added to improve the wear resistance of the CoCrCuFeNi HEA.

Table 4. Wear parameters.

Sample	Wear Rate [mm³/(Nm)] (×10⁻³)	Archard's Wear Coefficient [-] (×10⁻³)	Wear Resistance [Nm/mm³]
1	1.08	2.39	929.92
2	2.93	6.36	341.78
3	1.23	2.77	811.13
4	2.04	4.80	489.00
5	2.17	5.29	461.57
6	0.97	3.03	1032.16
7	0.85	2.66	1171.12
8	0.27	0.72	3645.47
9	0.14	0.43	7169.52
10	0.67	1.58	1484.68

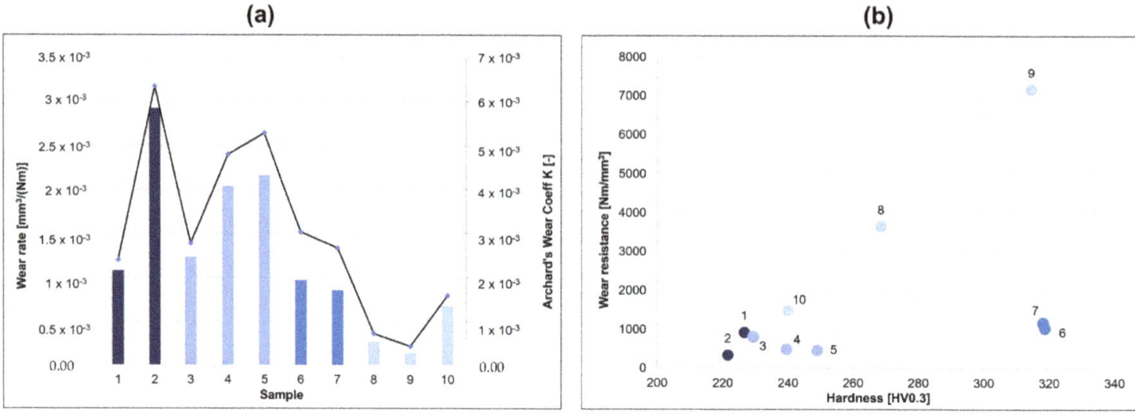

Figure 8. Wear rate (column) and Archard's coefficient (line) (**a**); wear resistance vs. hardness (**b**).

According to Archard's wear law, there is an inverse relation between microhardness and wear volume in a tribological test. This relation is confirmed for coatings, abrasive coatings [32], and for HEA coatings. In Figure 8 it can be observed that, for the investigated HEA coatings, the wear behavior increases with the average microhardness, thanks to better microstructural characteristics. The adhesive wear resistance is related to the wear coefficient. In fact, materials with high wear resistance require high energy to be removed by means of wear processes. In Table 4, the small wear coefficient is related to higher wear resistance. In the same Table 4, it can be observed that the wear coefficient decreases when the aluminum percent increases. If observing the relation between the aluminum content and wear coefficient, increasing the aluminum content five times results in a seven-times decrement in wear coefficient. For what concerns the relation between the wear coefficient and microhardness, according to [33], the wear coefficient of $Al_xCoCrCuFeNi$ HEA alloys is related to their hardness in transition from the FCC phase to BCC phase. In fact, in [33], it is reported that, generally, the wear resistance of materials is proportional to their Vickers hardness. The BCC phase is harder than the FCC phase and, as such, the wear resistance increases with increasing volume in the BCC phase, as can be observed in $Al_{0.5}$ HEA.

SEM images at the end of the wear tests, for each sample and the 100Cr6 disks, were also analyzed.

Figure 9 shows, as an example for samples 1 and 2, the abraded aspect of the coating surface, while in Figure 10 the superficial aspect of two 100Cr6 disks are illustrated. The same results for each sample are shown in the Supplementary Materials (Figure S4). For what concerns the SEM observation, obvious differences in wear surface morphology are seen as a function of aluminum content. With increasing aluminum content, the surface roughness and the degree of deformation decrease (compare these with Figures 8 and S2). In Figure S4, the SEM worn surfaces are reported. All of them show groove and plastic deformation along the wear grooves. A similar result is reported in [16] where the typical delamination wear mechanism of ductile materials is shown. According to [16], a higher aluminum content, as in samples 3, 4, 5, 6, and 7 compared to samples 1 and 2, gives a smoother worn surface. In the same paper [16], the oxidative wear is shown to be more effective for a higher content of Al. Samples 6 and 7 show coherently a higher wear resistance since the hard BCC phase resists to plastic deformation and the oxidized layer prevents delamination.

Figure 9. SEM images for samples 1 and 2.

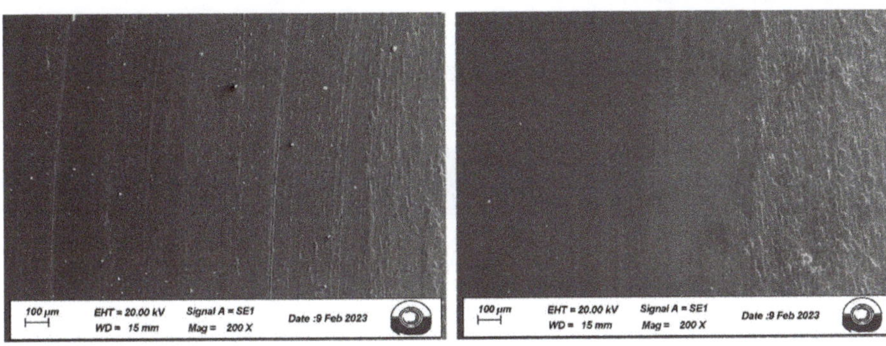

Figure 10. SEM images for 100Cr6 disks.

Moreover, the sliding wear behavior and mechanisms can be also analyzed. More in detail, observing Figure S2, samples 8 and 9, similar to what was found in [34], both the abrasive and adhesive wear mechanisms are observed in tribological tests between $AlCrCuFeNi_2$ and Si_3N_4. The shape of the grooves confirms this result, according to [34]. Similar to [17], wear particles can be observed along the sliding direction, as for example in Figure S2, samples 6, 8, 9, and 10, allowing the "three body abrasion" to take place. In Figure 9, sample 2, and in Figure S2, samples 4 and 5, a change in mechanism, according to [17], that is from abrasive to delamination and plastic deformation involving oxidation, is pointed out. This result is confirmed by the heating trend of samples during the wear tests (see Figure 11, sample 2, and Figure S3) and confirmed by the friction coefficients. For all specimens and tests, as one surface is curved, with increasing sliding distance the contact surface increases, and the phenomenon is larger for higher applied loads. With increasing sliding distance micro-scaled and material removal cutting can be observed, as in Figure S2. This phenomenon is also reported in [35]. The increase in applied load can increase the localized coating detachment, as reported in the literature.

Furthermore, for an HEA coating, wear cracks can also initiate and propagate in the coating surface and, therefore, the amount of wear debris can accordingly increase. The localized coating removal/detachment also can occur for higher wear loads due to the further propagation of surface cracks. These findings are in accordance with other investigations on the wear behavior of HEA coatings with a dual-phase BCC + FCC structure.

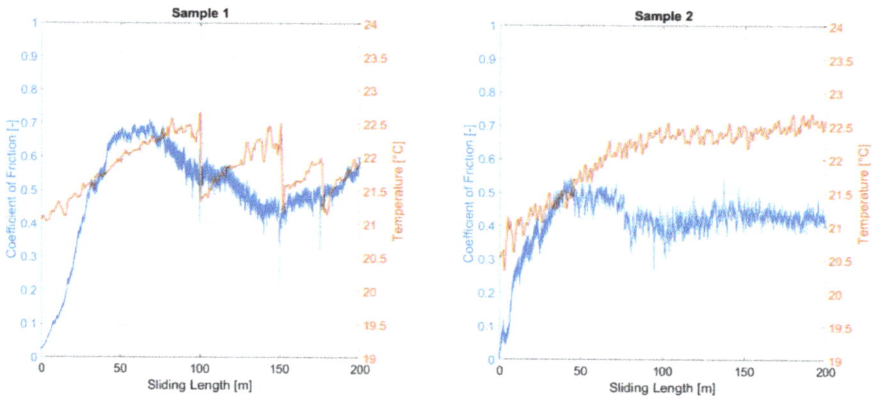

Figure 11. Coefficients of friction and temperature profiles.

Results obtained from the PT characterization are now presented. In particular, Figure 11 shows the COFs and thermal profiles, as an example for samples 1 and 2, for all the sliding lengths. The same results are shown for each sample in the Supplementary Materials (Figure S5). Moreover, a comparison between the corresponding maximum temperature increment and the mean value of the COF for each test is illustrated in Figure 12. On the basis of the results obtained, it can be emphasized that both the chemical composition and the shape of the coating influence heat propagation during tribological tests, making a direct comparison between the maximum temperature rise and the wear performance of the coatings impossible. Furthermore, the sliding distance and rotation speed, appropriately tuned for wear analysis, do not allow an appreciable temperature increment and do not keep it stable for a large part of the sliding distance. The COF curves show a typical trend: an initial rapid increase in COF to a peak value and then a gradual decrease to a steady value. In Figure S5, the COFs of all samples show this behavior. The initial peak value is justified by the curved surfaces of the samples that are then gradually abraded and become flat contact surfaces in the following part of the COF curve. The friction curves of the $Al_{0.5}CoCrCuFeNi$ HEA alloy show a peak value of COF of about 0.65 and a steady value of about 0.48 (Figures 11 and S5). The $Al_{0.5}CoCrCuFeNi$ HEA alloy exhibits a relatively large fluctuation in COF which can be caused by many phenomena. According to [16], the main ones are the periodic localized fracture of the coating surface and the accumulation of debris on the wear surface which is periodically eliminated. From the analysis of these results, it can be pointed out that in the COFs in Figures 11, 12 and S5, these phenomena are evident and moreover the relation of COF and temperature increment due to friction phenomena, plastic deformation, oxidation, and subsequent decrement in adhesive forces are reported. The conditions of the tribological test influence the generation of metal debris. As also illustrated in [36], the heat flow generated during the test increases with increasing loads and speed. These test conditions generate a higher contact surface temperature with a higher heat of friction. As a result, a faster oxidation film rate is generated. The temperature reached during the tribological tests also influences the adhesive forces. In fact, adhesive forces tend to decrease when the temperature of the contact interface reaches a critical value and, consequently, an overall temperature increment occurs. In addition, an increase in COF is observed and the wear rate also increases rapidly in accordance with the increase in temperature reached during tribological tests. It is important to note that the wear resistance is proportional to the hardness, as illustrated in Table 4. This behavior is also described in [17].

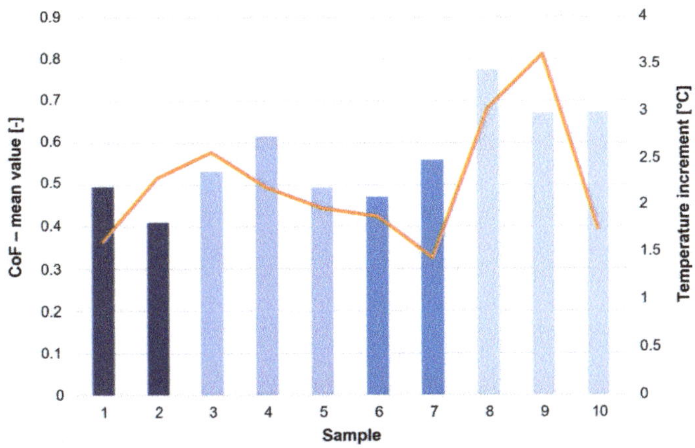

Figure 12. Coefficients of friction (mean value) and maximum temperature increments.

Results regarding the AT characterization are illustrated in Table 5, and Figures 13 and 14. The normalized relative radiance profiles for the before-wear and after-wear conditions, for each sample, are reported in the Supplementary Materials (Figure S6). Thermal area computations, according to Equation (3), are illustrated in Table 5, also with the corresponding before-wear and after-wear areas. From the analysis of Table 5, it can be observed that sample 10 generates a negative value of $\Delta A_{Thermal}$. This result can be assumed as 0 as it falls into the uncertainty of the measurement accuracy of the thermal measurement. In particular, the IR camera discriminates 0.08 °C. As illustrated in Figure S6 (see sample 10), the before-wear condition follows an opposite trend with respect to the other tested specimens. For what concerns sample 7, the $\Delta A_{cooling}$ is also negative and the same assumption is made.

Table 5. Thermal area results.

Sample	Before-Wear Area [s]	After-Wear Area [s]	$\Delta A_{Thermal}$ [s]	$\Delta A_{cooling}$ [s]
1	6.8080	6.9005	0.0925	0.0803
2	6.1009	6.5512	0.4503	0.4391
3	6.2781	6.4349	0.1568	0.1491
4	6.1092	6.2669	0.1577	0.1435
5	6.1268	6.2640	0.1372	0.0643
6	5.9957	6.1140	0.1183	0.117
7	6.2956	6.3206	0.0250	−0.0015
8	6.5883	6.5902	0.0019	0.2304
9	6.1339	6.3408	0.2069	0.23
10	6.1516	6.1018	−0.0498	−0.0368

Figure 13 compares the AT results with the classical tribological analysis results; that is, the linear regression models between $\Delta A_{Thermal}$ and the corresponding wear volume results ($\Delta Volume$ and abraded mass) are illustrated. The R^2 results are lower than 0.8 thus indicating a low linear correlation between the selected parameters if all the samples are analyzed together. In Figure 14, the same parameters are grouped related to each HEA coating composition. Again, no apparent relation between composition, process temperature, and wear testing temperature increment is observed. Nevertheless, it can be observed that for almost all samples the thermal cooling transition shows a positive change before and after wear test. This can be related to a change in surface composition, structure, and roughness.

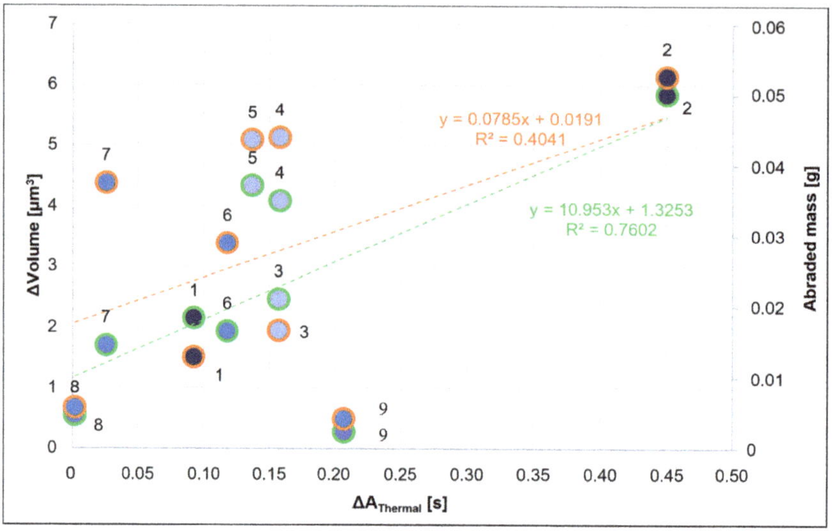

Figure 13. Δ*Volume* (orange) and abraded mass (green) vs. $\Delta A_{Thermal}$

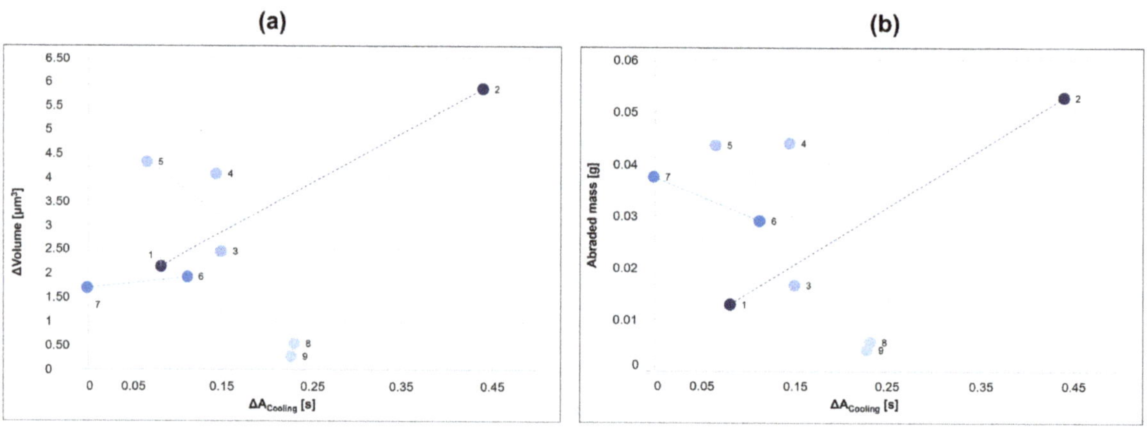

Figure 14. Δ*Volume* (**a**) and abraded mass (**b**) vs. $\Delta A_{Cooling}$

In the Supplementary Material section, the initial and final sample profiles are reported before and after the tribological tests. For all samples, the orientation of the profiles with respect to the center of rotation of the disk is preserved. It can be observed that a sort of anisotropic wear process took place thus possibly justifying this irregular behavior.

4. Conclusions

The following conclusions can be drawn from the presented results:

* The coating after the wear test shows microcracks and wear debris indicating HEA coating resistance to sliding stress.

* The HEA coatings show a mixed wear mechanism, both abrasive and adhesive. The temperature reached during the tests affects the adhesive forces and the COF.
* Passive thermography (PT) appears to be able to follow the transition between abrasive and adhesive wear processes where plastic deformation phenomena take place and can therefore be a useful tool for defining the type of friction as well as for estimating the COF.
* Active thermography (AT), however, to measure the abraded volume requires more in-depth studies and still remains a useful tool for distinguishing the different materials subjected to abrasive phenomena.

Furthermore, the two methods of data analysis provided acceptable and comparable results. To improve the accuracy of the method it is important to control the geometry of the sample.

Supplementary Materials: The following supporting information can be downloaded at: https://www.mdpi.com/article/10.3390/lubricants12060222/s1, Figure S1. Wear area comparison (left side) and coating surfaces (right side). Figure S2. Wear area comparison (left side) and coating surfaces (right side). Figure S3. Wear area comparison (left side) and coating surfaces (right side). Figure S4. SEM images for tested samples. Figure S5. Coefficient of frictions and thermal profiles. Figure S6. Normalized relative radiance profiles for the before-wear and after-wear conditions.

Author Contributions: Conceptualization: R.S., L.C. and N.S.; Supervision: R.L.; Methodology: L.C. and N.S.; Validation: R.S.; Writing—Original Draft Preparation: R.S., L.C. and S.Ö. All authors have read and agreed to the published version of the manuscript.

Funding: This research is funded by the European Commission, REA through its Marie Swklodowska Curie European Individual Fellowship to one of the authors (SÖ), Project No: 101018482 (November 2021–October 2023), which is gratefully acknowledged. This publication is part of the project PNRR-NGEU which has received funding from MUR—DM 352/2022.

Data Availability Statement: Data is contained within the article or supplementary material.

Conflicts of Interest: The authors declare no conflict of interest.

References

1. Patel, P.; Nair, R.B.; Supekar, R.; McDonald, A.; Chromik, R.R.; Moreau, C.; Stoyanov, P. Enhanced wear resistance of AlCoCrFeMo high entropy coatings (HECs) through various thermal spray techniques. *Surf. Coat. Technol.* **2024**, *477*, 130311. [CrossRef]
2. Dong, H. 3—Tribological properties of titanium-based alloys. In *Woodhead Publishing Series in Metals and Surface Engineering—Surface Engineering of Light Alloys*; Hanshan, D., Ed.; Woodhead Publishing: Sawston, UK, 2010; ISBN 9781845695378. [CrossRef]
3. Stoyanov, P.; Dawag, L.; Goberman, D.G.; Shah, D. Friction and wear characteristics of single crystal Ni-based superalloys at elevated temperatures. *Tribol. Lett.* **2018**, *66*, 47. [CrossRef]
4. Stoyanov, P.; Dawag, L.; Joost, W.J.; Goberman, D.G.; Ivory, S. Insights into the static friction behavior of Ni-based superalloys. *Surf. Coat. Technol.* **2018**, *352*, 634–641. [CrossRef]
5. Harrington, K.M.; Miller, E.C.; Frye, A.; Stoyanov, P. Tribological insights of Co-and Ni-based alloys in extreme conditions. *Wear* **2021**, *477*, 203827. [CrossRef]
6. Holmberg, K.; Erdemir, A. Influence of tribology on global energy consumption, costs and emissions. *Friction* **2017**, *5*, 263–284. [CrossRef]
7. Hong, S.; Mei, D.; Wu, J.; Lin, J.; Wu, Y.; Li, J.; Zheng, Y. Hydro-abrasive erosion and cavitation-silt erosion characteristics of HVOF sprayed WC-Ni cermet coatings under different flow velocities and sand concentrations. *Ceram. Int.* **2023**, *49*, 74–83. [CrossRef]
8. Stoyanov, P.; Harrington, K.M.; Frye, A. Insights into the Tribological characteristic of Cu-based coatings under extreme contact conditions. *JOM* **2020**, *72*, 2191–2197. [CrossRef]
9. Meghwal, A.; Anupam, A.; Murty, B.S.; Berndt, C.C.; Kottada, R.S.; Ang, A.S.M. Thermal spray high-entropy alloy coatings: A review. *J. Therm. Spray Technol.* **2020**, *29*, 57–893. [CrossRef]
10. Cantor, B.; Chang, I.T.H.; Knight, P.; Vincent, A.J.B. Microstructural development in equiatomic multicomponent alloys. *Mater. Sci. Eng. A* **2004**, *375*, 213–218. [CrossRef]
11. Patel, P.; Roy, A.; Sharifi, N.; Stoyanov, P.; Chromik, R.R.; Moreau, C. Tribological performance of High-Entropy Coatings (HECs): A review. *Materials* **2022**, *15*, 3699. [CrossRef]
12. Ranganathan, S. Alloyed pleasures: Multimetallic cocktails. *Curr. Sci.* **2003**, *85*, 1404–1406.

13. Bhaskaran Nair, R.; Supekar, R.; Morteza Javid, S.; Wang, W.; Zou, Y.; McDonald, A.; Mostaghimi, J.; Stoyanov, P. High-entropy alloy coatings deposited by thermal spraying: A review of strengthening mechanisms, performance assessments and perspectives on future applications. *Metals* **2023**, *13*, 579. [CrossRef]
14. Murty, B.S.; Yeh, J.-W.; Ranganathan, S.; Bhattacharjee, P.P. *High-Entropy Alloys*; Elsevier: Amsterdam, The Netherlands, 2019. [CrossRef]
15. Sesana, R.; Sheibanian, N.; Corsaro, L.; Ozbilen, S.; Lupoi, R.; Artusio, F. Cold spray HEA coating surface microstructural characterization and mechanical testing. *Results Mater.* **2024**, *21*, 100540. [CrossRef]
16. Wu, J.M.; Lin, S.J.; Yeh, J.W.; Chen, S.K.; Huang, Y.S.; Chen, H.C. Adhesive wear behavior of $Al_xCoCrCuFeNi$ high-entropy alloys as a function of aluminum content. *Wear* **2006**, *261*, 513–519. [CrossRef]
17. Chen, M.; Lan, L.; Shi, X.; Yang, H.; Zhang, M.; Qiao, J. The tribological properties of $Al_{0.6}CoCrFeNi$ high-entropy alloy with the sphase precipitation at elevated temperature. *J. Alloys Compd.* **2019**, *777*, 180–189. [CrossRef]
18. Hsu, C.Y.; Sheu, T.S.; Yeh, J.W.; Chen, S.K. Effect of iron content on wear behavior of$AlCoCrFe_xMo_{0.5}Ni$ high-entropy alloys. *Wear* **2010**, *268*, 653–659. [CrossRef]
19. Haghdadi, N.; Guo, T.; Ghaderi, A.; Hodgson, P.D.; Barnett, M.R.; Fabijanic, D.M. The scratch behaviour of $Al_xCoCrFeNi$ ($x = 0.3$ and 1.0) high entropy alloys. *Wear* **2019**, *428–429*, 293–301. [CrossRef]
20. Wu, D.-Y.; Zhou, X.; Li, L.P.; Zhai, G.F. Microstructure, mechanical properties and tribological behaviors of gold coating determined by surface quality. *Surf. Coat. Technol.* **2023**, *462*, 129463. [CrossRef]
21. Sesana, R.; Corsaro, L.; Sheibanian, N.; Özbilen, S. Application of NDT active thermography for the characterization of the cold spray process of high entropy alloys (Submission: 246712725). *submitted*.
22. Ma, R.; Dong, L.; Wang, H.; Guo, W. Research on the Contact Fatigue Failure of Thermal Sprayed Coating Based on Infrared Thermography. *Nondestruct. Test. Eval.* **2020**, *35*, 73–97. [CrossRef]
23. Piau, J.-M.; Bendada, A.; Maldague, X.; Legoux, J.G. Nondestructive testing of open microscopic cracks in plasma-sprayed coatings using ultrasound excited vibrothermography. *Nondestruct. Test. Eval.* **2008**, *23*, 109–120. [CrossRef]
24. Rachidi, R.; Elkihel, B.; Delaunois, F.; Deschuyteneer, D. Effectiveness of Phased Array Focused Ultrasound and Active Infrared Thermography Methods as a Nondestructive Testing of Ni-WC Coating Adhesion. *Nondestruct. Test. Eval.* **2019**, *34*, 205–220. [CrossRef]
25. *ASTM G99*; Standard Test Method for Wear Testing with a Pin-on-Disk Apparatus. ASTM International: West Conshohocken, PA, USA, 2006; Volume 5, pp. 1–6.
26. *ISO 18535:2016*; Diamond-Like Carbon Films—Determination of Friction and Wear Characteristics of Diamond-Like Carbon Films by Ball-on-Disc Method. International Organization for Standardization: Geneva, Switzerland, 2012; pp. 1–24.
27. Nasirpouri, F.; Alipour, K.; Daneshvar, F.; Sanaeian, M. Electrodeposition of Anticorrosion Nanocoatings. In *Micro and Nano Technologies*; Elsevier: Amsterdam, The Netherlands, 2020; pp. 473–497. [CrossRef]
28. *ASTM G133*; Standard Test Method for Linearly Reciprocating Ball-on-Flat Sliding Wear 1. ASTM International: West Conshohocken, PA, USA, 2011; Volume 5, pp. 1–10.
29. *ISO 4287:1997*; Geometrical Product Specifications (GPS)—Surface Texture: Profile Method—Terms, Definitions and Surface Texture Parameters. International Organization for Standardization: Geneva, Switzerland, 1997; pp. 1–38.
30. *ISO 18434*; Condition Monitoring and Diagnostics of Machine Systems Thermography, Part 2: Image Interpretation and Diagnostics. International Standard Organization: Geneva, Switzerland, 2019.
31. Verma, A.; Tarate, P.; Abhyankar, A.C.; Mohape, M.R.; Gowtam, D.S.; Deshmukh, V.P.; Shanmugasundaram, T. High temperature wear in $CoCrFeNiCu_x$ high entropy alloys: The role of Cu. *Scr. Mater.* **2019**, *161*, 28–31. [CrossRef]
32. Sesana, R.; Lupoi, R.; Pessolano Filos, I.; Yu, P.; Rizzo, S. Abrasion Power of Ti and Ni Diamond-Coated Coatings Deposited by Cold Spray. *Metals* **2022**, *12*, 1197. [CrossRef]
33. Khruschov, M.M. Principles of Abrasive Wear. *Wear* **1974**, *28*, 69–88. [CrossRef]
34. Gwalani, B.; Torgerson, T.; Dasari, S.; Jagetia, A.; Nartu, M.S.K.K.Y.; Gangireddy, S.; Pole, M.; Wang, T.; Scharf, T.W.; Banerjee, R. Influence of fine-scale B2 precipitation on dynamic compression and wear properties in hypo-eutectic $A_{l0.5}CoCrFeNi$ high-entropy alloy. *J. Alloys Compd.* **2021**, *853*, 157126. [CrossRef]
35. Löbel, M.; Lindner, T.; Mehner, T.; Lampke, T. Microstructure and Wear Resistance of AlCoCrFeNiTi High-Entropy Alloy Coatings Produced by HVOF. *Coatings* **2017**, *7*, 144. [CrossRef]
36. Ghadami, F.; Ghadami, S.; Davoudabadi, M.A. Sliding Wear Behavior of the Nanoceria-Doped AlCrFeCoNi High-Entropy Alloy Coatings Deposited by Air Plasma Spraying Technique. *J. Therm. Spray Technol.* **2022**, *31*, 1263–1275. [CrossRef]

lubricants

MDPI

Article

Tribological Properties of 7A04 Aluminum Alloy Enhanced by Ceramic Coating

Xiaobo Meng [1,†], Wei Zhang [1,†], Shizhong Wei [1,*], Kunming Pan [1,2], Xiaodong Wang [1], Tao Jiang [1], Xiran Wang [1,2], Changji Wang [1,2], Chong Chen [1,2], Feng Mao [1], Ziping Qiao [3], Jun Xue [3] and Cheng Zhang [1,2,*]

1 National Joint Engineering Research Center for Abrasion Control and Molding of Metal Materials, School of Materials Science and Engineering, Henan University of Science and Technology, Luoyang 471000, China; zw1987877362@163.com (W.Z.)

2 Longmen Laboratory, Luoyang 471003, China

3 Science and Technology on Transit Impact Laboratory, No.208 Research Institute of China Ordnance Industries, Beijing 102202, China; maryqiao@163.com (Z.Q.)

* Correspondence: wsz@haust.edu.cn (S.W.); zhangch06@126.com (C.Z.)

† These authors contributed equally to this work.

Abstract: The 7A04 Al alloy is a commonly used lightweight metal material; however, its low wear resistance limits its application. In this study, the wear resistance of this alloy was improved by preparing micro-arc oxidation (MAO) coatings, MAO/MoS_2 composite coatings, and hard-anodized (HA) coatings on its surface. The friction and wear behaviors of these three coatings with diamond-like coated (DLC) rings under oil lubrication conditions were investigated using a ring–block friction tester. The wear rates of the coatings on the block surfaces were determined using laser confocal microscopy, and the wear trajectories of the coatings were examined using scanning electron microscopy. The results indicated that, among the three coatings, the MAO/MoS_2 coating had the lowest coefficient of friction of 0.059, whereas the HA coating had the lowest wear rate of 1.47×10^{-6} mm/Nm. The MAO/MoS_2 coatings exhibited excellent antifriction properties compared to the other coatings, whereas the HA coatings exhibited excellent anti-wear properties. The porous structure of the MAO coatings stored lubricant and replenished the lubrication film under oil lubrication. Meanwhile, the introduced MoS_2 enhanced the densification of the coating and functioned as a solid lubricant. The HA coating exhibited good wear resistance owing to the dense structure of the amorphous-phase aluminum oxide. The mechanisms of abrasive and adhesive wear of the coatings under oil lubrication conditions and the optimization of the tribological properties by the solid–liquid synergistic lubrication effect were investigated. This study provides an effective method for the surface modification of Al alloys with potential applications in the aerospace and automotive industries.

Keywords: 7A04 Al alloy; wear resistance; surface strengthening; wear mechanism; ceramic coating; micro-arc oxidation (MAO)

1. Introduction

The 7A04 Al alloy is an important structural material in the aerospace [1] and automotive industries [2] owing to its low density, high specific strength [3], good ductility [4], and machinability [5]. However, it suffers from low hardness and poor wear resistance, which limit its application. To improve the wear resistance of Al alloys, a series of surface treatment techniques, such as ion implantation [6], laser treatment [7], and chemical conversion treatment [8], have been widely adopted in recent years. Although these technologies have made some progress, they fail to simultaneously provide the advantages of green environmental protection, low pretreatment requirements, simple processing, low environmental pollution, and excellent comprehensive performance of the film layer [9,10]. Therefore, in recent years, micro-arc oxidation (MAO), MAO/MoS_2, and hard-anodized

(HA) coatings have attracted attention as advanced surface treatment technologies. These ceramic coatings are generated in situ on the surfaces of Al alloys to enhance the wear and corrosion resistance of the alloys, compensate for the insufficient film–base bonding force of the traditional technology, and optimize the wear resistance of Al alloys.

Hard anodizing is an electrochemical surface treatment technique that is usually performed in an acidic electrolyte such as sulfuric acid. A thick and uniform aluminum oxide film is generated on the surface of an Al alloy substrate by applying a high voltage and low current density [11]. Although this technique can significantly enhance the hardness of a material, the limited heat resistance of the HA film may lead to cracking or peeling in high-temperature environments, limiting its use in high-temperature applications [12]. In addition, the use of acidic electrolytes can cause environmental pollution.

MAO is a surface-strengthening technique for the in situ formation of a ceramic oxide film on the surfaces of workpieces made of metals such as Al, Ti, and Mg and their alloys [13]. By applying a voltage to the workpiece, an electrochemical reaction with the electrolyte solution is stimulated, resulting in a micro-arc discharge that produces a film with high densities and good mechanical properties. Conventional MAO coatings typically exhibit irregular undulations and cracks, resulting in uneven hardness of the coating surface, which affects its friction properties [14]. However, increasing the lubricity of MAO coatings, which can reduce friction, is possible. MoS_2, as one of the many-layered solid lubricant materials, is particularly suitable for severe working conditions owing to its high chemical stability, low coefficient of friction (CoF), and high load-bearing capacity [15,16]. The porous nature of the MAO coating provides dispersion space for MoS_2 to form an effective lubrication film during friction, reducing the CoF. The metal-based self-lubricating composite ceramic layer prepared according to this principle contains uniformly dispersed, phase-stable, and strongly bonded lubricating phases both internally and on the surface, and its friction-reducing effect has been confirmed in various application scenarios. Lv et al. [17] synthesized a MAO/MoS_2 composite lubrication layer via a hydrothermal reaction on the surface of the MAO coating and investigated the MoS_2 lubrication layer, which had high purity and a unique micro-spherical structure. Comparison of the friction performance of MAO/MoS_2 and MAO coatings under different loads involving ball-on-disk wear revealed that the lower CoF of the MAO/MoS_2 composite coatings was due to the self-lubricating properties of MoS_2 particles in conjunction with the formation of MoO_3 by the wear process. Chen et al. [18] prepared MAO composite coatings by adding graphene particles to the MAO electrolyte of the 6064 Al alloy and found that the addition of graphene particles changed the breakdown voltage, promoted the MAO reaction, and increased the thickness of the MAO coatings, which increased the coating hardness. Currently, most ceramic surface coating lubrication tests employ the ball-on-disk wear method, which, despite its popularity, is more susceptible to external factors such as temperature and load variations compared to line contact under point contact conditions [19–23]. Line contact is prevalent in mechanical systems, including sliding guides, cam followers, and cylinder and piston systems. Although ring–block wear tests have been utilized to simulate these real-world conditions, there is a lack of comprehensive understanding regarding the lubrication effectiveness of coatings and their mechanisms in line contact scenarios, which restricts the broader application of ceramic surface coatings in industry. In this context, MAO composite coating technology predominantly enhances the hardness and wear resistance of coatings by incorporating hard particles like SiC and Cr_2O_3 [24]. However, there is a scarcity of research focused on enhancing the wear resistance of MAO through the addition of lubricating elements. As the demand for electric vehicles grows, it becomes increasingly important to optimize the wear resistance and lubricity of MAO coatings to mitigate wear on critical components, prolong their service life, and decrease energy consumption, thereby fostering the sustainable development of the automotive industry.

In this study, the friction and wear behaviors of a friction pair consisting of the 7A04 Al alloy, a MAO/MoS_2 composite coating, a MAO-coated block specimen, and a diamond-like carbon (DLC) ring were investigated. Scanning electron microscopy (SEM) and energy-dispersive X-ray spectroscopy (EDS) were used to analyze the wear surface and elemental changes in the block specimens. Additionally, Raman spectroscopy was used to analyze the wear surface and friction wear mechanism.

2. Specimen Preparation and Testing

2.1. Coating Preparation

A block specimen of the 7A04 Al alloy, which nominal composition is presented in Table 1, was employed in the experiment. After wire cutting a rectangular block of the size 19 mm × 12 mm × 12 mm, the specimen was sanded using 200#–1200# grit paper, rinsed with deionized water, ultrasonically cleaned in an acetone solution for 15 min at room temperature, and blown dry with cold air.

Table 1. Chemical composition of the 7A04 Al alloy (wt.%).

Element	Zn	Mg	Cu	Cr	Si	Mn	Al
Content	5–7	1.8–2.8	1.4–2	0.1–0.25	0.4	0.2–0.6	Balance

2.1.1. Micro-Arc Oxidation MAO

Prior to the MAO, the samples were treated with a NaOH and Na_3PO_4 pretreatment solution at 70–80 °C for 3–5 min. A basic alkaline electrolyte consisting of 10 g/L Na_2SiO_3, 5 g/L NaOH, and an appropriate amount of deionized water was used, and 4 g/L MoS_2 particles (grain size of 50 nm) were added to the basic electrolytes of the MAO/MoS_2 specimens. The added nanoparticles were prone to agglomeration; to prevent this, 4 g/L hydroxypropyl cellulose (HMC) was added to the base electrolyte and then ultrasonically stirred until it was uniformly suspended in the electrolyte. An asymmetric bipolar multifunctional alternating current pulse power supply was used, and the preparation parameters were set as follows: positive voltage, 460 V; positive frequency, 500 Hz; negative voltage, 60 V; duty cycle, 20%; oxidation time, 90 min; and current density, 3 A/dm^2. A constant current method was used to perform a MAO treatment on the surface of the Al alloy, and a cooling system was used to keep the temperature of the electrolyte below 45 °C. The electrolyte was then mixed via ultrasonic stirring until it was uniformly suspended in the electrolyte. During the MAO treatment, the sample was selected as the anode, and the electrolyzer was selected as the cathode. Following the treatment, the sample was rinsed with deionized water, blown dry with cold air, and placed in a sealed bag for subsequent testing and analysis. The detailed composition and parameters of the MAO electrolyte are presented in Table 2.

Table 2. Electrolyte composition and process parameters for the MAO process.

Component	Value
Positive voltage (V)	460
Negative voltage (V)	60
Positive frequency (Hz)	500
Duty cycle (%)	20
pH	12.3
Oxidation time (min)	90
Current density (A/dm^2)	3
Depositing temperature	≤45 °C
MAO Electrolyte	10 g/L Na_2SiO_3, 5 g/L NaOH, 4 g/L HMC
MAO/MoS_2 Electrolyte	10 g/L Na_2SiO_3, 5 g/L NaOH, 4 g/L MoS_2, 4 g/L HMC

2.1.2. Hard Anodizing—HA

HA coatings were performed in a properly prepared double-walled glass container. A direct current power supply with a constant current was used, with a current density of 2 A/dm^2. An aluminum block was used as the anode for the power supply, and stainless steel served as the cathode material. Ultrasonic stirring was employed to cool the electrolyte. The detailed composition and parameters of the HA electrolyte are presented in Table 3.

Table 3. Electrolyte composition and process parameters for the HA process.

Component	Value
Voltage (V)	120
Frequency (Hz)	500
pH	1.2
Current density (A/dm^2)	2
Oxidation time (min)	40
Depositing temperature	5 °C
HA Electrolyte	15% sulfuric acid

The ring specimens were made of 25Cr$_3$Mo$_3$NiNbZr mold steel and had an inner diameter of 42 mm, an outer diameter of 50 mm, and a width of 13 mm. The nominal composition of the material is presented in Table 4. The surface coating comprised DLC prepared using ionomer-enhanced chemical vapor deposition (PECVD), and the ring specimens were prepared using the same parameters. The chemical composition data in Tables 1 and 2 were provided by the material manufacturers.

Table 4. 25Cr$_3$Mo$_3$NiNbZr chemical composition (wt.%).

Element	C	Si	Mn	Cr	Mo	Ni	Nb	Zr	Fe
Content	0.28	<0.1	0.18	3.03	2.94	0.55	0.14	0.012	Balance

2.2. *Microstructural Characterization*

Field-emission SEM (JSM-IT800, JEOL Corporation, Tokyo, Japan) was used to examine the coating and cross-sectional morphologies of the specimen surfaces before and after wear. Elemental distribution analyses of the coating surface and wear marks were performed using EDS. The porosity of the coating surface under SEM was determined using ImageJ v1.54f software; three measurements were taken for each surface and averaged. The crystal structure was characterized using an X-ray diffractometer (Brux-D8, Bruker AXS, Karlsruhe, Germany) with Cu-Ka radiation at 40 kV and 40 mA. Cu was used as the target material, and the scanning range was 20°–80°, with a scanning speed of 6°/min. Hardness tests were performed using a microhardness tester (HVT-1000, Huayin Testing Instrument Factory, Laizhou City, China). This instrument was used to measure the microhardness of the surface at a certain depth with a load of 50 g and a loading time of 15 s. Owing to the surface roughness of the MAO film, the hardness test was performed after the surface of the sample was polished, the hardness of the structure of the uniform place of the 5-point hardness was tested, the mean value of the data points and the standard deviation of the data group was calculated, and a graph was made. A scratch tester (RST-300, Revetest, Peseux, Switzerland) was used to determine the coating and substrate bonding strength using Rockwell C-type diamond contacts (cone angle of 120° and contact radius of R = 200 μm). In the test, a diamond contact was applied to the coating surface under normal loads increasing from 0 to 100 N with a loading rate of 20 N/min and a scratch length of 5 mm. The critical loads were determined via specific damage observations and acoustic emission signal change analyses. The surface Raman spectra (InVia, Renishaw, London, UK) were examined using a confocal laser Raman spectrometer (LabRAM HR Evolution, laser wavelength of 532 nm).

2.3. *Tribological Properties*

A high-speed ring–block testing machine (MRH-3) was used under oil lubrication conditions at 25 °C with a normal load of 300 N, rotational speed of 200 rpm, rotation time of 20 min, and ring radius of 25 mm. The experiment was performed using the same substrate with three different coated block specimens. The ring was fabricated using the same process (DLC-coated $25Cr_3Mo_3NiNbZr$ mold steel). The blocks and rings were kept to maintain a smooth surface before the experiment, and lubricating oil was subsequently brushed onto their surfaces. A block-on-ring friction diagram is shown in Figure 1. At the end of the experiment, the blocks and rings were ultrasonically cleaned with anhydrous ethanol for 10 min and dried in an electric blast drying oven at 60 °C for 0.5 h. The worn-out specimens were then removed and weighed five times. Laser confocal microscopy (LCM) (0LS5100, Rui Ke System Integration Co., Ltd., Beijing, China) was used to measure the surface roughness of the coating and to model and analyze the wear surface topography. The wear rate was calculated via three-dimensional (3D) topographic modeling. Here, five cross-sections at equal distances from the abrasion mark trajectory were selected, the average cross-sectional area was calculated, and the average cross-sectional area multiplied by the length of the abrasion mark was used as the abrasion mark volume V. The abrasion rate was calculated using the following equation:

$$\text{Wear rate} = \frac{V}{Fn \times S}\left(\text{mm}^3/\text{Nm}\right), \quad (1)$$

where V represents the wear volume (mm^3), Fn represents the normal load (N), and S represents the total friction distance (m).

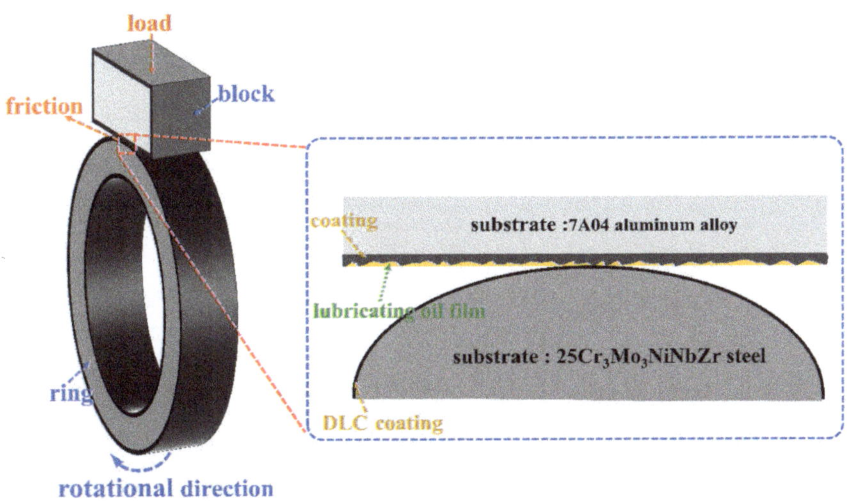

Figure 1. Friction diagram of block-on-ring.

3. Results and Discussion

3.1. Analysis of Microscopic Morphology Before Wear

Figure 2 presents the micromorphology of the surfaces of the MAO, MAO/MoS_2, and HA coatings on the 7A04 Al alloy before wear (backscattered electron (BSE) diagram). As shown in Figure 2a, the MAO coating surface mainly consisted of cracks, micropores, and raised cellular structures. Figure 2b indicated that, for the MAO/MoS_2 coating surface, cellular projections were the main features. The surface contained a few cracks, and the irregular cellular projections contained micropores. The formation of these micropores is related to the generation of gases from the molten oxide eruption and the generation of discharge channels. Comparing the microscopic morphologies of the MAO and MAO/MoS_2 coatings revealed that fewer cracks and micropores existed in the MAO/MoS_2 coatings. Additionally, the surface porosity of the MAO coatings was 10.221%, whereas that of the MAO/MoS_2 coating was 5.638%, as measured using ImageJ. The main reason for the large difference in surface porosity between the coatings is that the addition of MoS_2

nanoparticles increased the conductivity of the electrolyte, resulting in a more molten matrix ejected from the pores and oxidation. The micropores and cracks generated during the growth process were adsorbed on the coating pores and sealed with the diffusion and electrophoretic deposition of MoS_2 nanoparticles, which increased the surface density [25]. The surface of the HA coating (Figure 2c) exhibited small pits and a few holes and cracks on the surface, and the surface oxide film was less porous. The HA coating had the lowest porosity of 2.773%, as measured using ImageJ. Figure 2d shows the coating base material (7A04 Al alloy). A few cracks were observed on the surface. These cracks were due to the low plasticity of the 7A04 Al alloy, which led to the accumulation of internal residual stresses during processing. Table 5 presents the results of elemental analysis using the micromorphology energy spectra of the coatings shown in Figure 1. The ceramic coating was mainly composed of Al and O, the MAO coating surface contained Si and P, and a small amount of MoS_2 in the MAO/MoS_2 coating was deposited on the coating surface. The MAO coating surface has a high Si content, because the amorphous SiO_2 formed in the electrolyte cannot be dissolved in the electrolyte; thus, it was deposited on the coating surface as the coating grew [26].

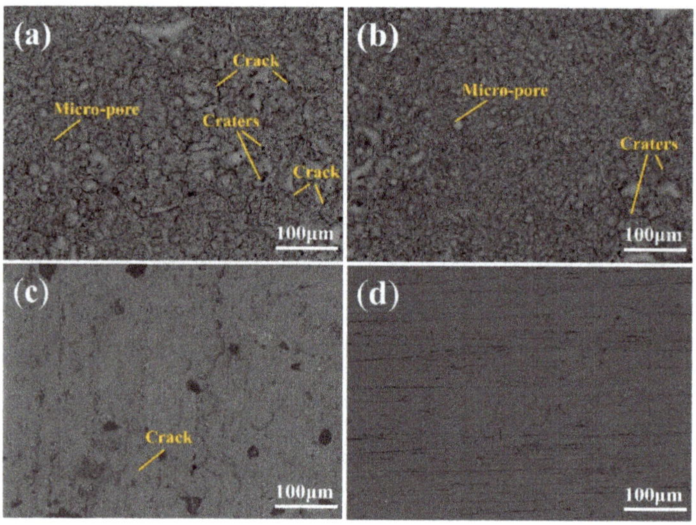

Figure 2. SEM image (500×) of the microscopic morphology of the coating surface: (**a**) MAO coating, (**b**) MAO/MoS_2 coating, (**c**) HA coating, and (**d**) 7A04 aluminum alloy.

Table 5. SEM images of the coated surfaces corresponding to elemental scans of (a) MAO coating (b) MAO/MoS_2 coating, and (c) HA coating.

Element	Content, wt.%		
	a	b	c
O	44.15	44.87	54.21
Al	32.08	32.16	40.21
P	2.03	2.39	-
Si	15.96	17.85	-
S	-	0.43	5.13
Mo	-	0.22	-
Others	4.62	2.08	0.53

Figure 3 presents cross-sectional views of the MAO, MAO/MoS_2, and HA coatings and the corresponding elemental line sweeps. The measured average thicknesses of the

MAO, MAO/MoS_2, and HA coatings were 60 ± 2.52 μm, 60 ± 3.79 μm, and 30 ± 1.12 μm, respectively. After polishing of the cross-section, as shown in Figure 3a,b for MAO and MAO/MoS_2, respectively, the undulation of the coating interfaces and surface contour lines were clearly observed, indicating that the coating grew inward and outward simultaneously and that the inward growth could increase the bonding force between the coating and the substrate [27]. In Figure 3b, the MAO/MoS_2 cross-section is denser, thicker, and has fewer cracks than MAO. The reason for the denser and thicker coating was related to the addition of MoS_2 to increase the discharge voltage and deposition [28]. The cross-sectional line sweeps of MAO and MAO/MoS_2 presented in Figure 3(a_1) and Figure 3(b_1), respectively, show a consistent rise and fall in the O and Al contents, respectively. The Si content gradually decreased from the outside to the inside, whereas the P content exhibited the opposite distribution, a phenomenon that was closely related to the pretreatment of the samples. Pretreatment with the phosphate electrolyte promoted the rapid formation of a dense ceramic film on the substrate surface. This step had an important influence on the initial formation stage of the coating, enhanced the bonding of the film layer with the substrate, and improved the densification of the coating. As shown in Figure 3c, the HA cross-section had a small number of holes and cracks, with no obvious cracks at the interface, and the contour line of the coating surface had a small undulation. As indicated by the three cross-sections, these coatings bonded well with the substrate. Table 6 summarizes the characteristics of the three coatings.

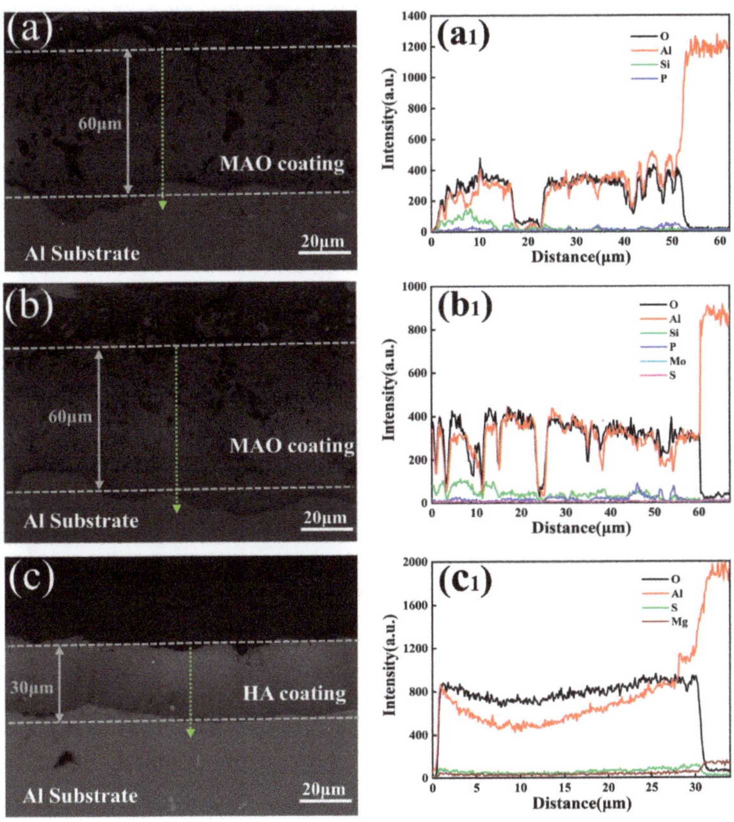

Figure 3. Cross-sectional morphology of the coatings and the corresponding elemental line sweeps: (**a**) MAO coating, (**b**) MAO/MoS_2 coating, and (**c**) HA coating. (**a_1–c_1** represents the corresponding element line).

Table 6. Characteristic properties of the three coatings.

Sample	Thickness/μm	Porosity/%
MAO	60 ± 2.52	10.221
MAO/MoS_2	60 ± 3.79	5.638
HA	30 ± 1.12	2.773

The phase structure of the elements in the coating was analyzed, and the results are presented in Figure 4. As shown, the three coatings can be detected in the high-intensity Al peaks, indicating that the high-intensity X-rays passed through the thick coating into the substrate. The diffraction peak shapes of the HA coating and the 7A04 Al alloy can only be observed in the Al peaks, indicating that the hard anodization of the oxide film layer after the generation of Al_2O_3 resulted in an amorphous phase. For the MAO and MAO/MoS_2 coatings, the common phases of α-Al_2O_3 and γ-Al_2O_3 were detected, of which the main phase was γ-Al_2O_3. The X-ray diffraction (XRD) pattern of the MAO/MoS_2 coatings exhibited an additional α-Al_2O_3 peak and a mullite peak. This formation was due to the gradual isolation of the initially formed dense layer from the electrolyte as the oxidation time increased, which reduced the cooling rate. The internal γ-Al_2O_3 was gradually converted into α-Al_2O_3 under the effect of arc heat [29]. Meanwhile, the growth of MAO coatings in the silicate electrolyte was mainly manifested by the deposition of SiO_2, which was accompanied by the mild oxidation of the Al substrate. With an increase in the voltage, the generated Al_2O_3 and SiO_2 layers gradually melted and sintered, forming a mullite structure. For the samples coated with MAO/MoS_2, the MoS_2 peaks were located at 40° and 59°, confirming that the MoS_2 particles participated in the MAO reaction and were successfully doped into the coating.

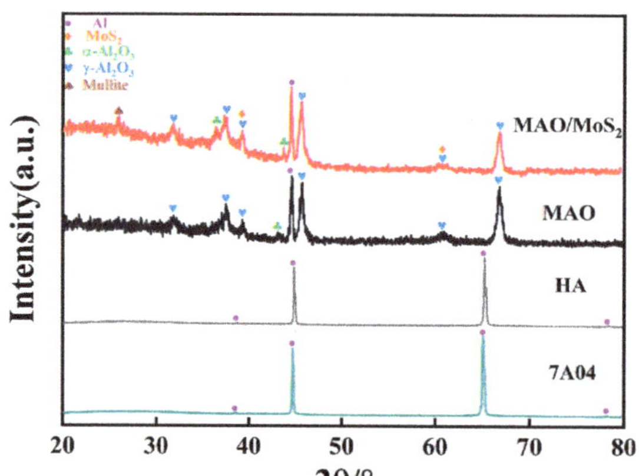

Figure 4. XRD plots of MAO, MAO/MoS_2, HA, and 7A04 aluminum alloys.

3.2. Mechanical Properties of the Coatings

Figure 5 shows the microhardness of the MAO, MAO/MoS_2, and HA coatings and the substrate. The hardness of the Al alloy was significantly increased after the formation of coatings on the surface. The hardness of the MAO coating was 483.7 HV, which was lower than that of the added MAO/MoS_2 coating. The average hardness of the MAO/MoS_2 coating was three times that of the 7A04 Al alloy. The HA coating had the highest average hardness, which was 4.1 times that of the substrate. The average hardness of the 7A04 Al alloy was the lowest (167.8 HV). The hardness of the coatings was affected by the phase composition and microstructure. The XRD results indicated that the MAO/MoS_2 coatings had a mullite phase and more α-Al_2O_3 phase, and the cross-sectional plots indicated that the MAO/MoS_2 coatings had a larger average thickness, which increased the average hardness of the coatings [30,31].

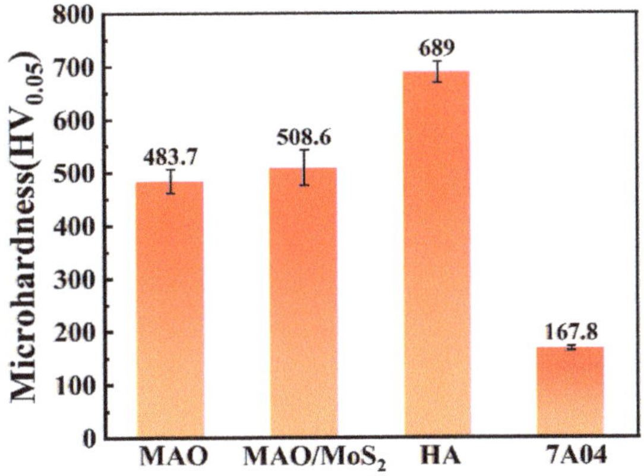

Figure 5. Hardness diagrams of MAO, MAO/MoS_2, HA, and 7A04 substrates.

The RST-300 instrument was used to determine the bond strength between the ceramic coatings and substrate. Figure 6a–d present the acoustic emission signal versus friction curves of the MAO, MAO/MoS_2, and HA coatings and the corresponding scratch morphologies. Owing to the ceramic particles in the MAO and HA coating layers and the influence of the surface quality, the testing process generated interfering signals that affected the accuracy of the critical loads [19,32]. Therefore, acoustic signals and friction curves were used in conjunction with scratch morphology maps (BSE images) to determine the critical load of the bonds (Lc_2). In the plots of the acoustic emission signals and friction profiles in Figure 6a–c, the friction profile of the MAO coating varied slowly, while the friction profile of the HA coating varied more rapidly. In Figure 6a,b, the MAO coating exhibited a few small peaks up to 70 N and started to exhibit significant peaks only above 70 N. Figure 6c shows a distinct peak corresponding to the HA coating acoustic signal at 25 N. To determine the exact Lc_2 value, the scratch morphology was examined using SEM. Figure 6d shows that the scratches are mainly divided into two parts: the grey ceramic layer in the first half of the abrasion mark and the silver substrate in the second half. In Figure 6d, the HA coating exhibited scratches in the grey–silver junction (2.3 mm; that is, a pro-bonding force of approximately 43 N). The acoustic signal at approximately 30 N appeared as an interfering signal that may be caused by the stripping of the membrane layer. The results indicate that the synthesis can be performed when the HA coating Lc_2 is 43 N. The acoustic signal bursts of the MAO and MAO/MoS_2 coating film layer failure matched the interface junction of the scratch micrograph in Figure 6d; thus, it was determined that the Lc_2 values of the MAO and MAO/MoS_2 coatings were 73 and 78 N, respectively. Observation of the edges of the scratched tracks did not reveal large-scale flaking. The resulting debris was scattered within the abrasion marks, and the coatings all exhibited good bonding strength.

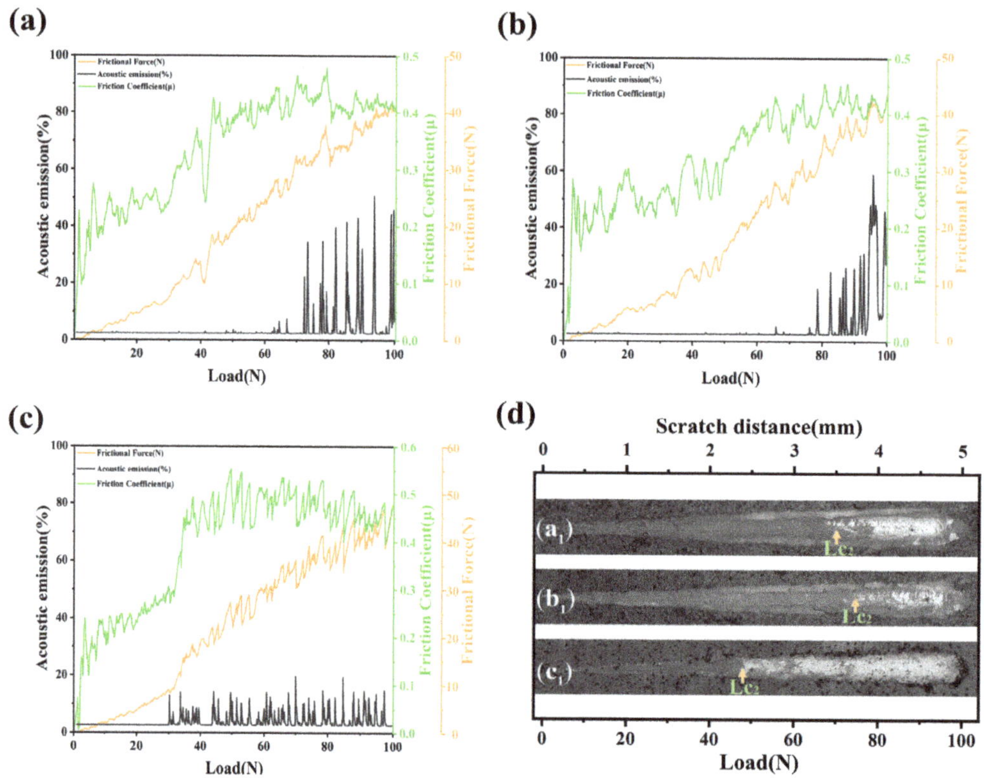

Figure 6. (**a**–**c**) Acoustic emission and friction profiles of MAO, MAO/MoS_2 and HA. (**d**) Scanning electron micrographs (BSE images) of the scratch trajectory (**a**$_1$–**c**$_1$ represents the corresponding BSE images).

3.3. Tribological Properties of the Coatings

The CoF versus sliding time curves for the MAO, MAO/MoS_2, and HA coatings and the 7A04 substrate under an applied load of 300 N are presented in Figure 7. As shown in Figure 7a, the initial CoF of the MAO coating was approximately 0.14. The CoF then decreased rapidly to 0.9, increased again, and gradually stabilized between 0.07 and 0.08 in 400 s. The fluctuation in the CoF between 50 and 200 s may have been due to the presence of large abrasive particles between the contact surfaces during the wear process. As the wear process proceeded, these abrasive particles were continuously extruded under a high contact pressure to form small particles that could eventually be stably embedded and stored between the contact surfaces. This dynamic change led to a corresponding fluctuation in the CoF. In Figure 7b, the MAO/MoS_2 and MAO coatings exhibited similar CoF change trends. The CoF at the beginning of the wear was approximately 0.13. After a relatively short period of time, to enter the stable wear stage, from the friction coefficient graphs, it could be observed in the wear periods of 20 and 100 s around the CoF fluctuation, which might be related to the abrasive particles, and then the stable wear stage was entered, and the CoF stabilized at approximately 0.5–0.6. Compared to the MAO coating, the addition of MoS_2 significantly reduced the wear time and resulted in a lower CoF in the stabilization stage, suggesting that the MoS_2 deposited on the MAO coating played a self-lubricating role. As indicated by Figure 7c, the CoF of the HA coating was 0.23 at the early stage of wear, and it stabilized at 0.12 after a short period of time and returned to 0.12 after a short fluctuation of approximately 700 s until the end of wear. Owing to the localized areas of the HA coating with bumps (e.g., Figure 7c), the oil film failed to cover the contact surface at the early stage of wear, resulting in a higher CoF at this stage. After the surface bumps were worn out, the surface morphology gradually became smooth, and the CoF stabilized. Figure 7d shows that the initial CoF of the Al alloy substrate was 0.15 and that the CoF decreased continuously until the end of wear.

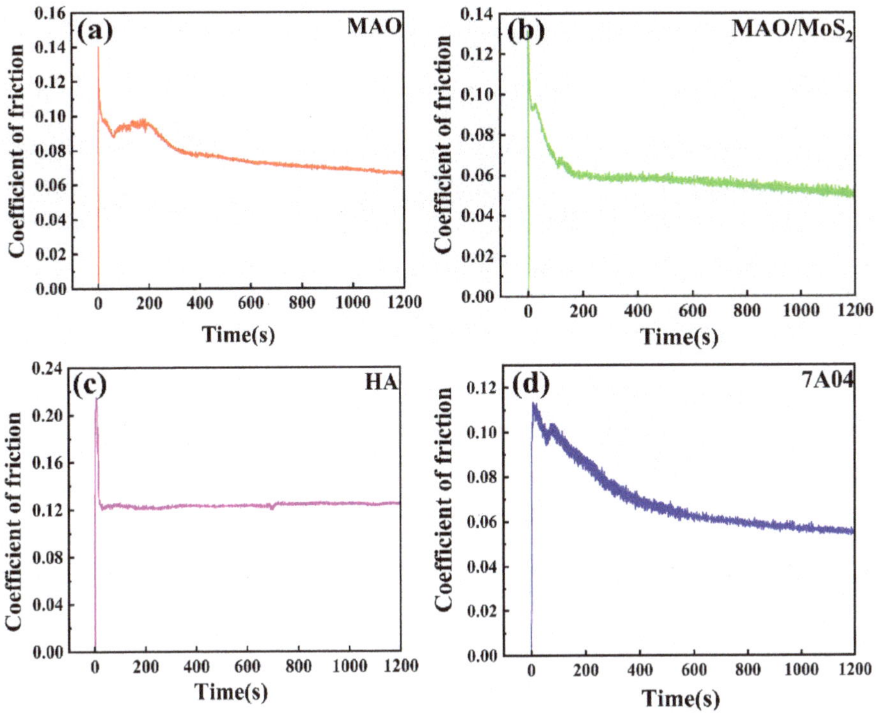

Figure 7. (**a–d**) The coefficient of friction versus time for MAO coating, MAO/MoS_2 coating, HA coating, and 7A04 substrate.

The wear rates and average CoFs of the MAO, MAO/MoS_2, and HA coatings and 7A04 substrate after wear are presented in Figure 8. In Figure 8a, the 7A04 substrate exhibited the highest wear rate, indicating that the coating had a protective effect on the substrate. The wear rate of MAO/MoS_2 was 13.73×10^{-6} mm^3/Nm lower than that of MAO (17.28×10^{-6} mm^3/Nm) owing to the self-lubricating effect of MoS_2. The HA coating exhibited the lowest wear rate. Notably, it also had the highest hardness and highest CoF owing to the generation of amorphous-phase aluminum oxide, which was relatively dense and had few cracks, thus exhibiting better wear resistance. The above experimental phenomena indicated that all the stabilized CoFs of the 7A04 Al alloy treated with MAO were low. Figure 8b shows that the average CoFs of the MAO, MAO/MoS_2, and HA coatings and the 7A04 substrate were 0.077, 0.059, 0.124, and 0.069, respectively. Owing to the porous nature of the MAO coating, the lubricant could be stored so that a stable oil film could be formed on the contact surface to protect the surface. The lowest CoF for the MAO/MoS_2 coating suggested that the MoS_2 slowed the increase in the roughness of the coating surface and formed a more stable oil film on the contact surface, resulting in a lower CoF, which reduced the wear rate [33,34].

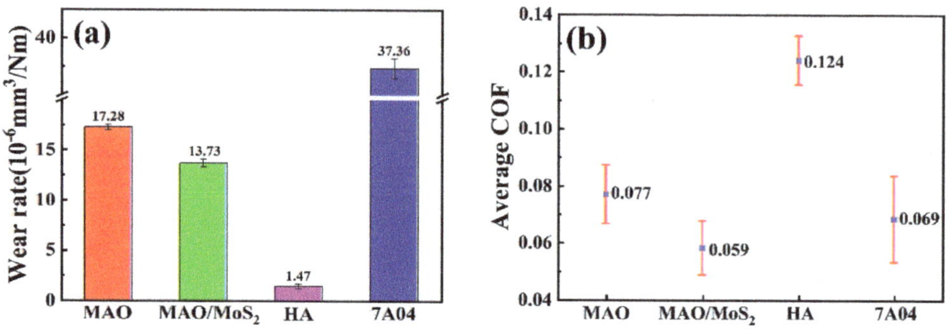

Figure 8. (**a**) Wear rate plot. (**b**) Average coefficient of friction plot for MAO coating, MAO/MoS_2 coating, HA coating, and 7A04 substrate after wear.

Figure 9a–d show the 3D morphology of the abrasion marks of the MAO, MAO/MoS_2, and HA coatings and the 7A04 substrate and the corresponding cross-sectional curves (Figure 9(a_1–d_1)). To examine the CoF changes during the break-in period, the surface and substrate roughness of the three coatings were measured using LCM, yielding the following results: MAO, 4.67 µm; MAO/MoS_2, 7.27 µm; HA, 2.06 µm; and 7A04 Al alloy, 0.977 µm. As shown, the widths of the abrasion marks of the MAO and MAO/MoS_2 coatings were divided into 4845.8 and 4252.5 µm, the HA coating had the narrowest abrasion mark (2004.4 µm), and the 7A04 substrate had the widest abrasion mark (4845.8 µm). In the cross-sectional diagrams (Figure 9(a_1–d_1)), the abrasion depth of the MAO coating was 95 µm, the abrasion depth of the MAO/MoS_2 coating was smaller (88 µm), and the HA coating had the smallest abrasion depth (18 µm). The maximum depth of the abrasion marks on the 7A04 substrate was 168 µm. The results indicated that all the coatings played a protective role with regards to the abrasion of the 7A04 substrate, and the wear resistance of the 7A04 Al alloy was increased. The HA coating exhibited the highest abrasion resistance, which was correlated with its surface roughness and densification.

3.4. Wear Surface Morphology Analysis

To investigate the wear mechanism of the coating, SEM characterization and the corresponding elemental analysis of the micromorphology of the abraded surface were performed, as shown in Figure 10. Figure 10a presents the grooves and black flaky bonded lumps on the wear scar surface. In Figure 10b, there were inconspicuous grooves and numerous black flaky bonding blocks on the surface of the abrasion marks, with localized coating detachment and obvious grooves on the substrate surface. Figure 10c shows the abrasion scar surface with evident grooves and a few black bonding blocks. Figure 10d shows the wear scar surface with obvious grooves, flaking, and cracks. In the EDS diagrams of the wear, the black bonding masses on the wear scar surfaces corresponded to the element C, which came from the DLC coating on the dyad. From the O elemental spectra, obvious parallel grooves were observed between the HA coating and the substrate. The grooves on all the wear marks were parallel to the sliding direction, indicating that no uneven load distribution existed during the wear process. According to the aforementioned characterization results for the wear marks, the main wear mechanism of all the specimens was a combination of abrasive and adhesive wears.

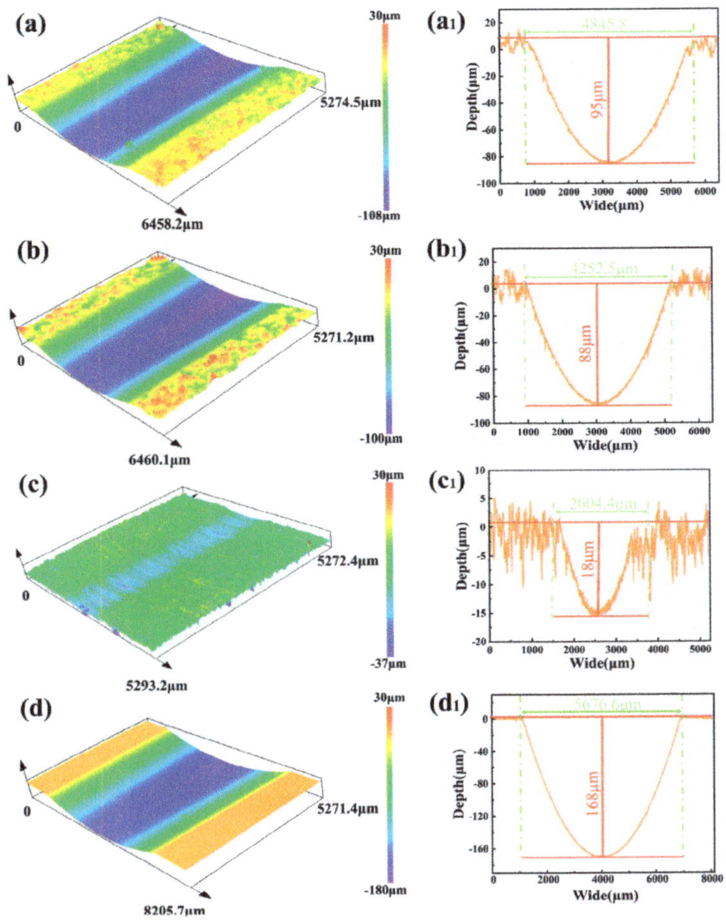

Figure 9. Three-dimensional morphology of the three coatings and the 7A04 substrate (**a–d**) and the corresponding cross-section curves (**a_1–d_1**): (**a,a_1**) MAO, (**b,b_1**) MAO/MoS_2, (**c,c_1**) HA, and (**d,d_1**) 7A04 substrate.

The wear morphology was further analyzed, and a comparison revealed that fewer grooves existed on the surface of the MAO coating, indicating that, in the early stage of ring wear, the MAO coating had a high roughness, direct contact between the friction partners occurred, and the actual contact involved a localized protrusion of the coating in contact with the opposing ring. Under a high load and cyclic shear stress, the hard debris wrapped around the coating was adsorbed on the surface of the coating with high hardness, which constituted three-body abrasive wear. When hard debris moved on the surface or inside the transfer film, it caused plastic deformation of the coating surface as it rolled. When the wear was further aggravated, the hard abrasive particles slid and cut the coating, resulting in deep furrows on the surface of the abrasion marks of the coating. As shown in Figure 10b, the surface of the abrasion marks had more transfer film and fewer abrasion marks, and the surface roughness of the MAO coating increased with the addition of MoS_2. Owing to its uneven surface properties, the contact area between the localized protrusions of the coating and the dyad was reduced during the initial stage of friction, significantly increasing the pressure at the contact point. This uneven microscopic contact surface led to the conversion of mechanical energy into thermal energy at the contact point, which increased the temperature of the contact area. The increased pressure and temperature of the contact surface may intensify the graphitization process on the DLC coating surface, increasing the tendency for adhesive wear to occur during the wear process. Jin et al. [35] also reported that severe adhesive wear occurred on the grinding balls with MAO/DLC coatings. Its surface energy spectrum indicated the presence of S, suggesting that MoS_2 played a lubricating role in the friction process and that forming a protective lubricating film with the dyadic parts under the action of shear force was easy, which played a lubricating role. This was reflected in the minimum CoF throughout the friction process (e.g., Figure 7b), and the surface of the wear marks had fewer grooves. Additionally, localized coating flaking was observed (Figure 10b). This may be due to microdefects in the MAO coating, which were prone to form crack sources under high loads. During the sliding test, these crack sources expanded, and spalling occurred under periodically applied high loads. In Figure 10c,d, the abrasion patterns were mainly plow grooves parallel to the sliding direction and a small amount of transfer film. The debris generated during the abrasion process acted on the contact surface to form the main abrasion patterns. A small amount of C was detected via EDS, as shown in Figure 10c, indicating that the stable wear phase accounted for a major part of the wear process.

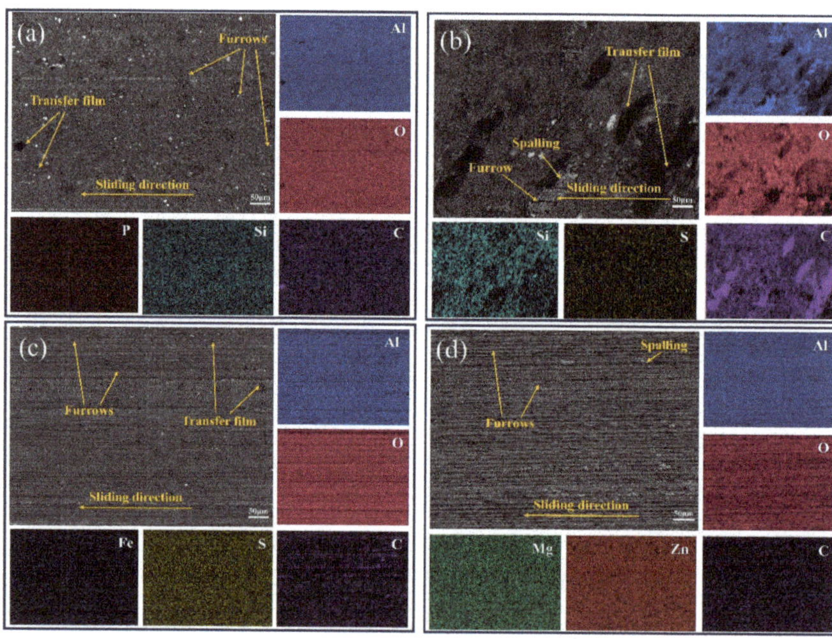

Figure 10. SEM images at the coating wear marks and their elemental scans of (**a**) MAO coating, (**b**) MAO/MoS_2 coating, (**c**) HA coating, and (**d**) 7A04 substrate.

Raman spectroscopy was performed on the wear marks to further analyze the influence of the friction process. Figure 11 shows the Raman spectra of the MAO-, MAO/MoS_2-, and HA-coated wear mark areas. A Gaussian function was fitted to the range of 400–2000 cm^{-1} to obtain the D peak associated with the graphite phase near 1360 cm^{-1} and the G peak associated with the sp^2 C–C bond near 1580 cm^{-1}. The variation in the I_D/I_G value can be used to analyze the effect of the transformation of DLC into a graphite structure during sliding in the friction process [36]. As shown in Figure 11d, the maximum I_D/I_G value was 1.08 for the MAO/MoS_2 coating. This indicated an increase in the transformation of the DLC ring surface in contact with the MAO/MoS_2 coating into the graphite structure. With an increase in the degree of graphitization, more C-based transfer films were generated, and the generated C-based transfer film entered the oil film and was stored together with the MoS_2 particles, which provided anti-friction and anti-wear effects.

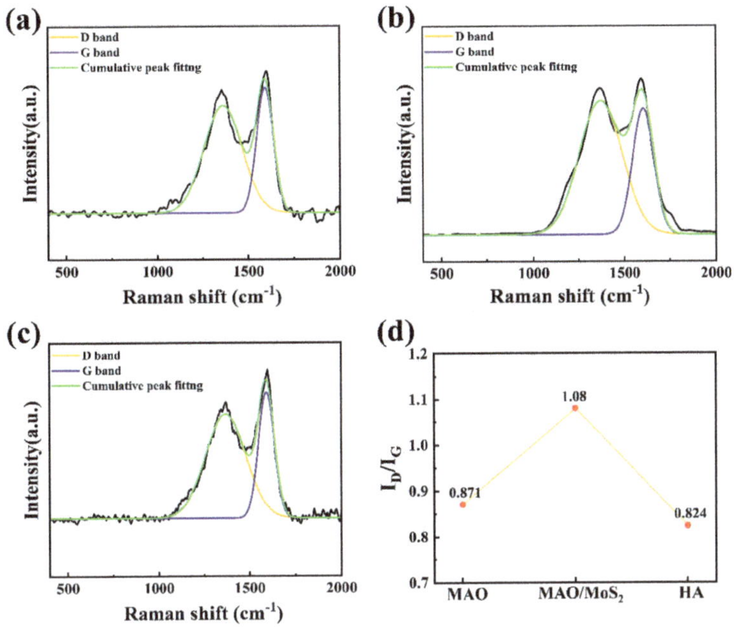

Figure 11. Raman spectra at the wear marks after the friction test: (**a**) MAO, (**b**) MAO/MoS_2, (**c**) HA, and (**d**) I_D/I_G.

MAO coatings can form thick oil films during wear owing to their inherent porosity. These oil films not only enhance the lubrication effect but also significantly improve the ability to capture exfoliated C-based transfer films [35]. In particular, larger flake transfer films are more easily captured and carried by oil films because of their larger surface areas and lower surface energies. As the wear process continues, the oil film decreases, making it difficult for large flake films to exit the system with the lubricant. Consequently, more C-based transfer films are observed to adhere to the wear surfaces of the MAO coatings, particularly the MAO/MoS_2 composite coating [37].

4. Conclusions

In this study, MAO, MAO/MoS_2, and HA coatings were prepared on the surface of the 7A04 Al alloy. The microstructures of the three coatings, as well as their tribological properties and wear mechanisms under oil lubrication conditions with DLC ring–block wear, were investigated. The following conclusions were drawn.

(1) The surface of the 7A04 Al alloy was effectively modified by the MAO, MAO/MoS_2, and HA treatments, which significantly improved its hardness and wear resistance.

(2) Among the coatings tested, the MAO/MoS_2 coating exhibited the lowest CoF, with superior anti-friction performance, owing to the self-lubricating properties of MoS_2. Meanwhile, the HA coating exhibited superior wear resistance owing to the dense structure of the amorphous-phase aluminum oxide.

(3) Under oil lubrication conditions, the main wear mechanisms of the MAO and MAO/MoS_2 coatings were abrasive and adhesive wear, whereas the wear mechanisms of the HA coatings and 7A04 Al alloy were dominated by abrasive wear, accompanied by slight adhesive wear. The surface micropores of MAO coatings store oil, which can replenish the lubricating film of the liquid and impede the movement of wear particles.

In addition, the MoS_2 particles and the solid lubrication provided by the graphitized layer work together to achieve a synergistic solid–liquid lubrication effect with the liquid lubrication film, optimizing the tribological properties.

Author Contributions: Conceptualization, Methodology, Investigation, and Data Curation, X.M. and W.Z.; Writing—Original Draft, X.M.; Software, S.W. and K.P.; Conceptualization and Writing—Review and Editing, X.W. (Xiaodong Wang), T.J., X.W. (Xiran Wang), C.W., and C.C.; Methodology, Study Design, and Writing—Review and Editing, F.M., Z.Q., J.X., and C.Z. All authors have read and agreed to the published version of the manuscript.

Funding: This project was supported by the State Key Lab of Advanced Metals and Materials (2022-Z17) and Frontier Exploration Projects of Longmen Laboratory (No. LMQYTSKT011).

Data Availability Statement: Data are contained within the article.

Acknowledgments: We wish to take this opportunity to thank the Provincial and Ministerial Coconstruction of Collaborative Innovation Center for Non-ferrous Metal New Materials and Advanced Processing Technology for their support.

Conflicts of Interest: The authors declare no conflict of interest.

References

1. Li, S.S.; Yue, X.; Li, Q.Y.; Peng, H.L.; Dong, B.X.; Liu, T.S.; Yang, H.Y.; Fan, J.; Shu, S.L.; Qiu, F.; et al. Development and applications of aluminum alloys for aerospace industry. *J. Mater. Res. Technol.* **2023**, *27*, 944–983. [CrossRef]
2. Hirsch, J. Recent development in aluminium for automotive applications. *Trans. Nonferrous Met. Soc. China* **2014**, *24*, 1995–2002. [CrossRef]
3. Song, Y.F.; Ding, X.F.; Xiao, L.R.; Zhao, X.J.; Cai, Z.Y.; Guo, L.; Li, Y.W.; Zheng, Z.Z. Effects of two-stage aging on the dimensional stability of Al-Cu-Mg alloy. *J. Alloys Compd.* **2017**, *701*, 508–514. [CrossRef]
4. Heinz, A.; Haszler, A.; Keidel, C.; Moldenhauer, S.; Benedictus, R.; Miller, W.S. Recent development in aluminium alloys for aerospace applications. *Mater. Sci. Eng. A* **2000**, *280*, 102–107. [CrossRef]
5. Vlach, M.; Cizek, J.; Kodetova, V.; Leibner, M.; Cieslar, M.; Harcuba, P.; Bajtosova, L.; Kudrnova, H.; Vlasak, T.; Neubert, V.; et al. Phase transformations in novel hot-deformed Al–Zn–Mg–Cu–Si–Mn–Fe(–Sc–Zr) alloys. *Mater. Des.* **2020**, *193*, 108821. [CrossRef]
6. Mändl, S.; Manova, D. Modification of metals by plasma immersion ion implantation. *Surf. Coat. Technol.* **2019**, *365*, 83–93. [CrossRef]
7. Xiong, L.; Wang, C.; Wu, W.; Xu, L.; Wang, C.; Deng, H.; Cheung, C.F. The surface softening mechanism of AlN ceramic by laser treatment. *Surf. Interfaces* **2024**, *46*, 104023. [CrossRef]
8. Fockaert, L.I.; Pletincx, S.; Boelen, B.; Hauffman, T.; Terryn, H.; Mol, J.M.C. Effect of zirconium-based conversion treatments of zinc, aluminium and magnesium on the chemisorption of ester-functionalized molecules. *Appl. Surf. Sci.* **2020**, *508*, 145199. [CrossRef]
9. Sun, Y. The use of aluminum alloys in structures: Review and outlook. *Structures* **2023**, *57*, 105290. [CrossRef]
10. Dai, Y.; Yan, L.; Hao, J. Review on Micro-Alloying and Preparation Method of 7xxx Series Aluminum Alloys: Progresses and Prospects. *Materials* **2022**, *15*, 1216. [CrossRef]
11. Kikuchi, T.; Takenaga, A.; Natsui, S.; Suzuki, R.O. Advanced hard anodic alumina coatings via etidronic acid anodizing. *Surf. Coat. Technol.* **2017**, *326*, 72–78. [CrossRef]
12. Mora-Sanchez, H.; del Olmo, R.; Rams, J.; Torres, B.; Mohedano, M.; Matykina, E.; Arrabal, R. Hard Anodizing and Plasma Electrolytic Oxidation of an Additively Manufactured Al-Si alloy. *Surf. Coat. Technol.* **2021**, *420*, 127339. [CrossRef]
13. Zhang, J.; Dai, W.; Wang, X.; Wang, Y.; Yue, H.; Li, Q.; Yang, X.; Guo, C.; Li, C. Micro-arc oxidation of Al alloys: Mechanism, microstructure, surface properties, and fatigue damage behavior. *J. Mater. Res. Technol.* **2023**, *23*, 4307–4333. [CrossRef]
14. Yerokhin, A.L.; Nie, X.; Leyland, A.; Matthews, A.; Dowey, S.J. Plasma electrolysis for surface engineering. *Surf. Coat. Technol.* **1999**, *122*, 73–93. [CrossRef]
15. Hongmei, X.; Bin, J.; Changping, T.; Jiahong, D.; Xin, Q.; Fusheng, P. Experimental evaluation of MoS_2 nanosheets as lubricant additive in different types of base oil for magnesium alloy/steel pairs. *Mater. Res. Express* **2019**, *6*, 066575. [CrossRef]
16. Mutyala, K.C.; Singh, H.; Fouts, J.A.; Evans, R.D.; Doll, G.L. Influence of MoS2 on the Rolling Contact Performance of Bearing Steels in Boundary Lubrication: A Different Approach. *Tribol. Lett.* **2016**, *61*, 20. [CrossRef]
17. Lv, X.; Zou, G.; Ling, K.; Yang, W.; Mo, Q.; Li, W. Tribological properties of MAO/MoS_2 self-lubricating composite coating by microarc oxidation and hydrothermal reaction. *Surf. Coat. Technol.* **2021**, *406*, 126630. [CrossRef]
18. Chen, Q.; Jiang, Z.; Tang, S.; Dong, W.; Tong, Q.; Li, W. Influence of graphene particles on the micro-arc oxidation behaviors of 6063 aluminum alloy and the coating properties. *Appl. Surf. Sci.* **2017**, *423*, 939–950. [CrossRef]
19. Li, Z.; Di, S. Preparation and properties of micro-arc oxidation self-lubricating composite coatings containing paraffin. *J. Alloys Compd.* **2017**, *719*, 1–14. [CrossRef]

20. Gecu, R.; Yurekturk, Y.; Tekoglu, E.; Muhaffel, F.; Karaaslan, A. Improving wear resistance of 304 stainless steel reinforced AA7075 aluminum matrix composite by micro-arc oxidation. *Surf. Coat. Technol.* **2019**, *368*, 15–24. [CrossRef]
21. Fei, C.; Hai, Z.; Chen, C.; Yangjian, X. Study on the tribological performance of ceramic coatings on titanium alloy surfaces obtained through microarc oxidation. *Prog. Org. Coat.* **2009**, *64*, 264–267. [CrossRef]
22. Ouyang, J.-H.; Wang, Y.-H.; Liu, Z.-G.; Wang, Y.-M.; Wang, Y.-J. Preparation and high temperature tribological properties of microarc oxidation ceramic coatings formed on Ti 2 AlNb alloy. *Wear* **2015**, *330–331*, 239–249. [CrossRef]
23. Nie, Z.; Lu, H.; Liu, Q.; Chai, G.; Ding, Y.; Xu, G.; Guo, J. Effect of copper introduction on the properties of micro-arc oxidation coating on powder metallurgy aluminum disk. *Surf. Coat. Technol.* **2024**, *479*, 130520. [CrossRef]
24. Zhao, J.; Zhang, H.; Yang, X.; Gu, Y.; Liu, Y. Local Electrochemical Corrosion of 6061 Aluminum Alloy with Nano-SiO_2/MAO Composite Coating. *Materials* **2023**, *16*, 6721. [CrossRef] [PubMed]
25. Ma, C.; Cheng, D.; Zhu, X.; Yan, Z.; Fu, J.; Yu, J.; Liu, Z.; Yu, G.; Zheng, S. Investigation of a self-lubricating coating for diesel engine pistons, as produced by combined microarc oxidation and electrophoresis. *Wear* **2018**, *394–395*, 109–112. [CrossRef]
26. Zhai, D.; Qiu, T.; Shen, J.; Feng, K. Growth kinetics and mechanism of microarc oxidation coating on Ti−6Al−4V alloy in phosphate/silicate electrolyte. *Int. J. Miner. Metall. Mater.* **2022**, *29*, 1991–1999. [CrossRef]
27. Erfanifar, E.; Aliofkhazraei, M.; Nabavi, H.F.; Rouhaghdam, A.S. Growth kinetics and morphology of microarc oxidation coating on titanium. *Surf. Coat. Technol.* **2017**, *315*, 567–576. [CrossRef]
28. Li, W.; Yan, Z.; Shen, D.; Zhang, Z.; Yang, R. Microstructures and tribological properties of MoS2 overlayers on MAO Al alloy. *Tribol. Int.* **2023**, *181*, 108348. [CrossRef]
29. Jahnichen, T.; Carstens, S.; Franz, M.; Laufer, O.; Wenzel, M.; Matysik, J.; Enke, D. Towards High Surface Area alpha-Al_2O_3-Mn-Assisted Low Temperature Transformation. *Materials* **2023**, *16*, 3047. [CrossRef]
30. krishna, L.R.; Purnima, A.S.; Wasekar, N.P.; Sundararajan, G. Kinetics and Properties of Micro Arc Oxidation Coatings Deposited on Commercial Al Alloys. *Metall. Mater. Trans. A* **2007**, *38*, 370–378. [CrossRef]
31. Li, H.; Kong, S.; Liu, Z.; Wang, Z.; Geng, Y. Hardness Distribution and Growth Behavior of Micro-Arc Oxide Ceramic Film with Positive and Negative Pulse Coordination. *Nanomaterials* **2024**, *14*, 842. [CrossRef] [PubMed]
32. Ma, H.; Gu, Y.; Liu, S.; Che, J.; Yang, D. Local corrosion behavior and model of micro-arc oxidation HA coating on AZ31 magnesium alloy. *Surf. Coat. Technol.* **2017**, *331*, 179–188. [CrossRef]
33. Zhang, D.; Du, X.; Bai, A.; Wang, L. The synergistic effect of MAO-treated and PAO–graphene oil on tribological properties of Ti6Al4V alloys. *Wear* **2022**, *510–511*, 204494. [CrossRef]
34. Wu, H.; Yin, S.; Du, Y.; Wang, L.; Wang, H. An investigation on the lubrication effectiveness of MoS_2 and BN layered materials as oil additives using block-on-ring tests. *Tribol. Int.* **2020**, *151*, 106516. [CrossRef]
35. Jin, L.; Li, Y.; Liu, C.; Fan, X.; Zhu, M. Friction mechanism of DLC/MAO wear-resistant coatings with porous surface texture constructed in-situ by micro-arc oxidation. *Surf. Coat. Technol.* **2023**, *473*, 130010. [CrossRef]
36. Rajak, D.K.; Kumar, A.; Behera, A.; Menezes, P.L. Diamond-Like Carbon (DLC) Coatings: Classification, Properties, and Applications. *Appl. Sci.* **2021**, *11*, 4445. [CrossRef]
37. Liu, Y.; Erdemir, A.; Meletis, E.I. A study of the wear mechanism of diamond-like carbon films. *Surf. Coat. Technol.* **1996**, *82*, 48–56. [CrossRef]

lubricants

MDPI

Article

Influence of Molybdenum Addition on the Structure, Mechanical Properties, and Cutting Performance of AlTiN Coatings

Tao Yang [1,2,3], Jun Yin [1,4], Puyou Ying [1,2,3,*], Changhong Lin [1,2,3], Ping Zhang [1,2,3], Jianbo Wu [1,2,3], Alexander Kovalev [1,2,3], Min Huang [5], Tianle Wang [1,3], Andrei Y. Grigoriev [6], Dmitri M. Gutsev [6] and Vladimir Levchenko [1,2,3,*]

1 International Joint Institute of Advanced Coating Technology, Taizhou University, Taizhou 318000, China; yangtaotzc@tzc.edu.cn (T.Y.); yj15240271225@163.com (J.Y.); lin201191@163.com (C.L.); zhangp03@tzc.edu.cn (P.Z.); wujb@tzc.edu.cn (J.W.); alexander.v.kovalev@gmail.com (A.K.); wtl0203@tzc.edu.cn (T.W.)
2 Wenling Research Institute of Taizhou University, Taizhou 318000, China
3 Zhejiang Provincial Key Laboratory for Cutting Tools, Taizhou University, Taizhou 318000, China
4 College of Biological and Chemical Engineering, Zhejiang University of Science and Technology, Hangzhou 310000, China
5 Civil Aviation Department, Zhejiang Institute of Communication, Hangzhou 311112, China; mhuang@tzc.edu.cn
6 Metal-Polymer Research Institute, National Academy of Sciences of Belarus, 246050 Gomel, Belarus; aygri@tut.by (A.Y.G.); gucevd@muai.ru (D.M.G.)
* Correspondence: ypu@tzc.edu.cn (P.Y.); vladlev@yahoo.com (V.L.)

Abstract: Though AlTiN coating has been intensively studied, there is still a need to develop AlTiN coating to meet the growing demand of industrial machining. One effective way to improve the performance of AlTiN coating is by adding alloying elements. In this study, AlTiN and AlTiMo coatings were deposited using multi-arc ion plating to investigate the influence of molybdenum addition on the structure, mechanical properties, and cutting performance of AlTiN coatings. Spherical droplets formed on the surfaces of both coatings, with the AlTiMoN coating exhibiting more surface defects than the AlTiN coating. The grazing incidence X-ray diffraction results revealed the formation of an (Al,Ti)N phase formed in the AlTiN and AlTiMoN coatings. Molybdenum doping in the AlTiMoN coating slightly reduced the grain size. Both coatings exhibited excellent adhesion to the substrate. The hardness (H), elastic moduli (E), H/E, and H^3/E^2 ratios of the AlTiMoN coating were higher than those of the AlTiN coating. The improvement in the mechanical properties was attributed to grain refinement and solution strengthening. Molybdenum doping improved the tribological properties and cutting performance of the AlTiN coatings, which was ascribed to the formation of MoO_3 as a solid lubricant. These results show a path to increase the performance of AlTiN coating through molybdenum addition and provide ideas for the application of AlTiMoN coatings for cutting tools.

Keywords: AlTiN and AlTiMoN coatings; mechanical properties; tribological properties; cutting performance

1. Introduction

Metal nitride coatings have been widely used as hard protective coatings in various industries, especially in machining and forming processes. Among them, TiN coatings were developed as the first generation of metal nitride coatings for protecting various tool materials. However, TiN coatings have been gradually replaced because of their limited mechanical and tribological properties [1]. Furthermore, TiN coatings are unstable and rapidly oxidize above 500 °C [2].

Combining TiN coatings with Al significantly improves the hardness, wear resistance, thermal stability, and oxidation resistance of AlTiN coatings [3]. The improved hardness of AlTiN coatings is attributed to an increase in the internal strain of the lattice owing to the substitution of Ti atoms with Al atoms [1]. AlTiN coatings exhibit hardnesses of up to 30 GPa [4]. AlTiN coatings exhibit more effective wear and oxidation resistance than TiN coatings because an Al_2O_3 top layer is formed during friction, which prevents further oxidation and reduces the coefficient of friction (COF) [5]. AlTiN coatings remain stable in the single-phase cubic structure at 800 °C/1 h with only a 250-nanometer-thick surface layer oxidation [6], indicating that the coating has high thermal stability and is suitable for high-speed cutting operations.

AlTiN wear-resistant coatings exhibit serious disadvantages, namely a high COF exceeding 0.7 at room and high temperatures [7]. High COF increases the tool temperature and accelerates the wear process during cutting. The mechanical and tribological properties and cutting performance of AlTiN coatings must be further improved. The development of AlTiN coatings via the addition of alloying elements has gained extensive attention [8,9].

Molybdenum is a promising alloying element for AlTiN coatings because its introduction improves the mechanical performance, tribological properties, and oxidation resistance of AlTiN coatings [10–14]. In particular, the inclusion of Mo in AlTiN coatings results in the formation of a MoO_3 phase during friction. MoO_3, known as the Magnéli phase, is regarded as a solid–lubricant oxide [15]. MoO_3 is composed of layers bonded by van der Waals forces that easily deform under shearing force, effectively reducing the COF [16–18]. These coatings have attracted extensive attention owing to their excellent comprehensive properties. Tavares investigated the influence of the N_2 flow, Ar partial pressure, and bias voltage on the structure, mechanical properties, and thermal stability of multilayered AlTiMoN coatings [19–24]. Thereafter, several studies have been conducted on the effect of the deposition parameters (bias voltage [17,25], N_2 pressure [17,25], and rotation speed [18,26–28]) on the structure and properties (smoothness, hardness, fracture toughness, wear behavior, and tribological properties) of AlTiMoN coatings. Several studies investigated the influence of the Mo content on the microstructure, wettability, mechanical performance, corrosion resistance, thermal stability, and tribological properties of the AlTiMoN coating system. However, few studies have focused on the cutting performance of AlTiMoN coatings [10,12,13,16,29–34]. AlTiN-based hard coatings are often used as protective coatings on cutting tools; therefore, it is necessary to study the cutting performance of AlTiMoN coatings.

AlTiN and AlTiMoN coatings were prepared in this study using multi-arc ion plating, and the effect of Mo addition on the structure, mechanical properties, and cutting performance of the AlTiN coatings was investigated. This study aimed to develop an AlTiMoN coating for cutting tools and provide ideas for the application of AlTiMoN coatings prepared using multi-arc ion plating in machining and forming applications.

2. Materials and Methods

AlTiN and AlTiMoN coatings were deposited on two different substrates, namely YG8 cemented carbide sheets (16 mm × 16 mm × 4.5 mm) and YT15 carbide blades. The coatings prepared on the carbide sheets were used for structural characterization and mechanical performance testing, while the coatings prepared on the carbide blades were employed for evaluating the cutting performance. The substrates were ultrasonically cleaned before use. Ultrasonic cleaning was first performed in an acetone solution at 45 °C for 20 min, followed by ultrasonic cleaning in anhydrous ethanol at 45 °C for 20 min.

The coatings were deposited using a multi-arc ion plating apparatus (SP-0806ASI, Beijing Powertech Technology Co., Ltd., Beijing, China). Three Φ100 mm × 30 mm cylindrical targets, namely Ti, $Al_{67}Ti_{33}$, and $Al_{60}Ti_{30}Mo_{10}$ targets, with 99% purity were employed.

Before deposition, the temperature and pressure of the vacuum chamber were set to 400 °C and 9×10^{-3} Pa, respectively. The substrate and sample holder rotations were initiated, and the rotation speeds were set to 2 and 6 rpm, respectively. Argon (50 sccm) was injected into the cavity, and the substrate was etched using Ti plasma for 5 min to eliminate pollutants. The bias voltage, arc current, frequency, and pulse width of the etching process were −800 V, 60 A, 20 KHz, and 10 μs, respectively. After etching, a Ti interlayer was deposited at a bias voltage, arc current, frequency, and pulse width of −400 V, 60 A, 10 KHz, and 40 μs, respectively. The Ti layer deposition time was set to 30 min. Finally, AlTiN and AlTiMoN coatings were deposited at a bias voltage, arc current, frequency, and pulse width of −150 V, 60 A, 75 KHz, and 10 μs, respectively. The N_2/Ar gas flow rates for the AlTiN and AlTiMoN coatings were 320/20 and 200/20 sccm, respectively. Both coatings were deposited for 180 min.

The hardness (H) and elastic moduli (E) of the AlTiN and AlTiMoN coatings were evaluated using a nano-indenter (Nano Test Vantage, Micro Materials, Wrexham, UK). The international standard ISO 14577 was adopted [35]. The hardness evaluations were performed using a Berkovich diamond indenter with a maximum indentation load of 30 mN. The loading, peak load holding, and unloading times were 30 s, 60 s, and 30 s, respectively. The Oliver–Phar method was adopted to calculate the hardness and elastic moduli of the coatings [36]. Poisson's ratios of the indenter and coatings were 0.07 and 0.18, respectively. Ten indentations were performed on each sample to reduce measurement errors.

The friction and wear behaviors were tested using a reciprocating friction and wear testing machine (MFT-5000, Rtec Instruments, San Jose, CA, USA). The standard ASTM G137 was adopted [37]. Friction testing was conducted at 25 °C for 60 min at a relative humidity of approximately 60%. The tribotest was performed using a 6-milimeter-diameter Si_3N_4 ball. The applied load, sliding distance of a single stroke, and test frequency were 10 N, 5 mm, and 2 Hz, respectively. Two samples were used in the tribological test in order to reduce experimental error. After the tribotests, white-light interferometry (Lambda, Rtec Instruments, San Jose, CA, USA) was used to observe the wear. The wear rate (W) was calculated according to Ref. [38]:

$$W = V / (F \cdot L) \tag{1}$$

where V (mm^3) is the wear volume of the coating, F (N) is the applied load, and L (m) is the sliding distance. V was obtained by analyzing the wear tracks using MountainsMap Imaging Topography 9 software. Two wear tracks were selected to calculate the respective wear rates, and the average value was taken as the wear rate.

The coating adhesion was qualitatively investigated using Rockwell-C indentation according to the JB/T 11442-2013 [39]. The Rockwell-C indentation tests employed a load of 588 N (60 kgf) and a diamond cone indenter with an angle of 120°. Optical microscopy (ZEISS AXIO Scope.A1, Oberkochen, Germany) was used to obtain indentation images after the Rockwell-C indentation tests.

A Computer Numerical Control machine tool (CK6140S, Zhejiang Kaida Machine Tool Co., Ltd., Zhuji, China) was used for the cutting experiment according to the GB/T 16461-2016 [40]. The workpiece material was a No. 45 steel round rod with a diameter of 50 mm. The cutting parameters have a remarkable influence on cutting quality and surface finish, and the assignment of cutting parameters should be considered comprehensively. According to the recommendations of cutting tool manufacturer, existing industry practice, and research references, the common ranges of the cutting speed, cutting depth, and feed rate for steel are 40–250 m/min, 0.2–0.5 mm, and 0.1–0.3 mm/r. Therefore, within the range of the above process parameters, the cutting speed, cutting depth, and feed rate in this study were set at 80 m/min, 0.5 mm, and 0.25 mm/r, respectively. A dynamometer (9257 B) and thermal imager (A655SC) were employed to analyze the cutting force and cutting temperature, respectively.

Scanning electron microscopy (SEM; S4800, Hitachi, Tokyo, Japan) was used to examine the surface and cross-sectional morphologies of the coatings. The coating composition was analyzed using energy-dispersive X-ray spectroscopy (EDS). Three different regions were selected for EDS analysis, and the average value of the result was taken as the element component. The phase structures of the AlTiN and AlTiMoN coatings were characterized using grazing incidence X-ray diffraction (GIXRD, D8 Advance, Bruker, Saarbruecken, Germany) with Cu-Ka radiation (40 kV, 100 mA) in a grazing mode (0.1°). The scanning range, step size, and scanning speed of the GIXRD analyses were 20°~80°, 0.02°, and 0.03°/s, respectively. The crystallite sizes of the AlTiN and AlTiMoN coatings were calculated using the Scherrer formula (Equation (2)) [10]:

$$D = k\lambda / (\beta \cdot \theta) \quad (2)$$

where k is the shape factor (k = 0.94), λ is the X-ray wavelength (λ_{Cu} = 0.154 nm), β is the full width–half maximum, and θ is the Bragg angle.

3. Result and Discussion

3.1. Structural Characterization of the AlTiN and AlTiMoN Coatings

Figure 1 shows the surface and cross-sectional morphologies of the AlTiN and AlTiMoN coatings. As shown in Figure 1, spherical droplets [41] formed on the surfaces of both coatings. Figure 1a,b show that the number of droplets on the surface of the AlTiMoN coating was significantly higher than that on the surface of the AlTiN coating. Additionally, the droplets on the AlTiMoN-coated surface were coarse. These results indicate that the AlTiMoN coating had more surface defects than the AlTiN coating. This can be attributed to two factors. First, the thermal conductivity of Mo is higher than that of Ti (a higher thermal conductivity coefficient implies that the cooling of microdroplets is faster, which indicates that the probability of solidification of Mo droplets before contacting the surface is higher than that of Ti droplets [41]). Additionally, Mo has a higher melting point than Al and Ti; therefore, Mo is more difficult to ionize during deposition. Research suggests that, owing to fast evaporation during cathode ion plating, the un-ionized atoms of the target material reach the substrate surface and form droplet particles [26].

Figure 1. Scanning electron microscopy (SEM) images of the surface and cross-sectional morphologies of the (**a**,**c**) AlTiN and (**b**,**d**) AlTiMoN coatings.

Table 1 lists the elemental compositions of the AlTiN and AlTiMoN coatings determined using EDS; the Al, Ti, and N contents in the AlTiMoN coating, as compared with those in the AlTiN coating, were slightly reduced, and the Mo was 5.1%.

Table 1. Elemental composition of the AlTiN and AlTiMoN coatings (at.%).

	Al	Ti	Mo	N
AlTiN	27.4 ± 0.9	15.8 ± 0.7	—	56.8 ± 1.5
AlTiMoN	24.9 ± 1.0	14.2 ± 0.9	5.1 ± 0.6	55.8 ± 1.2

Figure 1c,d show that both coatings consisted of two layers, namely, an outer AlTiN or AlTiMoN layer and an inner Ti layer. The thicknesses of the AlTiN and AlTiMoN coatings were 3.41 and 2.75 μm, respectively. The thickness of the thicker AlTiN coating was expected owing to the higher total pressure in the chamber during the AlTiN coating deposition process. The probability of ions colliding with gas atoms increases with an increasing total pressure in the chamber; thus, the ion mobility, impinging energy, and resputtering efficiency decrease, increasing the deposition rate [42]. Both the AlTiN and AlTiMoN coatings exhibited typical columnar structures arranged perpendicular to the substrate. Belgroune reported that Mo addition of up to 16% in AlTiN coatings densified the coatings without the formation of a columnar structure and that the coating structure changed from zone II to zone T based on the Anders's structural zone model [30]. In this study, approximately 5.1% Mo was added to the coating, indicating that the low Mo addition used in this study cannot change the coating structure.

Figure 2 shows the GIXRD patterns of the AlTiN and AlTiMoN coatings. A diffraction peak was detected in both coatings at a 2θ of approximately 40°, which corresponds to the Ti layer in both coatings. Figure 2 also shows diffraction peaks of the substrate (mainly tungsten carbide). The remaining peaks were attributed to metallic nitrides, which were the main phase components. The peak positions of the (Al,Ti)N phase in the AlTiN coating were located at 37.1°, 43.3°, 63.0°, and 75.8°, which correspond to the (111), (200), (220), and (311) planes, respectively. The GIXRD pattern of the AlTiN coating shows that the (Al,Ti)N phase has a face-centered cubic (fcc) B1-NaCl structure. Compared to the standard TiN diffraction peaks, the peak positions of the (Al,Ti)N phase shifted by a small angle, which indicates that part of the Ti in the cubic TiN lattice was replaced by Al with a smaller atomic radius. This result was similar to the XRD results reported in Ref. [28]. The formation of the (Al,Ti)N phase was also observed in the AlTiMoN coating. The peaks of the (Al,Ti)N phase were located at 37.1°, 43.2°, 62.7°, and 75.4°, corresponding to the (111), (200), (220), and (311) planes, respectively. The addition of Mo to the AlTiMoN coating shifted the peak positions of the (Al,Ti)N phase to lower angles (Figure 2b), owing to the substitution of Mo atoms [12,30]. No peaks corresponding to pure metallic Mo were observed, which may be related to the low content of pure metallic Mo in the AlTiMoN coating. The diffraction angles of the corresponding peaks of the (Al,Ti)N and Mo_2N phases were close to each other, and it is difficult to distinguish from the GIXRD results whether the Mo_2N phase exists. The Mo_2N phase was found in the AlTiMoN coating with 8% Mo addition in some studies [17,25,34]. XPS may be needed to further determine whether the MoN_2 phase exists in the AlTiMoN coating with a 5% Mo addition in this study.

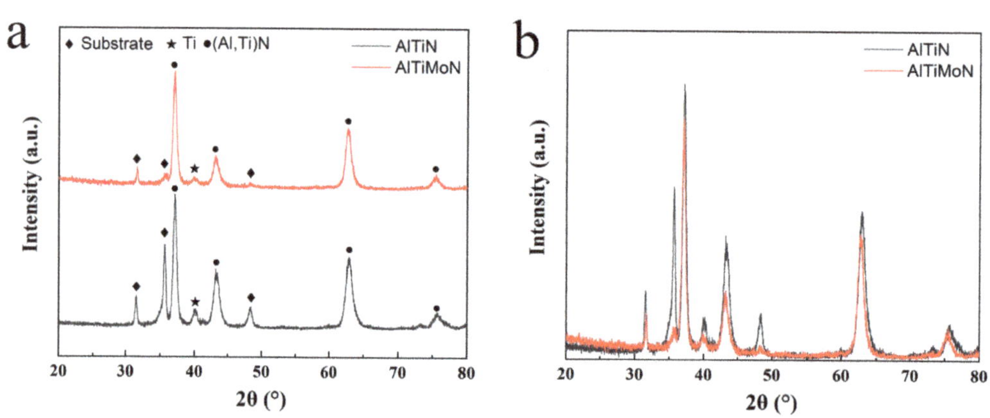

Figure 2. Grazing incidence X-ray diffraction (GIXRD) patterns of the AlTiN and AlTiMoN coatings. (**a**) is the GIXRD pattern of the two coatings for phase analysis, (**b**) is the GIXRD pattern of the two coatings after overlapping, which is used to compare the peak positions of the (Al,Ti)N phase.

Based on the Debye–Scherrer equation, the peaks of (Al,Ti)N (111), (200), (222), and (311) were selected, and the average grain sizes of the AlTiN and AlTiMoN coatings were calculated using GIXRD data analysis. The grain sizes of the AlTiN and AlTiMoN coatings were approximately 14 and 11 nm, respectively. Thus, the addition of Mo slightly reduced grain size, which is consistent with previously reported results [10,30]. Gao reported that the addition of 8% Mo reduced the grain size of AlTiN coatings from 30 to 28 nm [10]. Belgroune reported that the addition of 4% Mo to an AlTiN coating reduced the grain size of the coating from 58 to 55 nm [30].

3.2. Mechanical Properties of the AlTiN and AlTiMoN Coatings

The adhesion properties of the AlTiN and AlTiMoN coatings on the cemented carbide substrate were evaluated using Rockwell-C indentation. Figure 3 shows images of the indentations after the indentation tests. Only a few tiny cracks are visible around the indentation, as shown by the red arrow in Figure 3. No indication of spallation was observed. Radial cracking without chipping indicates that both coatings exhibited excellent adhesion (HF1 level) and that Mo did not significantly influence the adhesion properties of the AlTiN coatings. This may be because of the favorable effect of the deposited Ti interlayer on improving the adhesion properties between the coating and the substrate. The Ti interlayer can reduce the difference in the physical and mechanical properties between the nitride coating and substrate; therefore, the stress gradient at the interface is lowered [12]. Both coatings exhibited excellent adhesion to the substrate.

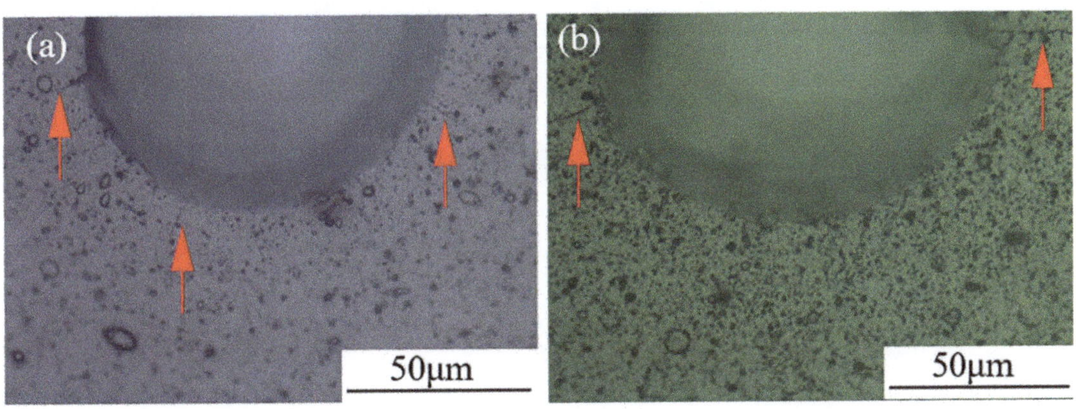

Figure 3. Optical micrographs of the (**a**) AlTiN and (**b**) AlTiMoN coatings after the Rockwell-C indentation tests. Red arrows indicate cracks.

Table 2 lists the H and E of the AlTiN and AlTiMoN coatings. The H and E of the AlTiN coating were 26 and 318 GPa, respectively. The H and E of the AlTiMoN coating increased to 30 and 341 GPa, respectively. The addition of Mo clearly increased the H and E values of the AlTiN coating. Furthermore, the AlTiMoN coating exhibited higher H/E and H^3/E^2 ratios.

Table 2. H and E of the AlTiN and AlTiMoN coatings.

Coating	Hardness (GPa)	Elastic Moduli (GPa)	H/E	H^3/E^2
AlTiN	26 ± 2	318 ± 9	0.082	0.174
AlTiMoN	30 ± 3	341 ± 12	0.088	0.232

The AlTiMoN coating exhibited a higher hardness than the AlTiN coating because (1) the grain size of the AlTiMoN coating is finer than that of the AlTiN coating; thus, the increase in hardness is related to grain refinement strengthening [30]; (2) the increased hardness is related to the increased solution-strengthening effect from the substitution of Ti atoms with Mo atoms [10,12].

The increase in the elastic modulus of the coating owing to the Mo addition can be attributed to the substitution of Mo atoms. The Ti atoms are replaced by Mo atoms, resulting in a smaller lattice. Smaller interatomic spaces result in greater binding forces, which facilitates improvements in the elastic modulus [43].

The values of H/E and H^3/E^2 indicate the wear resistance of coatings and can be used to predict their tribological behavior [44,45]. The H/E and H^3/E^2 ratios of the AlTiMoN coating with added Mo were higher than those of the AlTiN coating, which should result in better wear resistance and tribological behavior.

3.3. Tribological Properties of the AlTiN and AlTiMoN Coatings

Figure 4 shows the COFs of the AlTiN and AlTiMoN coatings under 10 N with a sliding duration of 3600 s, as well as the COF of the substrate. As shown in Figure 4, the COF of the substrate is stable and low at approximately 0.45. The COF of the AlTiMoN coating is stable and exhibits a progressively increasing trend. The COF of the AlTiMoN coating is approximately 0.60. The COF of the AlTiN coating exhibits two distinct stages, that is, it first increased and then decreased, and then finally stabilized. The first stage of the curve lasts for approximately 430 s, during which the COF is approximately 0.85. Subsequently, the COF decreases rapidly and then fluctuates at approximately 0.45. Notably, the second stage of the COF curve is similar to that of the substrate, which suggests the COF reduced from 0.85 to 0.45 because the AlTiN coating was abraded. Therefore, the COF of 0.85 is the COF of the AlTiN coating during the first stage, and the COF of 0.45 is the COF of the substrate during the second stage. The COF of the AlTiMoN coating is lower than that of the AlTiN coating.

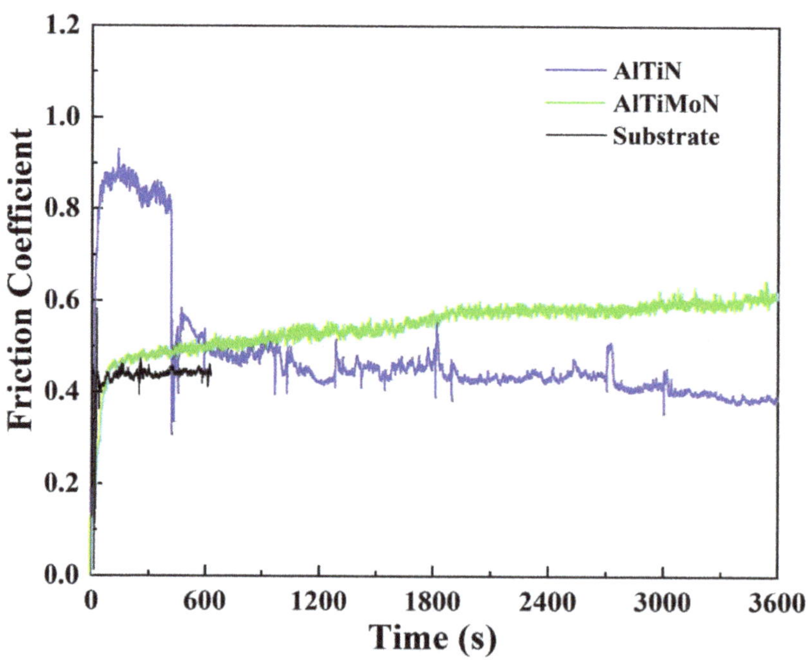

Figure 4. Coefficient of friction dynamics of the AlTiN and AlTiMoN coatings.

Figure 5 shows the SEM topographies of the wear tracks on the AlTiN and AlTiMoN coatings. The metal substrate is exposed in Figure 5a, indicating that the AlTiN coating was completely abraded. In contrast, the AlTiMoN coating in Figure 5b does not exhibit peeling. The worn surface of the AlTiMoN coating is smooth and clean, except for small holes observed on the wear track. Table 3 lists the elemental compositions of the wear tracks on the AlTiN and AlTiMoN coatings. Tungsten and Co, originating from the cemented carbide substrate, appeared in the wear track of the AlTiN coating, indicating that the AlTiN coating was completely abraded. However, no W or Co was found in the wear track of the AlTiMoN coating, indicating that the AlTiMoN coating was not completely abraded, which is consistent with the SEM observations shown in Figure 5.

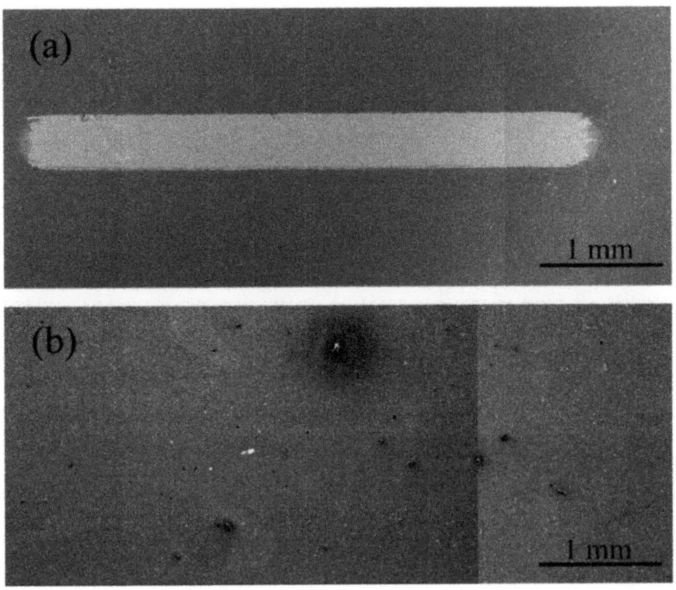

Figure 5. SEM topography of the wear tracks of the (**a**) AlTiN and (**b**) AlTiMoN coatings.

Table 3. Elemental composition of the wear tracks of the AlTiN and AlTiMoN coatings (at.%).

	Al	Ti	Mo	N	W	C	Co
AlTiN	—	—	—	—	36.1 ± 0.8	61.3 ± 1.3	2.6 ± 0.7
AlTiMoN	23.1 ± 0.6	14.8 ± 0.5	5.1 ± 0.5	57.0 ± 1.2	—	—	—

Figure 6 shows the wear rates of the AlTiN (5.24×10^{-6} $mm^3/(N \cdot m)$) and AlTiMoN coatings (1.67×10^{-6} $mm^3/(N \cdot m)$). The wear rate of the AlTiMoN coating is lower than that of the AlTiN coating. The results indicate that the addition of Mo increased the wear resistance of the AlTiN coating.

The heat generated by sliding friction can sufficiently increase the temperature of the contact surface for oxidation to occur. TiO_2 and Al_2O_3 phases were formed in the wear track during dry friction of the AlTiN coating. These phases have low ionic potentials, and the cations between them can strongly interact to form strong ionic or covalent bonds, making them difficult to shear [46]. Accordingly, the formation of an oxide layer or wear debris patches of such oxides does not provide any lubricious effect, resulting in a high COF [44]. Additionally, the TiO_2 and Al_2O_3 oxidation phases are hard and act as abrasive particles during the wear tests, causing abrasive wear [10]. Leyland and Tsui reported that lower H/E and H^3/E^2 values indicate diminished wear resistance [44,45]. Thus, the AlTiN coating failed due to abrasion during sliding, which caused the coating to delaminate, thereby exposing the substrate. The coating failed owing to the low H/E and H^3/E^2 values and high COF.

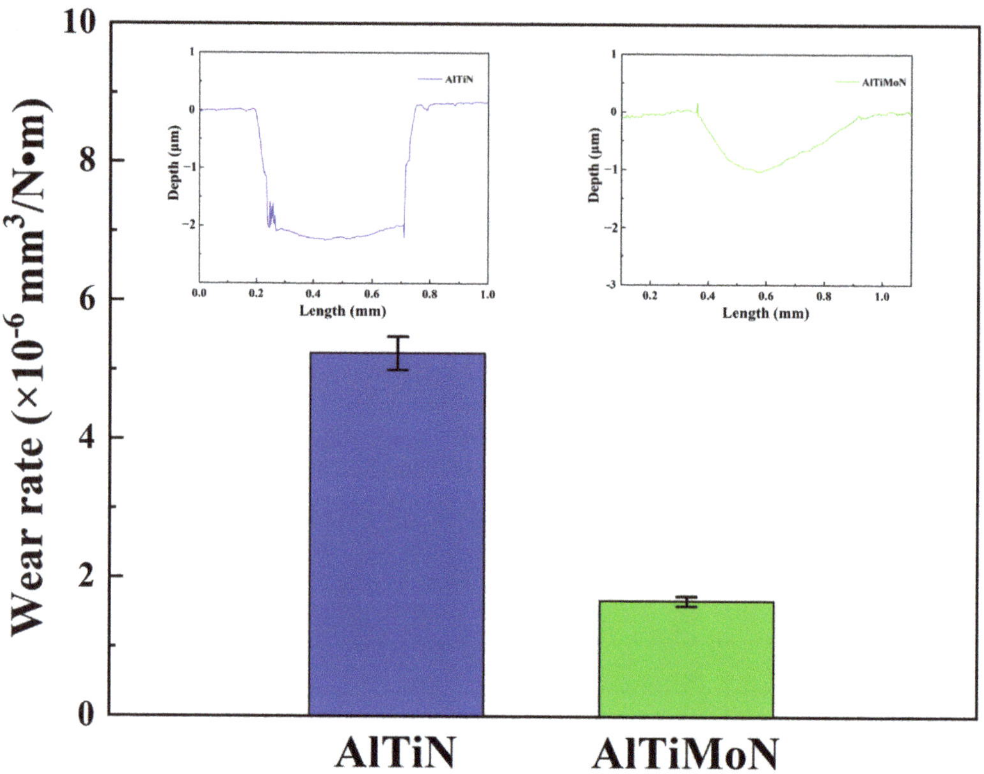

Figure 6. Wear rates of the wear tracks on the AlTiN and AlTiMoN coatings. The width and depth profiles of the wear tracks on both coatings are also shown.

Molybdenum doping enhanced the tribological properties and wear resistance of the AlTiN coatings. The COF decreased from 0.85 (AlTiN coating) to 0.60 (AlTiMoN coating), and the wear rate of the AlTiMoN coating was reduced to one-third of that of the AlTiN coating. In addition to the TiO_2 and Al_2O_3 phases, the MoO_3 phases were also formed on the wear track of the AlTiMoN coating during dry friction. The layered structure of MoO_3 is bonded by weak van der Waals forces; therefore, MoO_3 is easily sheared layer by layer. Soft and non-abrasive MoO_3 can be used as a solid lubricant to reduce the friction between friction pairs. The wear resistance improvement of the AlTiMoN coating is also attributed to the beneficial contribution of the MoO_3 phases, which exhibit self-lubricating properties and facilitate the sliding processes. Moreover, the improved wear resistance of the AlTiMoN coating is related to its higher H/E and H^3/E^2 ratios [10–13,16–18,26,28–32,47].

3.4. Cutting Performance of the AlTiN and AlTiMoN Coatings

Figure 7 shows a schematic diagram illustrating the cutting force direction of the cutting tool. The cutting forces can be divided into three directions, namely, the F_x (feed force) direction is opposite to the direction of feed movement, the F_y (tangential force) direction is horizontally outward, and the F_z (radial thrust force) direction is vertically downward. Figure 8 shows the cutting forces on the AlTiN- and AlTiMoN-coated tools measured using a force meter during cutting. Figure 8 clearly shows that the cutting force in the F_z direction is more stable than that in the other two directions for both coatings. Specifically, the average cutting forces on the AlTiMoN-coated tool were F_x = 197.4 N, F_y = 302.4 N, and F_z = 195.4 N. The average cutting forces on the AlTiN-coated tool were F_x = 239.2 N, F_y = 345.7 N, and F_z = 233.8 N. The experimental results showed that the forces on the AlTiMoN-coated tool in all three directions were smaller than those on the AlTiN-coated tool, indicating that the addition of Mo resulted in the AlTiN-coated tool exhibiting lower cutting forces.

Figure 7. Schematic diagram illustrating the cutting force directions of the cutting tool.

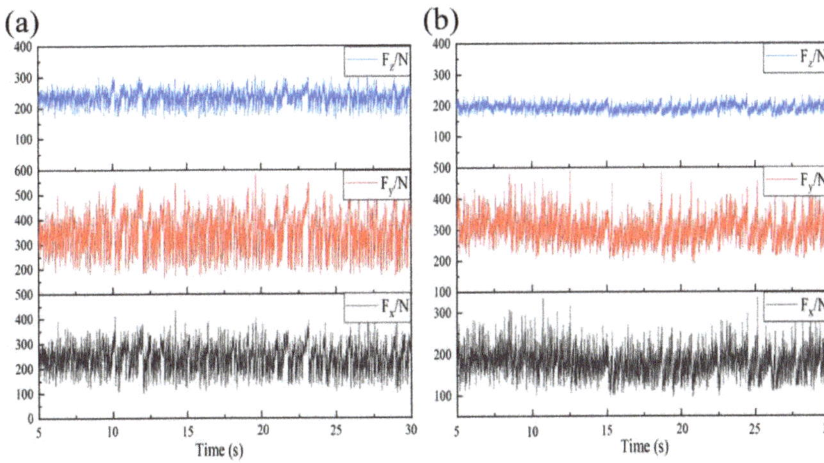

Figure 8. Cutting forces on (**a**) the AlTiN- and (**b**) AlTiMoN-coated tools.

The cutting temperatures of the AlTiMoN- and AlTiN-coated tools during the cutting process were measured using an A655sc thermal imager. Table 4 shows that the highest cutting temperatures of the AlTiMoN- and AlTiN-coated tools were 577 and 611 °C, respectively. Additionally, the average temperatures of the AlTiMoN- and AlTiN-coated tools at which cutting was stable were approximately 560 and 596 °C, respectively. Therefore, the highest cutting temperature of the AlTiMoN-coated tool was 34 °C lower than that of the AlTiN-coated tool, and the average cutting temperature at which cutting was stable was 36 °C lower than that of the AlTiN-coated tool. Therefore, it can be concluded that the AlTiMoN coating was more effective than the AlTiN coating in reducing the cutting temperature of cutting tools during the cutting process.

Table 4. Cutting temperatures of the AlTiN- and AlTiMoN-coated tools.

	Highest Temperature During Cutting (°C)	Average Temperature at Which Cutting Is Stable (°C)
AlTiN	611	596
AlTiMoN	577	560

Cutting tools produce unavoidable heat during the cutting process. Therefore, cutting heat is the primary reason for the increase in the cutting temperature. A high cutting temperature oxidizes or even causes the coating on the tool surface to fail, which directly impacts the machining accuracy and surface quality of the workpiece.

TiO_2 and Al_2O_3 phases formed during the cutting process of the AlTiN coating. The formation of the TiO_2 and Al_2O_3 phases by a possible reaction during the cutting process may be as follows:

$$2TiN + 2O_2 = 2TiO_2 + N_2 \quad (3)$$

$$2AlN + 3/2O_2 = Al_2O_3 + N_2 \quad (4)$$

These oxides have low ionic potentials, and the cations between them can strongly interact to form strong ionic or covalent bonds. As a result, these oxides are not easily deformed under shearing force and do not have any lubricating effect [46]. What is worse, the TiO_2 and Al_2O_3 oxides are hard and act as abrasive particles, which are detrimental to the cutting process. Owing to the above reasons, the cutting force and temperature of the AlTiN-coated tools were high during the cutting process.

The oxidation temperature of Mo nitride films is approximately 350–400 °C [12], and the melting point for the lubricious oxide MoO_3 is 795 °C [48]. Consequently, the solid–lubricant MoO_3 phase formed during the cutting process of the AlTiMoN coating, though the TiO_2 and Al_2O_3 phases were also formed. The formation of the MoO_3 phase by tribo-chemical reaction during the cutting process may be as follows [30]:

$$Mo_2N + 2O_2 = 2MoO_2 + 1/2N_2 \quad (5)$$

$$MoO_2 + 1/2O_2 = MoO_3 \quad (6)$$

or

$$Mo + 3/2O_2 = MoO_3 \quad (7)$$

MoO_3 phase is a solid-lubricant oxide, which is composed of layers bonded by van der Waals forces that easily deform under shearing force. The existence of the soft and non-abrasive MoO_3 phase minimized friction between the tool and chip interfaces, thereby reducing the cutting force and heat generated during machining. Xing [49] and Li [50] found that the formation of a lubricating phase with a low shear strength between the tool and chip interfaces contributes to the improvement of the cutting performance. Low cutting forces and temperatures result in low wear and a long tool life.

In summary, Mo addition improves the mechanical performance, tribological properties, wear resistance, and cutting performance of AlTiN coatings. This study shows a path to increase the performance of AlTiN coating through molybdenum addition and provides ideas for the application of AlTiMoN coatings for cutting tools.

4. Conclusions

AlTiN and AlTiMo coatings were deposited in this study using multi-arc ion plating. The influence of Mo addition on the structure, mechanical properties, and cutting performance of the AlTiN coating was investigated. The main conclusions drawn from this study are as follows:

1. Spherical droplets formed on the surfaces of both coatings, with the AlTiMoN coating exhibiting more surface defects than the AlTiN coating. The GIXRD results for the

AlTiN and AlTiMoN coatings showed that an (Al,Ti)N phase was formed. Doping of the AlTiMoN coating with Mo slightly reduced the grain size.
2. Both coatings exhibited excellent adhesion to the substrate. The hardness, elastic modulus, and H/E and H^3/E^2 ratios of the AlTiMoN coating were higher than those of the AlTiN coating. The improvement in the mechanical properties can be attributed to grain refinement and solution strengthening.
3. The tribological properties and cutting performance of the AlTiN coating improved with the Mo doping, which can be attributed to the formation of MoO_3 as a solid lubricant.

The presented research focuses the Mo addition on the structure, mechanical properties, and cutting performance of the AlTiN coating. However, the effect of Mo composition on the structure and properties of the AlTiMoN coating is not involved, especially on the cutting performance. This is what our next research will focus on. In the field of advanced coating technologies for cutting tools, the cutting performance and tool life of coated tools can be further improved by using nano-layer and nano-composite coating. Future research can be focused on depositing nanolayered and nanocomposite AlTiMoN coatings to further improve the performance of AlTiMoN coatings for machining applications.

Author Contributions: Methodology, T.Y. and V.L.; investigation, J.Y. and T.Y.; resources, V.L.; data curation, J.Y. and P.Y.; writing—original draft preparation, P.Y.; writing—review and editing, T.Y., C.L., P.Z., J.W., A.K., M.H., A.Y.G. and D.M.G.; supervision, V.L.; project administration, T.W. All authors have read and agreed to the published version of the manuscript.

Funding: This research was funded by the Zhejiang Provincial Natural Science Foundation of China, grant numbers LQ22E010007 and LTZ20E020001; Wenling Key Research and Development Project, grant number 2023G00007; and Taizhou Science and Technology Plan Project, grant number 22gya09.

Data Availability Statement: The data presented in this study are available on request from the corresponding author. The data are not publicly available due to privacy.

Conflicts of Interest: The authors declare no conflicts of interest.

References

1. Avila, R.F.; Mancosu, R.D.; Machado, A.R.; Vecchio, S.D.; Silva, R.B.d.; Vieira, J.M. Comparative analysis of wear on PVD TiN and $(Ti_{1-x}Al_x)N$ coatings in machining process. *Wear* **2013**, *302*, 1192–1200. [CrossRef]
2. Chauhan, K.V.; Rawal, S.K. A review paper on tribological and mechanical properties of ternary nitride based coatings. *Procedia Technol.* **2014**, *14*, 430–437. [CrossRef]
3. PalDey, S.; Deevi, S.C. Single layer and multilayer wear resistant coatings of (Ti,Al)N: A review. *Mater. Sci. Eng. A* **2003**, *342*, 58–79. [CrossRef]
4. Musil, J.; Hrubý, H. Superhard nanocomposite $Ti_{1-x}Al_xN$ films prepared by magnetron sputtering. *Thin Solid Films* **2000**, *365*, 104–109. [CrossRef]
5. Kale, A.N.; Ravindranath, K.; Kothari, D.C.; Raole, P.M. Tribological properties of (Ti,Al) N coatings deposited at different bias voltages using the cathodic arc technique. *Surf. Coat. Technol.* **2001**, *145*, 60–70. [CrossRef]
6. Richthofen, A.; Cremer, R.; Witthaut, M.; Domnick, R.; Neuschütz, D. Composition, binding states, structure, and morphology of the corrosion layer of an oxidized $Ti_{0.46}Al_{0.54}N$ film. *Thin Solid Films* **1998**, *312*, 190–194. [CrossRef]
7. Hardell, J.; Prakash, B. Tribological performance of surface engineered tool steel at elevated temperatures. *Int. J. Refract. Met. Hard Mater.* **2010**, *28*, 106–114. [CrossRef]
8. Sousa, V.F.C.; Silva, F.J.G.D.; Pinto, G.F.; Baptista, A. Characteristics and wear mechanisms of TiAlN-based coatings for machining applications: A comprehensive review. *Metals* **2021**, *11*, 260. [CrossRef]
9. He, J.; Liu, G.; Ren, H.X.; Liu, J.; Wang, H.; Qiu, L.S.; Hu, X.G. Investigations on the structure, mechanical property, wear and oxidation resistances of TiAlSiN coatings by tailoring Al/(Al + Ti) ratios. *Adv. Eng. Mater.* **2024**, *26*, 2400625. [CrossRef]
10. Gao, T.; Yu, L.H.; Lu, G.Y.; Ju, H.B.; Xu, J.H. Influence of Mo content on properties of Ti-Al-Mo-N films. *Surf. Eng.* **2021**, *37*, 519–526. [CrossRef]
11. Sergevnin, V.S.; Blinkov, I.V.; Belov, D.S.; Volkhonskii, A.O.; Krupin, A.Y.; Chernogor, A.V. Hardness, adhesion strength, and tribological properties of adaptive nanostructured ion-plasma vacuum-arc coatings (Ti,Al)N-Mo_2N. *Russ. J. Non-Ferr. Met.* **2016**, *57*, 572–579. [CrossRef]
12. Yang, K.; Xian, G.; Zhao, H.B.; Fan, H.Y.; Wang, J.; Wamg, H.; Du, H. Effect of Mo content on the structure and mechanical properties of TiAlMoN films deposited on WC-Co cemented carbide substrate by magnetron sputtering. *Int. J. Refract. Met. Hard Mater.* **2015**, *52*, 29–35. [CrossRef]

13. Sergevnin, V.S.; Blinkov, I.V.; Belov, D.S.; Smirnov, N.I.; Volkhonskii, A.O.; Kuptsov, K.A. Wear and erosion of arc-PVD multilayer Ti-Al-Mo-N coatings under various conditions of friction and loading. *Int. J. Adv. Manuf. Technol.* **2018**, *98*, 593–601. [CrossRef]
14. Sangiovanni, D.G.; Chirita, V.; Hultman, L. Toughness enhancement in TiAlN-based quarternary alloys. *Thin Solid Films* **2012**, *52*, 4080–4088. [CrossRef]
15. Gassner, G.; Mayrhofer, P.H.; Kutschej, K.; Mitterer, C.; Mitterer, M. Magnéli phase formation of PVD Mo-N and W-N coatings. *Surf. Coat. Technol.* **2006**, *201*, 3335–3341. [CrossRef]
16. Tomaszewski, Ł.; Gulbinski, W.; Urbanowicz, A.; Suszko, T.; Lewandowski, A.; Gulbinskin, W. TiAlN based wear resistant coatings modified by molybdenum addition. *Vacuum* **2015**, *121*, 223–229. [CrossRef]
17. Sergevnin, V.S.; Blinkov, I.V.; Volkhonskii, A.O.; Belov, D.S.; Kuznetsov, D.V.; Gorshenkov, M.V.; Skryleva, E.A. Wear behaviour of wear-resistant adaptive nano-multilayered Ti-Al-Mo-N coatings. *Appl. Surf. Sci.* **2016**, *388*, 13–23. [CrossRef]
18. Yousaf, M.I.; Pelenovich, V.O.; Yang, B.; Liu, C.S.; Fu, D.J. Effect of bilayer period on structural and mechanical properties of nanocomposite TiAlN/MoN multilayer films synthesized by cathodic arc ion-plating. *Surf. Coat. Technol.* **2015**, *282*, 94–102. [CrossRef]
19. Tavares, C.J.; Rebouta, L.; Alves, E.; Cavaleiro, A.; Goudeau, P.; Riviere, J.P.; Declemy, A. A structural and mechanical analysis on PVD-grown (Ti,Al)N/Mo multilayers. *Thin Solid Film.* **2000**, *377–378*, 425–429. [CrossRef]
20. Tavares, C.J.; Rebouta, L.; Riviere, J.P.; Pacaud, J.; Garem, H.; Pischow, K.; Wang, Z. Microstructure of superhard (Ti,Al)N/Mo multilayers. *Thin Solid Film.* **2001**, *398–399*, 397–404. [CrossRef]
21. Tavares, C.J.; Rebouta, L.; Andritschky, M.; Guimarães, F.; Cavaleiro, A. Mechanical and surface analysis of $Ti_{0.4}Al_{0.6}N$/Mo multilayers. *Vacuum* **2001**, *60*, 339–346. [CrossRef]
22. Tavares, C.J.; Rebouta, L.; Alves, E.; Barradas, N.P.; Pacaud, J.; Riviere, J.P. Study of roughness in $Ti_{0.4}Al_{0.6}N$/Mo multilayer structures. *Nucl. Instrum. Methods Phys. Res. B* **2002**, *188*, 90–95. [CrossRef]
23. Tavares, C.J.; Rebouta, L.; Riviere, J.P.; Girardeau, T.; Goudeau, P.; Alves, E.; Barradas, N.P. Atomic environment and interfacial structural order of TiAlN/Mo multilayers. *Surf. Coat. Technol.* **2004**, *187*, 393–398. [CrossRef]
24. Tavares, C.J.; Vidrago, C.; Rebouta, L.; Riviere, J.P.; Bourhis, E.L.; Denanot, M.F. Optimization and thermal stability of TiAlN/Mo multilayers. *Surf. Coat. Technol.* **2005**, *200*, 288–292. [CrossRef]
25. Sergevnin, V.S.; Blinkov, I.V.; Belov, D.S.; Volkhonskii, A.O.; Skryleva, E.A.; Chernogor, A.V. Phase formation in the Ti-Al-Mo-N system during the growth of adaptive wear-resistant coatings by arc PVD. *Inorg. Mater.* **2016**, *52*, 735–742. [CrossRef]
26. Yousaf, M.I.; Abudouwufu, T.; Yang, B.; Tolstoguzov, A.; Fu, D.J. Influence of bilayer thickness on mechanical and tribological properties of (Ti-Al)N/MoN nanostructured hard coatings deposited by cathodic arc ion plating. *Coatings* **2023**, *13*, 1654. [CrossRef]
27. Yousaf, M.I.; Pelenovich, V.O.; Yang, B.; Liu, C.S.; Fu, D.J. Influence of substrate rotation speed on the structure and mechanical properties of nanocrystalline AlTiN/MoN coatings synthesized by cathodic arc ion-plating. *Surf. Coat. Technol.* **2015**, *265*, 117–124. [CrossRef]
28. Wang, Z.S.; Tian, C.X.; Tolstogouzov, A.; Liang, F.; Zou, C.W.; Li, S.Q.; Gusev, S.I.; Yousaf, M.I.; Pelenovich, V.; Zuo, W.B.; et al. Microstructure and Rutherford backscattering spectrometry of hard/lubricant Mo-Ti-Al-N multilayered coatings prepared by multi-arc ion plating at low substrate rotation. *Coatings* **2020**, *10*, 101. [CrossRef]
29. Jirón-Lazos, U.; Rodil, S.E.; Mazón-Montijo, D.A.; Pérez-Higareda, J.R.; Torres, D.; Garay-Tapia, A.M.; Montiel-González, Z. Microstructural behavior of the Ti-Al-Mo-N system controlled by Mo content: Impact on the performance as hard coatings. *J. Mater. Sci.* **2023**, *58*, 11771–11787. [CrossRef]
30. Belgroune, A.; Aissani, L.; Alhussein, A.; Zaabat, M.; Obrosov, A.; Rtimi, S. In vitro corrosion and wear investigation of multifunctional TiAlMoN sputtered coatings on cold-sprayed SS316L. *ACS Appl. Eng. Mater.* **2024**, *2*, 345–359. [CrossRef]
31. Glatz, S.A.; Moraes, V.; Koller, C.M.; Riedl, H.; Bolvardi, H.; Kolozsvári, S.; Mayrhofer, P.H. Effect of Mo on the thermal stability, oxidation resistance, and tribo-mechanical properties of arc evaporated Ti-Al-N coatings. *J. Vac. Sci. Technol. A* **2017**, *35*, 061515. [CrossRef]
32. Glatz, S.A.; Koller, C.M.; Bolvardi, H.; Kolozsvári, S.; Riedl, H.; Mayrhofer, P.H. Influence of Mo on the structure and the tribomechanical properties of arc evaporated Ti-Al-N. *Surf. Coat. Technol.* **2017**, *311*, 330–336. [CrossRef]
33. Moreno, M.; Andersson, J.M.; Johansson-Jöesaar, M.P.; Friedrich, B.E.; Boyd, R.; Schramm, I.C.; Johnson, L.J.S.; Odén, M.; Rogström, L. Wear of Mo- and W-alloyed TiAlN coatings during high-speed turning of stainless steel. *Surf. Coat. Technol.* **2022**, *446*, 128786. [CrossRef]
34. Demirov, A.P.; Blinkov, I.V.; Sergevnin, V.S.; Volkhonskii, A.O.; Chernogor, A.V.; Shchetinin, I.V. Thermal stability and electrochemical behavior of nanostructured Ti-Al (~1%(at.))-Mo-N coatings deposited using the Arc-PVD method. *J. Alloys Compd.* **2022**, *929*, 167269. [CrossRef]
35. *ISO 14577*; Metallic Materials-Instrumented Indentation Test for Hardness and Materials Parameters. ISO Central Secretariat: Geneva, Switzerland, 2002.
36. Wang, H.M.; Shi, P.J.; Yu, H.L.; Zhang, W.; Xu, B.S. Applications of nanoindentation techniques in the field of surface coatings. *Key Eng. Mater.* **2008**, *373–374*, 802–805. [CrossRef]
37. *ASTM G137*; Standard Test Method for Ranking Resistance of Plastic Materials to Sliding Wear Using a Block-on-Ring Configuration. ASTM International: West Conshohocken, PA, USA, 2017.

38. Ying, P.Y.; Sun, H.Y.; Zhang, P.; Lin, C.H.; Yang, T.; Wu, J.B.; Huang, M.; Wang, T.L.; Lian, Z.B.; Levchenko, V. Preparation and tribological properties of WS_2 solid lubricating coating with dense structure using HiPIMS. *J. Mater. Res. Technol.* **2024**, *32*, 530–540. [CrossRef]
39. *JB/T 11442-2013*; Coated Solid-Carbide Tools. Detection Methods. Machinery Industry Press: Beijing, China, 2013.
40. *GB/T 16461-2016*; Tool-Life Testing with Single-Point Turning Tools. China Standards Press: Beijing, China, 2016.
41. Grigoriev, S.; Vereschaka, A.; Milovich, F.; Sitnikov, N.; Andreev, N.; Bublikov, J.; Sotova, C.; Sadov, I. Investigation of the influence of microdroplets on the coatings nanolayer structure. *Coatings* **2020**, *10*, 1204. [CrossRef]
42. Wang, L.; Zhang, S.H.; Chen, Z.; Li, J.L.; Li, M.X. Influence of deposition parameters on hard Cr-Al-N coatings deposited by multi-arc ion plating. *Appl. Surf. Sci.* **2012**, *258*, 3629–3636. [CrossRef]
43. Ju, H.B.; Yu, L.H.; Yu, D.; Asempah, I.; Xu, J.H. Microstructure, mechanical and tribological properties of TiN-Ag films deposited by reactive magnetron sputtering. *Vacuum* **2017**, *141*, 82–88. [CrossRef]
44. Leyland, A.; Matthews, A. On the significance of the H/E ratio in wear control: A nanocomposite coating approach to optimised tribological behaviour. *Wear* **2000**, *246*, 1–11. [CrossRef]
45. Tsui, T.Y.; Pharr, G.M.; Oliver, W.C.; Bhatia, C.S.; White, R.L.; Anders, S.; Anders, A.; Brown, I.G. Nanoindentation and nanoscratching of hard carbon coatings for magnetic disks. *Mater. Res. Soc. Symp. Proc.* **1995**, *383*, 447. [CrossRef]
46. Erdemir, A. A crystal chemical approach to the formulation of self-lubricating nanocomposite coatings. *Surf. Coat. Technol.* **2005**, *200*, 1792–1796. [CrossRef]
47. Yang, Q. Wear resistance and solid lubricity of molybdenum-containing nitride coatings deposited by cathodic arc evaporation. *Surf. Coat. Technol.* **2017**, *322*, 283–295. [CrossRef]
48. Biksa, A.; Yamamoto, K.; Dosbaeva, G.; Veldhuis, S.C.; Fox-Rabinovich, G.S.; Elfizy, A.; Wagg, T.; Shuster, L.S. Wear behavior of adaptive nano-multilayered AlTiN/Me_xN PVD coatings during machining of aerospace alloys. *Tribol. Int.* **2010**, *43*, 1491–1499. [CrossRef]
49. Xing, Y.Q.; Deng, J.X.; Li, S.P.; Yue, H.Z.; Meng, R.; Gao, P. Cutting performance and wear characteristics of Al_2O_3/TiC ceramic cutting tools with WS_2/Zr soft-coatings and nano-textures in dry cutting. *Wear* **2014**, *318*, 12–26. [CrossRef]
50. Li, X.M.; Deng, J.X.; Lu, Y.; Duan, R.; Ge, D.L. Machinability investigation and sustainability assessment of dry cutting AISI1045 steel using tools configured with shark-skin-inspired structures and WS_2/C coatings. *Int. J. Precis. Eng. Manuf.-Green Technol.* **2022**, *9*, 83–106. [CrossRef]

lubricants

MDPI

Article

Lubrication-Enhanced Mechanisms of Bentonite Grease Using 2D MoS_2 with Narrow Lateral Size and Thickness Distributions

Shaoyicheng Zhu [1], Shuo Xiang [1,*], Xue Yang [1], Xin Yang [1], Hebin Bao [1], Hao He [1], Xin Du [1], Qinhui Zhang [1], Junjie Zhang [1], Kai Ma [2,*], Yi Cao [2], Yuehao Liu [2], Liangyi Peng [2], Zhitao Li [2] and Yunhong Fan [2]

1 Army Logistics Academy of PLA, Chongqing 401331, China; 15025173173@163.com (S.Z.)
2 Synthetic Lubes Branch, Sinopec Lubricant Co., Ltd., Chongqing 400039, China
* Correspondence: xslaplace@163.com (S.X.); mak.lube@sinopec.com (K.M.)

Abstract: 2D MoS_2 with narrow lateral size and thickness distributions was introduced to promote the anti-friction and anti-wear properties of the bentonite grease (BG) in a state of boundary lubrication. Optical microscopy (OM), and 3D optical profilers (3D OP), Raman spectrometry (Raman), scanning electron microscope, energy dispersion spectrum (SEM-EDS), and X-ray photoelectron spectroscopy (XPS) were applied to characterize the wear surface of the GCr15 bearing steel/GCr15 bearing steel contact. It is found that the average friction coefficient (AFC), wear scar diameter (WSD), surface roughness and average wear scar depth of BG + 1.2 wt.% 2D MoS_2 were effectively reduced by approximately 22.15%, 23.14%, 55.15%, and 21.1%, respectivlely, compared with BG under the working condition of 392N, 75 °C, 1 h, and 1200 rpm. Raman, EDS and XPS results jointly demonstrated that a stable adsorbed film and a robust tribochemical film composed of Fe_2O_3, $FeSO_4$, $Fe_2(SO_4)_3$, $FeSO_3$, FeS, FeO and MoO_3, which further contributes to the enhancement of lubrication performance.

Keywords: bentonite grease; lubricating performance; friction modifiers additives; 2D MoS_2

1. Introduction

Friction exists extensively in nature and industry, which is confirmed to be the main cause of energy consumption and equipment life shortening. Utilizing lubricants properly can not only effectively reduce the energy required to operate machinery and equipment, but also prolong the life of machinery and equipment, which can minimize the necessity for energy-intensive maintenance and repair [1]. Lubricants act as a barrier, preventing direct contact between surfaces in relative mutual motion, and reduce both the frictional force and wear. Modern lubricants are normally divided into liquid (the most important), gaseous and solid lubricants based on their physical state. In general terms, a lubricating grease is defined as a kind of lubricating fluid thickened to a solid or semi-fluid product by means of a thickener, which are particularly suitable for applications where a continuous oil supply is not feasible or frequent lubrication is impractical. A lubricating grease is formulated from a base oil, a thickener and various performance-improving additives [2]. Although the thickener accounts for a comparatively small proportion in a grease, which does regulate the foundamental properties of grease, such as the dropping point, consistency, water resistant and thermal stability. There are many different types of thickeners [3], which can be classified into two major types soap (lithium soap, sodium soap, aluminum soap, barium soap, calcium soap, lithium complex soap, complex sodium soap, complex aluminum soap, complex barium soap, complex calcium soap, etc.) and non-soap (polyurea, organophilic bentonite, modified attapulgite, fumed silica, fluoropolymers solid hydrocarbons, etc.), however, lithium soap is the most common thickener of greases due to their superior performance and ease of manufacturing over comparable technologies. Currently, the future of lithium greases is at a crossroads due to the unprecedented growth in electric vehicles (EVs), which results in a sharp increase in the price of lithium.

Bentonite grease (BG) is a type of clay thickened grease, which can be a potential replacement for lithium greases due to its significant advantages, such as temperature change resistance, great wear and tear protection, exceptional water tolerance, good mechanical or shear stability, and impressive adhesiveness [4]. Friction modifiers (FM) additives are components added to lubricans that help prevent metal-to-metal contact when loads are extremely high, which work chemically with the metal surfaces to form protective films that shield them from scuffing, welding, and wear [5]. Traditional FM additives containing S, P, and Cl elements have been used to enhance the lubricating performance of lubricants, for example, chlorinated paraffin [6], sulfurized isobutylene [7], tricresyl phosphate [8], amine phosphate [9], zinc dialkyl dithiophosphate [10] and thiophosphate amine salt [11]. However, traditional FM additives are environmentally unfriendly and are harmful to humans [12], for example, chlorine-containing additives have been prohibited by many countries due to their toxicity and corrosive problems and sulfur can result in sulfur pollution to the environment, whereas phosphorus can lead to eutrophication of water bodies. Due to the characteristics of outstanding high specific surface area, nanometer dimension effect, and unique physical and chemical properties [13,14], two-dimensional (2D) materials have been comprehensively investigated in the field of FM additives, for instance, graphene [15], black phosphorus (BP) [16], hexagonal boron nitride [17], transition metal dichalcogenides (TMDs) [18], two-dimensional molybdenum carbide (MXene) [19], layered double hydroxides (LDHs) [20] and covalent–organic frameworks (COFs). Top-down exfoliation is suitable for scale industrial production. The as-prepared 2D materials have wide distributions in lateral size and thickness, which is a dilemma for some applications requiring controlled nanosheet sizes: 2D material-assembled thin films [21–24], active site exposure of 2D material catalysts [25–27], functionalization of 2D material sheets [28–31] as well as the field-induced alignment order of 2D material liquid crystals. As indeed suitable candidates for traditional FM additives, achieving homodisperse in both lateral size and thickness of 2D materials, which is of great importance to maximize the potential of their extreme pressure and lubrication performance in lubricants.

In the present article, lubrication-enhanced mechanisms of bentonite grease using 2D MoS_2 with narrow lateral size and thickness distributions were systematically investigated by a four-ball tribometer. Optical microscopy (OM) and 3D optical profilers (3D OP) methods were adopted to characterize the worn surface. The composition and microstructure of the physical adsorption film and tribo-chemical film were analyzed by Raman spectrometry (Raman), scanning electron microscope (SEM), energy dispersion spectrum (EDS), and X-ray photoelectron spectroscopy (XPS). The lubrication-enhanced mechanisms of bentonite grease using 2D MoS_2 with narrow lat-eral size and thickness distributions was discussed.

2. Materials and Methods

2.1. Materials

As reported in our previous paper [26], 2D MoS_2 had been prepared by the ultrasound-assisted liquid-phase exfoliation method and the atomic force microscopy (AFM) study revealed a mean thickness of 22 nm, as shown in Figure 1. 150BS base oil (Typical characteristics showed in Table 1) was commercially obtained from the PetroChina Karamay Petrochemical Company (Karamay, China). The commercial surface-modified organo-bentonite modified by HTMAB was purchased from ZheJiang AnJi Tianlong OrganicBentonite Co., Ltd. (Huzhou, China). GCr15 bearing steel balls made of GCr15 bearing steel with a diameter of 12.7 mm and an HRC59-61 were obtained from the SINOPEC Research Institute of Petroleum Processing Co., Ltd. (Beijing, China).

Table 1. Typical physico-chemical properties of the 150BS base oil.

Test Description	Result	Method
Kinematic viscosity (mm^2/s) (40 °C)	490.7	ASTM D445
Kinematic viscosity (mm^2/s) (100 °C)	31.75	ASTM D445
Viscosity Index	95	ASTM D2270
Appearance	Clear to bright	Visual
Colour (ASTM) (Quantitative)	L2.0	ASTM D1500
Density (kg/m^3) (15 °C)	0.9012	ASTM D4052
Density (kg/m^3) (30 °C)	0.8917	ASTM D4052
Refractive index (20 °C)	1.46	ASTM D1218
Pour point (°C)	−6	ASTM D5950
Flash point (°C) (PMcc)	316	ASTM D92
Specific Gravity (60/60 °F)	0.9017	ASTM D4052
Total acid number (mgKOH/g)	0.01	ASTM D664
Cartxon residue (micro method) (wt.%)	0.41	ASTM D4530
Sulphur content (wt.%)	0.536	ASTM D4294
Water content (vol.%)	Nill	ASTM D95

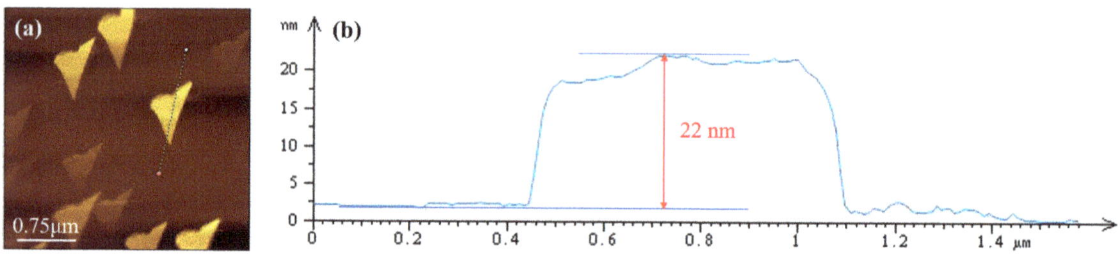

Figure 1. AFM images (**a**) of 2D MoS_2 with the corresponding height profiles (**b**).

2.2. Preparation and Characterization of Bentonite Grease with 2D MoS_2

The bentonite greases were prepared in accordance with the literature [3]. Firstly, the 150BS base oil was poured into the vessel and next stirred. Secondly, organo-bentone was added to the 150BS base oil little by little, being stirred vigorously. When the 150BS base oil and the organo-bentonite had been well dispersed, acetone, at approximately 50 wt.% of the entire quantity of the thickener, was introduced to guarantee that the organo-bentone can be thoroughly dispersed throughout the 150BS base oil. After continuous stirring for 30 min, acetone was removed from the mixture via heating for 30 min at 80 °C. The 2D MoS_2 of specified mass (0, 0.3, 0.6, 0.9, 1.2, 1.5, 1.8 and 2.1 wt.%) were put into the mixture. Thirdly, when the mixture was cooled naturally to room temperature and then the mixture was ground three times using a three-roll mill to gain the samples, which were labeled as BG (BG-0), BG + 0.3 wt.% 2D MoS_2 (BG-0.3), BG + 0.6 wt.% 2D MoS_2 (BG-0.6) , BG + 0.9 wt.% 2D MoS_2 (BG-0.9), BG + 1.2 wt.% 2D MoS_2 (BG-1.2) , BG + 1.5 wt.% 2D MoS_2 (BG-1.5), 2D MoS_2 BG + 1.8 wt.% 2D MoS_2 (BG-1.8) and BG + 2.1 wt.% 2D MoS_2 (BG-2.1) for short, respectively. Schedule of all prepared grease samples are illustrated in Figure 2.

The consistency, thermal stability, colloid stability, evaporation loss and corrosive properties of the propared grease samples were evaluated on the basis of the GB/T 269, GB/T 3498, NB/SH/T 0324, GB/T 7325 and GB/T 7326 standards, respectively.

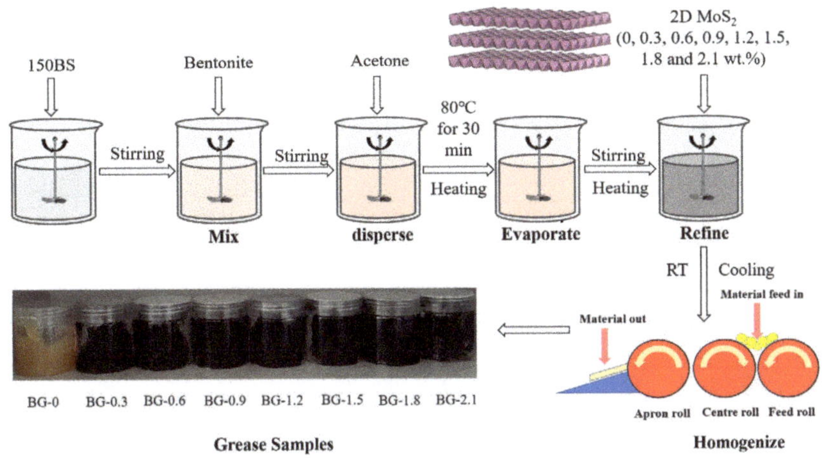

Figure 2. Schedule of preparation of bentonite grease containing 2D MoS_2.

2.3. Tribology Tests and Analysis

The lubrication performance of eight grease samples is carried out via the four-ball tester using the SH/T 0204, which is similar to the ASTM D2266. The tribometer is displayed schematically in Figure 3, and a series of friction tests were conducted under atmospheric conditions with a relative humidity (RH) of about 75%–80%.

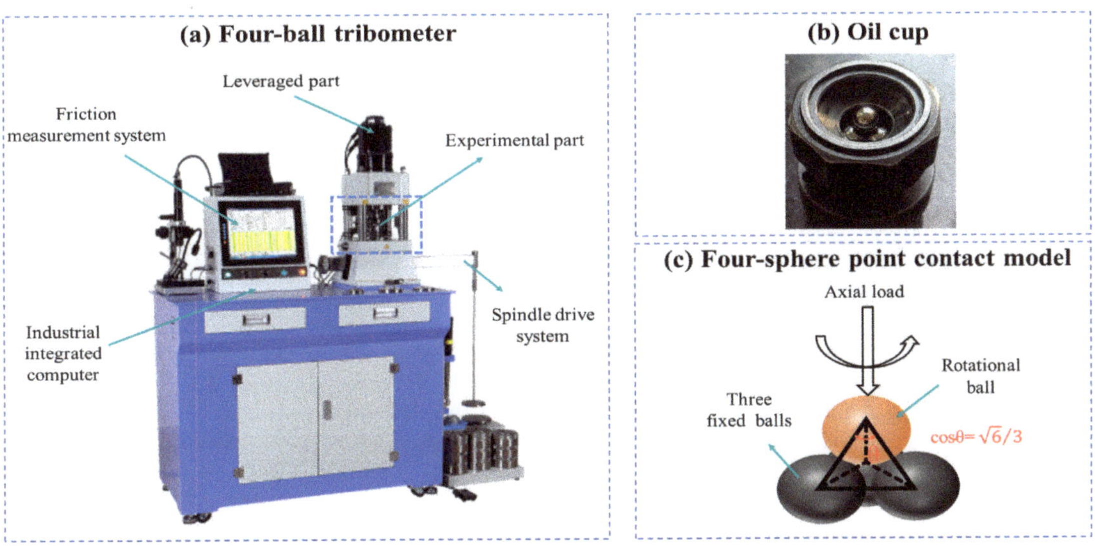

Figure 3. The schematic diagram of four-ball tribometer. (**a**) Four-ball tribometer, (**b**) Oil cup, (**c**) Four-sphere point contact model.

The friction coefficient (COF) was automatically measured and recorded in real time by the computer and the wear scar diameter (WSD) of the three fixed steel balls were scanned by optical microscope. To ensure the repeatability and accuracy of the data, friction tests were performed three times under the identical experimental conditions, and the average value was calculated. The parameters of the tested GCr15 steel balls are shown in Table 2.

Table 2. Experimental conditions and basic properties of the steel balls utilized.

Parameter	BG-0	BG-0.3	BG-0.6	BG-0.9	BG-1.2	BG-1.5	BG-1.8	BG-2.1
Speed				1200 rpm				
Load				392 N				
Temerature				75 °C				
Test Duration				60 min				
Component	Elastic modulus (MPa)		Poisson ratio	Diameter	Rockwell harness (HR)		Surface roughness (Ra)	
GCr15	2.085×10^5		0.3	12.7 mm	60 ± 1		0.256 μm	

For spherical contact, the contact pressure can be estimated according to the Hertzian point contact theory [32]:

$$q_{max} = \frac{3p}{2\pi a^2} \tag{1}$$

$$a = (\frac{3}{4} \times \frac{pR'}{E^*})^{\frac{1}{3}} \tag{2}$$

$$w = 3p\cos\theta\left(\cos\theta = \sqrt{6}/3\right) \tag{3}$$

$$\frac{1}{R'} = \frac{1}{R_{above}} + \frac{1}{R_{below}} \tag{4}$$

$$E^* = (\frac{1 - v_{above}^2}{E_{above}} + \frac{1 - v_{below}^2}{E_{below}})^{-1} \tag{5}$$

Herein, q_{max} is the maximum contact pressure between the four GCr15 steel balls, p is the effective load, a is the Hertzian contact radius, E^* is the equivalent Young's modulus, R' is the comprehensive radius, w is the total load (392N), R_{above} is the radius of the rotating upper ball, R_{below} are the radii of the lower three stationary balls, v_{above} and E_{above} are Poisson's ratio and Young's modulus of the rotating upper ball, v_{below} and E_{below} are Poisson's ratio and Young's modulus of the lower three stationary balls, respectively.

The λ was estimated on the basis of the Dowson and Hamrock minimum film thickness formula [33]:

$$h_{min} = 3.63\frac{G^{0.49}U^{0.68}R'}{W^{0.073}}\left(1 - 0.61e^{-0.68k}\right) \tag{6}$$

$$\lambda = \frac{h_{min}}{\sqrt{\sigma_1^2 + \sigma_2^2}} \tag{7}$$

where, h_{min} was the minimum oil film thickness, $G = \alpha E'$, $U = \eta_0 u/E'R'$, $W = p/E'R'^2$, $k = 1.03(R_y/R_x)^{0.64} = 1.03$ were dimensionless material parameter, dimensionless speed parameter, dimensionless load parameter, ellipticity parameter, respectively. α and η_0 were the viscosity-pressure coefficient and the dynamic viscosity, respectively. u (0.461 m/s) was the relative sliding velocity of the two friction pairs, k was the ellipticity parameter, $E' = 2E^*$ was the effective modulus of elasticity, σ_1 (2.1 μm) and σ_2 (2.1 μm) were the surface roughness of the worn area of the upper rotating ball and the lower stationary ball, respectively. Calculated from the formula, the maximum contact pressure was 2.34 GPa and the lambda ratio is 0.45 for these test conditions, which indicating that the contact area was firmly in a state of boundary lubrication under the four-ball tribometer.

After testing, the morphology of the worn surface was observed using a JEOL JSM-6610LV scanning electron microscope (JEOL, Tokyo, Japan) and a 3D optical profilers (Sensofar, Terrassa, Spain), and the elemental distribution and composition of the worn surface was identified and quantified using an Oxford X-Max 20 mm^2 energy dispersive X-ray spectrometer (Oxford Instruments, Oxford, UK). The Raman spectra of worn surfaces were measured using Raman spectrometry (LabRAM HR Evolution, HORIBA, Longjumeau, France) with a laser wavelength of 532 nm. For the sake of exploring the elemental composition and chemical state of the films on the worn surfaces, X-ray photoelectron spectroscopy (XPS) tests were performed using an ESCALAB 250Xi X-ray photoelectron spectrometer (Bruker, Karlsruhe, Germany) to probe the deposition of the ternary films.

3. Results and Discussion

3.1. Physico-Chemical Properties

The consistency, thermal stability, colloid stability, evaporation loss and corrosive properties of the BG containing 2D MoS_2 are presented in Table 3. As the addition of 2D MoS_2 in BG was increased, the thermal stability, colloid stability and evaporation loss values increased first, then decreased and increased in the end. The 2D MoS_2 introduced to BG could effectively enhance the holding performance of bentonite grease structural networks for base oil owing to its high adsorption property. The consistency of samples increased after adding 2D MoS_2, which means that 2D MoS_2 has an has a considerable thickening effect on the BG due to its big specific surface area and excellent dispersity. Additionally, 2D MoS_2 is proved to be uninfluential on the corrosion of BG.

Table 3. Effect of 2D MoS_2 on penetration, dropping point, oil separation, evaporation loss and copper corrosion of BG.

Parameter	BG-0	BG-0.3	BG-0.6	BG-0.9	BG-1.2	BG-1.5	BG-1.8	BG-2.1	Method
Penetration/0.1 mm	285	244	261	292	297	301	306	310	GB/T 269
Dropping point	278	292	290	285	286	287	287	291	GB/T 3498
Oil separation	0.93	0.54	1.46	1.22	0.92	0.91	1.42	1.64	NB/SH/T 0324
Evaporation loss	0.20	0.41	0.47	0.40	0.39	0.43	0.45	0.52	SH/T 0661
Copper corrosion	1a	1a	1a	1a	1a	1a	1a	1a	GB 7326

3.2. Friction and Wear Performance

Figure 4a shows the curves of the COF variation with time of BG with different mass fraction of 2D MoS_2. Figure 4b displays the AFC and WSD variations with 2D MoS_2. As shown in Figure 4a, the tendency of COF among different samples can be distinctly discovered that the three periods turned up during the friction process. At the initial stage of test, COF value rised sharply indicating that there is severe friction, because of the inevitable surface wear occuring during the running-in process, which leads to pollution of the grease by the wear debris. The COF value has been changed to be more mild after continuing for around 300 s. When 1.2 wt.% 2D MoS_2 is introduced, COF is obvious lower than that of BG-0. Figure 4b reveals that AFC initially decreases and subsequently increases with the introduction of 2D MoS_2, implying that an appropriate addition of these nanosheets to BG can provide a certain anti-friction effect. Among them, the AFC and WSD of BG-1.2 are the smallest, reducing by 22.15%, 23.14%, respectively, which exhibits the best performance. When the addition amount is less than 1.2 wt.%, 2D MoS_2 easily penetrate and distribute evenly between the friction pairs, while the anti-friction effect is not as effective as BG-1.2 due to their insufficient quantity. On the other hand, when the addition is more than 1.2 wt.%, high specific surface area and strong surface energy of 2D MoS_2 leads to irreversible accumulation and agglomeration into abrasive particles, exacerbating wear on these particles and weakening their friction-reducing effect.

For purpose of further distinctly comparing the anti-wear properties of BG with different concentrations of 2D MoS_2, the 3D OP is used in Figure 5. In comparison with the surface roughness of wear surface lubricated by BG-0, the wear surface lubricated by BG-1.2 seems to have the smaller surface roughness of about 2.097 μm, reducing by 55.15%, as displayed in Figure 5a,b. After a testing time of 1 h, the worn surface lubricated by BG-0 are seriously worn and show considerable deep furrows and rough scratches along the sliding direction. Meanwhile, the surface roughness and the average wear depth was remarkably reduced (21.1%) after the introduction of layered structure 2D MoS_2 into the BG, which is consistent with the change in WSD, as illustrated in Figure 5c,d. Due to the their small size and sheet shape, the 2D MoS_2 in BG can lightly permeate into the friction surfaces and fill the gaps on surfaces to decrease surface roughness and generate a physical absorption film in the central worn surface ,which can avoid severe wear of the uneven peak of the steel balls as much as possible, resulting in promising anti-wear properties on account of the sliding effect between the two sliding surfaces by reason of the shear stress. This requires to be further testified by a Raman characterization.

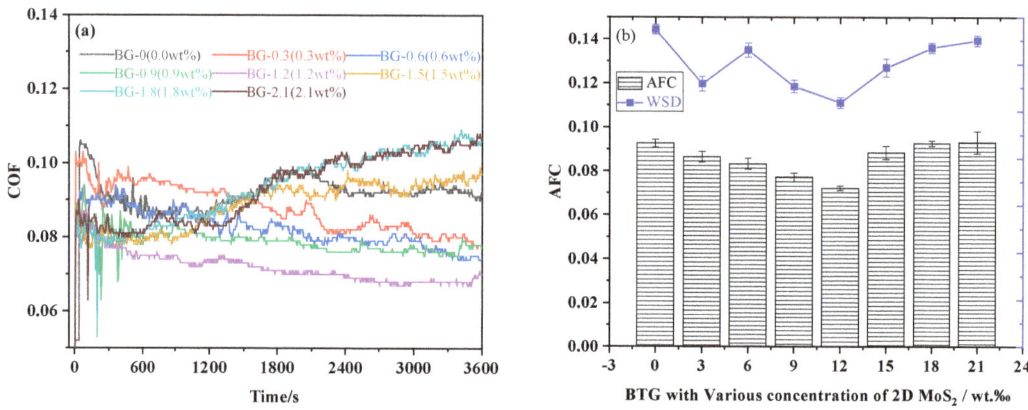

Figure 4. The COF-Time curve (**a**), AFC and WSD (**b**) of BG with different concentrations of 2D MoS_2 (1200 rpm, 392 N, 60 min, 75 °C).

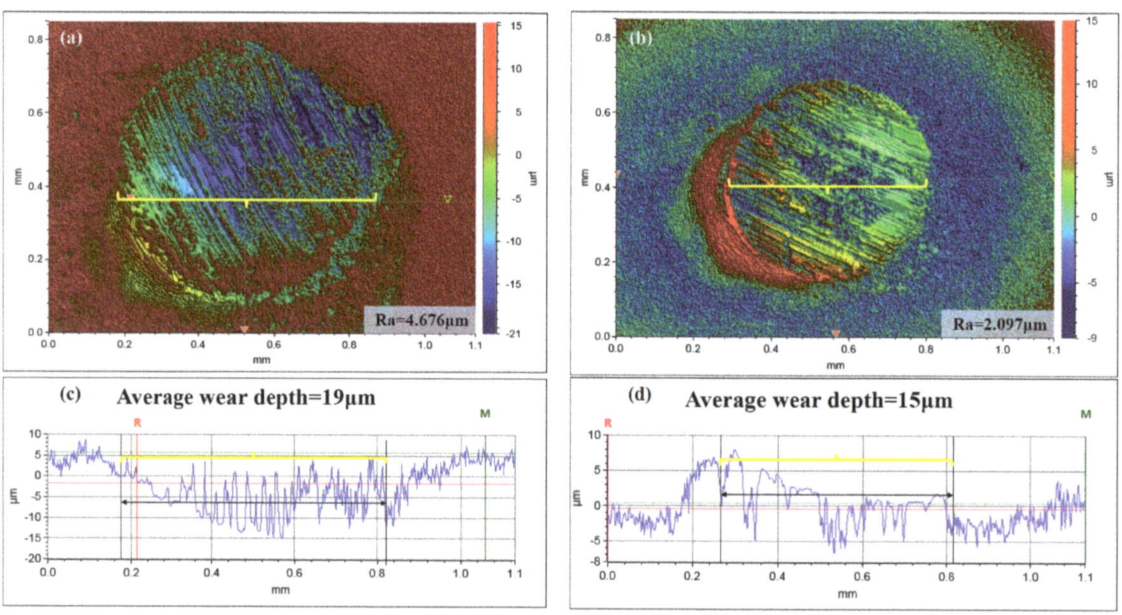

Figure 5. The 3D surface profiles and average wear depth of worn surfaces under the lubrication of BG-0 (**a**,**c**) and BG-1.2 (**b**,**d**), respectively.

3.3. Worn Surface Analysis

In order to detect the physical absorption film of 2D MoS_2 on the worn surface, Raman spectroscopy was utilized to anatomize the worn surfaces on the GCr15 bearing steel balls after the tribotest. The generation of 2D MoS_2 on the wear surface confirmed by the Raman results. The Raman spectra were recorded at the as prepared 2D MoS_2 additive, worn surface lubricated by BG-0 and BG-1.2, respectively, as shown in Figure 6. The notable fact is that the representative peaks at about 380 and 404 cm^{-1} for the worn surface lubricated by BG-1.2, which are $E^1{}_{2g}$ and A_{1g} modes were observed. The presence of the as prepared 2D MoS_2 additive on the worn surface verified by the appearance of similar patterns located at the same frequency as in the Raman spectra of the 2D MoS_2 additive itself. It demonstrates that 2D MoS_2 in BG can smoothly slide with oil into the point contact of the steel balls to avoid impetuous collision of the coarse peak of the steel balls. However, the physical absorption film may fracture due to the harsh working conditions. At the same time, a new film with superior mechanical properties could stand up to the scuffing and prevent the steel ball surfaces from severe crashes, which will be further investigated.

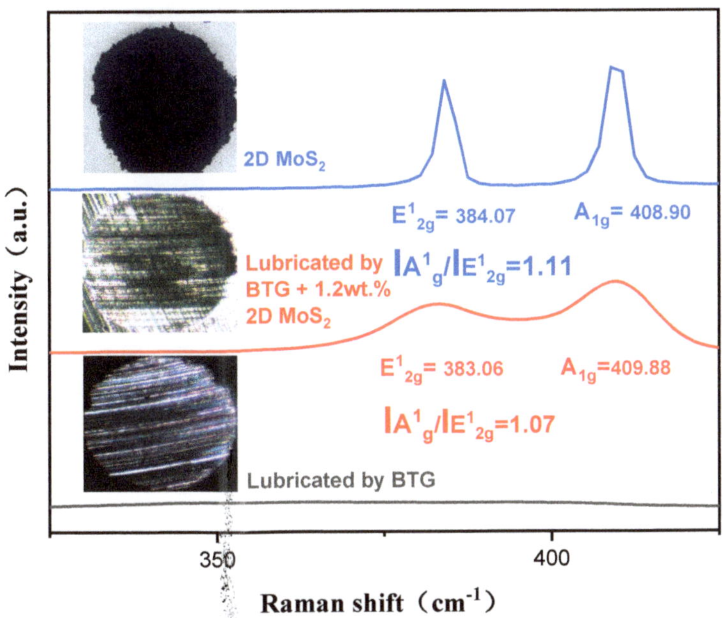

Figure 6. Raman spectra of 2D MoS_2, wear surface on the GCr15 bearing steel ball under the lubrication of BG-1.2, BG-0.

In order to elucidate the lubrication effect of 2D MoS_2 on the surface of the GCr15 bearing steel ball and preliminarily evaluate the appearance of the tribo-chemical film, the EDS and element mapping analyses on the wear surface lubricated by BG-0 and BG-1.2 were measured and the corresponding results are displayed in Figures 7 and 8, respectively. The deep grooves and large pits can ba observed on the wear surface under the lubrication of BG-0, as illustrated in Figure 7. Five main elements were measured on the worn surface, i.e., C, O, Al, Si, and Fe. The detected Fe elements were primarily from the GCr15 bearing steel. However, the detected C, O, Al, Si elements originated from the 150BS base oil and organo-bentone, respectively.

After adding the 2D MoS_2 into BG-0, the pits and grooves on the worn surface lubricated by BG-1.2 were more smoother and smaller, as shown in Figure 8. For example, the mapping illustrated the generation of C, O, Al, Si, Fe, Mo, S, and the like. Even if the proportions of Mo and S elements account for 0.14% and 0.10% owing to the deficient concentration required for the detection limits for EDS, which given evidence of the 2D MoS_2 do work during the friction process and 2D MoS_2 made a large contribution to the significant improvement of lubricantion performance. Compared with worn surface lubricated by BG-0, two new elements Mo and S occured on the wear surface. It could be readily acquired that the Mo and S elements were enriched in the area of contact between the worn surface and GCr15 matrix, which was in accord with to the distribution of Hertz contact stress. Furthermore, the oxidation of the GCr15 bearing steel ball surfaces was visibly cut down in the middle area, where the thickness of the lubrication film was the least. It was shown that the fierce frictional process continually updated sliding GCr15 ball surfaces in the central area, which were hard to completely participate in the tribochemical reaction. In the mean time, the 2D MoS_2 as additives could play a role in avoiding direct crash of asperities during the friction period. Hence, a comparatively thorough tribofilm, abundant in Mo and S elements, which were mainly generated on the aregion of the GCr15 bearing steel ball surfaces with comparatively low contact stress.

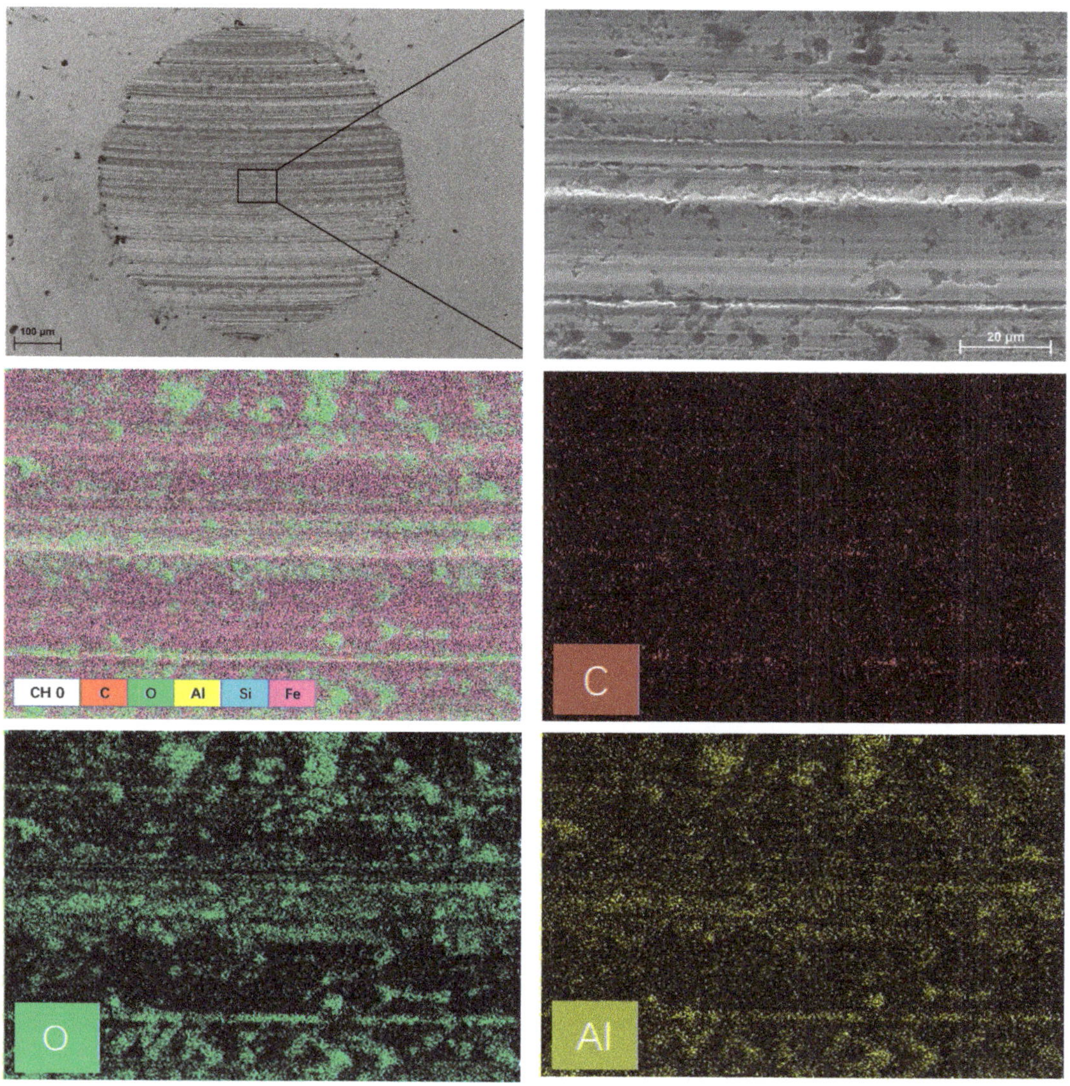

Figure 7. *Cont.*

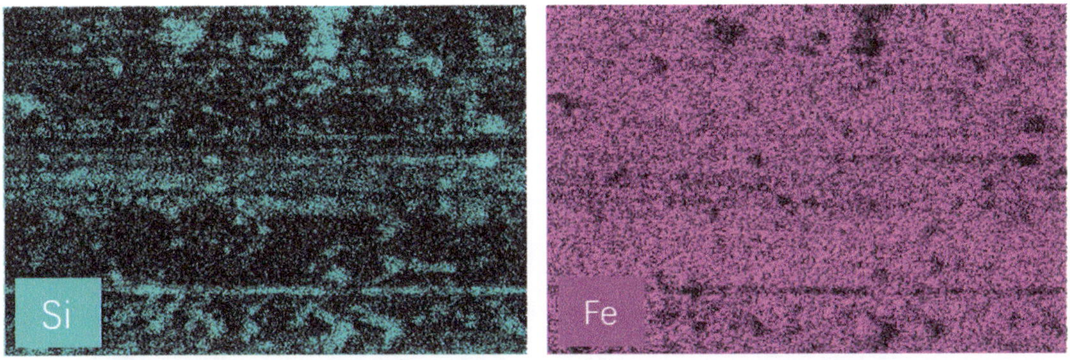

Figure 7. SEM images and C, O, Al, Si and Fe distribution of worn surface lubricated by BG-0.

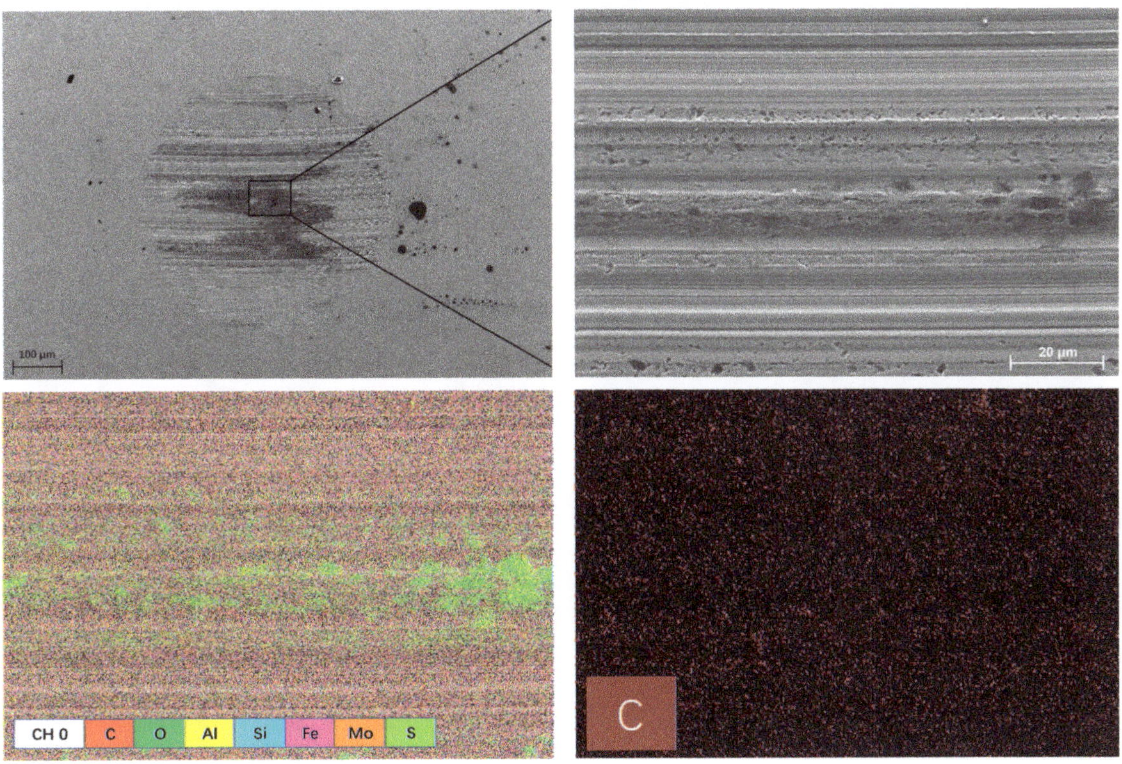

Figure 8. *Cont.*

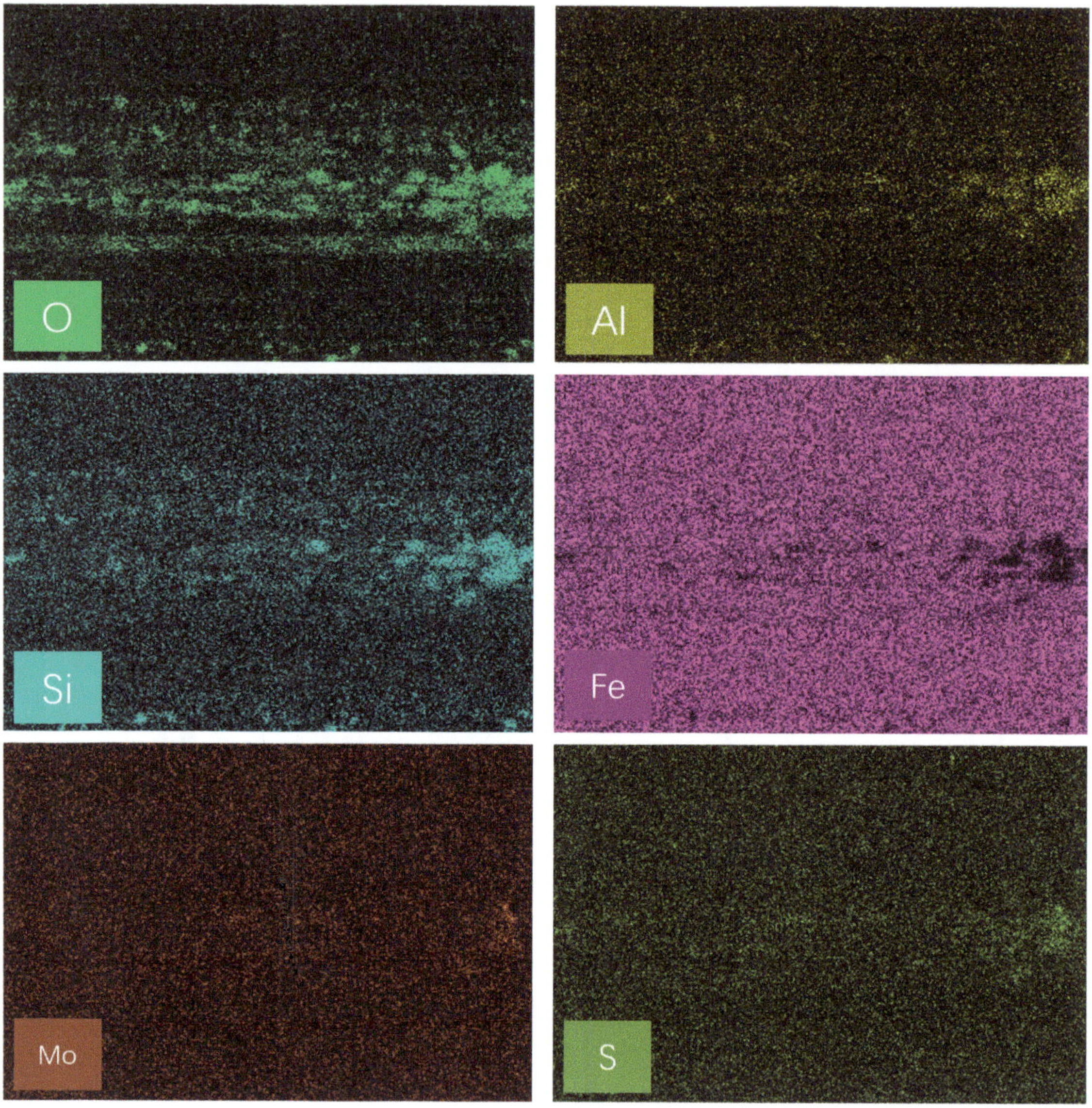

Figure 8. SEM images and C, O, Al, Si, Fe, Mo and S distribution of worn surface lubricated by BG-1.2.

For the sake of exploring the detailed chemical tribo-film formation mechanisms on the steel ball lubricated by BG-0 and BG-1.2, the chemical components on the worn surfaces were detected. The C 1s of 284.8 eV was utilized to calibrate the whole high-resolution XPS photoelectron spectra and subsequently the Gaussian–Lorentzian fitting was used to separate the peaks. Figure 9 presents the XPS survey of worn surface with BG-0 lubrication and showed the patterns of C 1s, O 1s, Si 2p, Al 2p, Fe 2p. As illustrated in Figure 9b, the greatest characteristic peak at 284.80 eV attributes to the C-C bond, the middle characteristic peak at 286.00 eV belongs to the C-O bond, and the smallest characteristic peak at 288.99 eV corresponds to the C=O bond, which originates from the organic matter in 150BS base oil or tribo-chemical film formed by the BG-0 during friction test [34]. As displayed in Figure 9c, the peak of 530.43 eV corresponds to metal oxides, and the characteristic peak near 533.38 eV belongs to the C–O bond [35]. Meanwhile, the peak of 102.29 eV (Figure 9d) and 74.14 eV (Figure 9e) both to refer to the Aluminosilicate which proves that the prepared bentonite grease is relatively stable, and the internal structural skeleton has not been significantly damaged. The peaks located at 711.37.6 and 725.63 eV (Figure 9d) ascribed to the Fe2p1/2 and Fe2p3/2 of –Fe(III)-O- of Fe_2O_3, and there were also two characteristic peaks at 714.14 eV and 728.11 eV on the worn surface of steel ball ascribed to the Fe2p1/2 and Fe2p3/2 of –Fe(II)-O- of FeO [36], respectively, indicating that a stable and robust tribo-film generated on the wear surface.

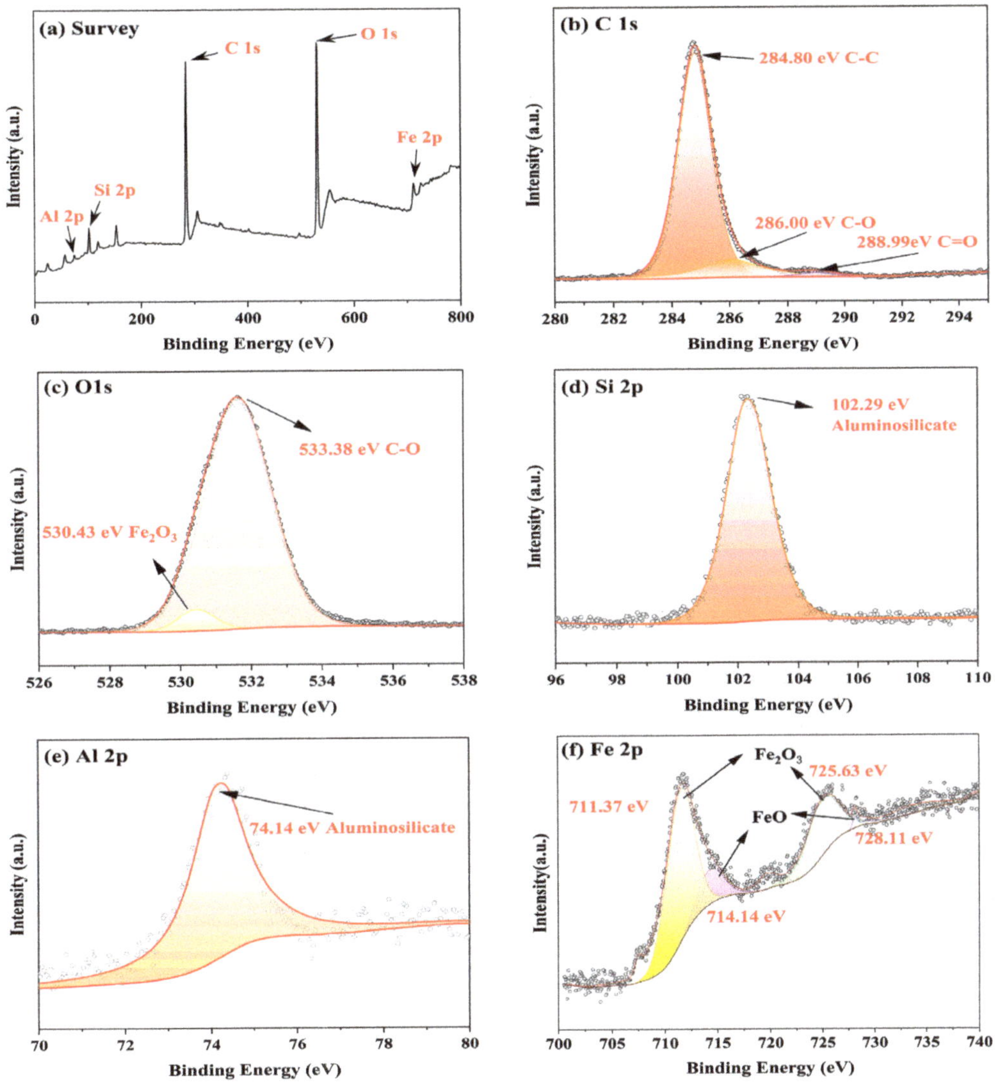

Figure 9. XPS spectra of worn surface with BG-0 lubrication, (**a**) survey, (**b**) C 1s fine spectrum, (**c**) O 1s fine spectrum, (**d**) Si 2p fine spectrum, (**e**) Al 2p fine spectrum, (**f**) Fe 2p fine spectrum.

Figure 10 dispalys the XPS survey of wear surface with BG-0 lubrication and the spectra of C 1s, O 1s, Si 2p, Al 2p, Fe 2p, Mo 3d and S 2p are at around 284.8 eV, 531.84 eV, 102.08 eV, 56.08 eV, 712.08 eV, 233.08 eV and 168.51 eV, respectively, as shown in Figure 10a. Similar to BG-0, the three characteristic peak at 284.80 eV, 286.00 eV and 288.93 eV belong to the C-C bond, the C-O bond and the C=O bond, respectively, as illustrated in Figure 10b. As shown in Figure 10c, the peaks of O1s spectra at 530.33 eV, 531.84 eVand 533.38 eV could be attributed to the metal oxides, the C-O bond and sulfate. Fe elements participate in the metal oxidation reaction. In addition, the spectral peak of 533.38eV is $FeSO_4$, which can be reflected by the increase of the peak of Fe2p3/2 of Fe(II) near 712.47eV in Figure 10f and the appearance of 168.51 eV and 169.66 eV of $SO4^{2-}$ spectral peaks in Figure 10h. Figure 10g is Mo 3d spectrum of worn surface. The peak at 288.97 eV and 232.02 eV ascribe to the Mo^{4+} in MoS_2 [37], which indicates the occurrence of 2D MoS_2 on the worn surface. The peaks located at nearby 232.80 and 235.98 eV correspond to Mo 3d 3/2 and Mo 3d 5/2, respectively, indicating that the the formation of the possible oxidative product molybdenum trioxide (MoO_3) in the tribo-chemical film [38]. Figure 10h is the S 2p spectra of wear surface. The characteristic peaks at 161.78 and 162.93 eV belong to S^{2-} in MoS_2 and FeS [39]. The characteristic peak vicinity 168.51 and 169.66 eV belong to S^{6+}, which confirms that there is not only Fe oxide, but also ferrous sulfate ($FeSO_4$) or ferric sulfate ($Fe_2(SO_4)_3$) on the worn surface of the steel ball. In other words, S is oxidized during the friction test. However, the existence of weak peaks of 165.57 eV and 166.72 eV refering to $SO_3{}^{2-}$ shows that the oxidation reaction on the worn surface of the steel ball is incomplete, and also indirectly confirms the existence of the intermediate $FeSO_3$. In Figure 10e, apart from the peak of aluminosilicate, a distinct peak at 99.22 eV is observed, indicating the presence of monatomic silicon. This can be ascribed to the layered structure of montmorillonite crystal within bentonite grease, which provides ample space for Mo elements from 2D MoS_2 to insert and form chemical bonds, leading to reduction of silicon element in the aluminosilicate. Conversely, no additional peaks are observed in the spectrum of Al 2p (Figure 10e) except for its aluminosilicate peak at 74.30 eV, suggesting that aluminum element does not participate in tribological reactions on the wear surface of GCr15 bearing steel ball. According to the above analysis, a stable adsorption film and a robust tribochemical film composed of Fe_2O_3, $FeSO_4$, $Fe_2(SO_4)_3$, $FeSO_3$, FeS, FeO and MoO_3 generated on the wear urface during frictional test, which account for the enhancement of lubricantion performance.

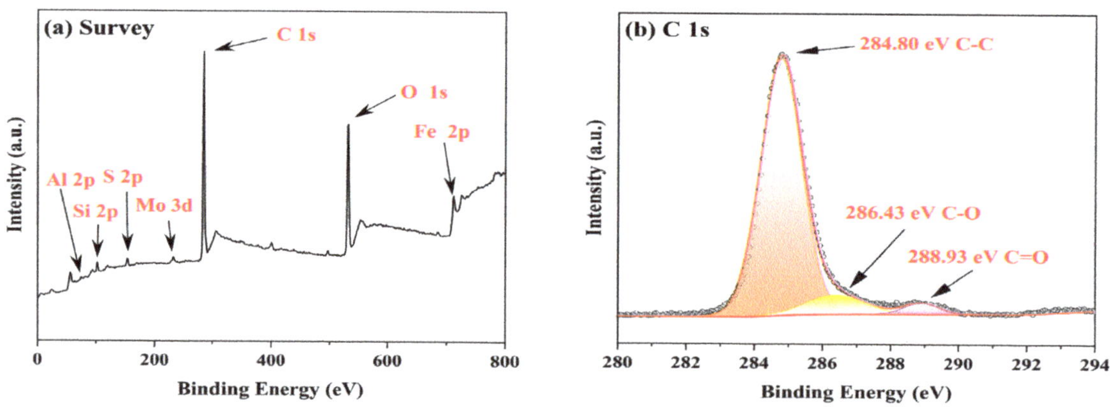

Figure 10. *Cont.*

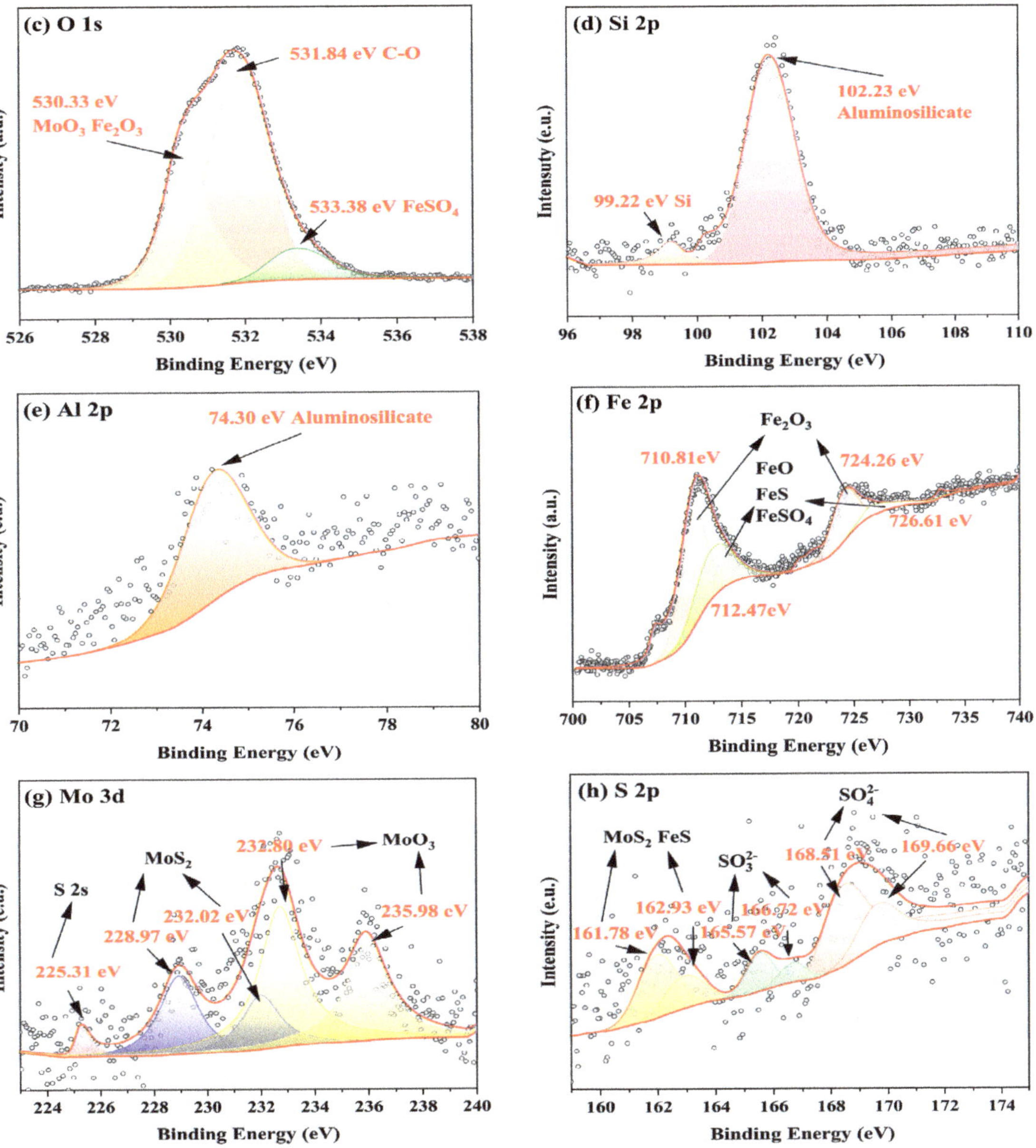

Figure 10. XPS spectra of worn surface with BG-1.2 lubrication, (**a**) survey, (**b**) C 1s fine spectrum, (**c**) O 1s fine spectrum, (**d**) Si 2p fine spectrum, (**e**) Al 2p fine spectrum, (**f**) Fe 2p fine spectrum, (**g**) Mo 3d fine spectrum, (**h**) S 2p fine spectrum.

3.4. Lubrication Mechanism of 2D MoS_2

Based on the above, the schematic illustration of sliding process of GCr15 bearing steel/GCr15 bearing steel friction couples in bentonite grease with 2D MoS_2 has been shown in Figure 11. During the early stages of friction test, 2D MoS_2 can readily penetrate between the sliding friction couples and fill the gaps or adsorb to the worn surface owing to the advantages of the nanometer-sized effect. At the moment, the actual contact region between GCr15 bearing steel balls is much less than the cleavage strength of interatomic covalent bond (S-Mo-S) in 2D MoS_2 sheets is adequately more than actual contact pressure, exceeding 30 GPa [40], which indicates that 2D MoS_2 could bear extremely high Hertz contact pressure between asperities and alleviate the abrasion phenomenon, promoting the anti-wear and friction-reducing performance, which confirmed by Raman analysis.As the sliding proceeds, the lubrication state in the contact area changing to boundary lubrication is inevitable, leading to the physical adsorption film would rupture. However, the heat, plastic distortion and defects generated on the wear surface provided a suitable condition for the formation of subsequent tribo-chemical reactions, resulting in a new and stable tribofilm with superior mechanical properties could withstand plastic deformation and protect GCr15 bearing steel ball surface from serious abrasion. The above-mentioned tribo-film mainly composes of Fe_2O_3, $FeSO_4$, $Fe_2(SO_4)_3$,$FeSO_3$, FeS, FeO, and MoO_3 confirmed by XPS analysis. Therefore, 2D MoS_2 were more easily adsorbed on the GCr15 bearing steel/GCr15 bearing steel surfaces to generate a stable adsorption film and a robust tribochemical film composed of Fe_2O_3, $FeSO_4$, $Fe_2(SO_4)_3$,$FeSO_3$, FeS, FeO, and MoO_3, which is typically more ductile than the GCr15 bearing steel substrate and protect the GCr15 bearing steel/GCr15 bearing steel substrate from severe wear by avoiding direct metal-to-metal contact.

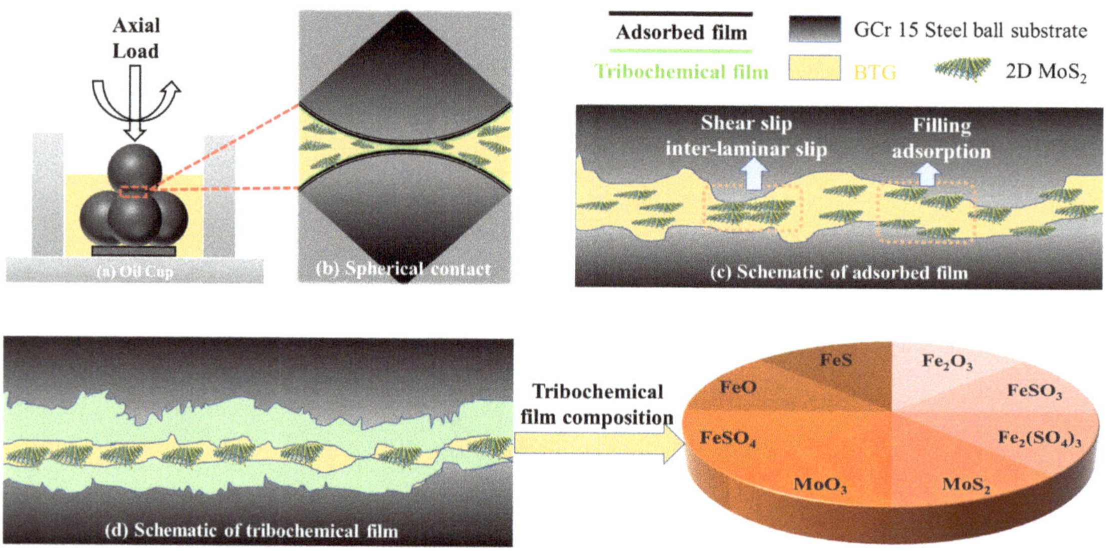

Figure 11. Schematic diagram of the lubricating mechanisms under BG with 2D MoS_2.

4. Conclusions

In summary, 2D MoS_2 with narrow lateral size and thickness distributions was introduced to enhance the friction-reducing and anti-wear performance of the bentonite grease. The relevant tribological mechanisms were illustrated.

a The 2D MoS_2 as lubricating additives utilized in the bentonite grease have significant effects on its penetration, dropping point, oil separation, evaporation, copper corrosion and friction-reducing and antiwear properties.

b The COF, WSD, surface roughness and wear scar depth of BG + 1.2 wt.% 2D MoS_2 were effectively reduced by approximately 22.15%, 23.14%, 55.15%, and 82.64%, respectilvely, in comparison with that of BG. In addition, the contact region was firmly in a state of boundary lubrication under the four-ball tribometer according to the calculation of the Dowson and Hamrock minimum film thickness formula.

c Raman, EDS and XPS results collectively showed that a stable adsorption film and a robust tribochemical composed of Fe_2O_3, $FeSO_4$, $Fe_2(SO_4)_3$,$FeSO_3$, FeS, FeO, and MoO_3, which is typically more ductile than the GCr15 bearing steel substrate and protect the GCr15 bearing steel/GCr15 bearing steel substrate from severe wear by avoiding direct metal-to-metal contact.

Author Contributions: Investigation, S.Z., S.X., X.Y. (Xue Yang), J.Z., K.M. and Y.F.; resources, S.X., X.Y., H.H., Q.Z., Y.C. and Z.L.; methodology and validation, S.X., X.Y., H.B., K.M. and Y.L.; visualization and formal analysis, S.X., X.Y. (Xue Yang), H.B., Q.Z. and L.P.; supervision, X.Y., H.H., X.D., Q.Z. and Y.L.; writing—original draft preparation, S.Z., S.X., X.Y. (Xue Yang) and Y.C.; writing—review and editing, X.Y., X.D., J.Z., L.P. and Z.L. All authors have read and agreed to the published version of the manuscript.

Funding: This research was funded by Science and Technology Research Program of Chongqing Municipal Education Commission, grant number KJZD-K202212905 and Natural Science Foundation of Chongqing, grant number cstc2019jcyj-msxmX0453.

Data Availability Statement: The original contributions presented in the study are included in the article, further inquiries can be directed to the corresponding author.

Conflicts of Interest: Auhtors Kai Ma, Yi Cao, Yuehao Liu, Liangyi Peng, Zhitao Li and Yunhong Fan were employed by Sinopec Lubricant Co., Ltd. The remaining authors declare no conflicts of interest.

References

1. Luo, J.; Min, L.; Ma, L. Origin of friction and the new frictionless technology—Superlubricity: Advancements and future outlook. *Nano Energy* **2021**, *86*, 106092. [CrossRef]
2. Chen, C.; Liu, Y.; Tang, Q.; Xu, H.; Tang, M.; Li, X.; Liu, L.; Dong, J. Tribological and rheological performance of lithium grease with poly-α-olefin and alkyl-tetralin as base oils. *Chin. J. Chem. Eng.* **2023**, *56*, 180–192. [CrossRef]
3. Zhou, C.; Ren, G.; Fan, X.; Lv, Y. Probing the effect of thickener microstructure on rheological and tribological properties of grease. *J. Ind. Eng. Chem.* **2022**, *111*, 51–63. [CrossRef]
4. Calderon Salmeron, G.; Leckner, J.; Schwack, F.; Westbroek, R.; Glavatskih, S. Greases for electric vehicle motors: Thickener effect and energy saving potential. *Tribol. Int.* **2022**, *167*, 107400. [CrossRef]
5. Li, H.; Zhang, Y.; Li, C.; Zhou, Z.; Nie, X.; Chen, Y.; Cao, H.; Liu, B.; Zhang, N.; Said, Z.; et al. Extreme pressure and antiwear additives for lubricant: Academic insights and perspectives. *Int. J. Adv. Manuf. Technol.* **2022**, *120*, 1–27. [CrossRef]
6. Vetter, W.; Sprengel, J.; Krätschmer, K. Chlorinated paraffins—A historical consideration including remarks on their complexity. *Chemosphere* **2022**, *287*, 132032. [CrossRef]
7. Niu, W.; Yuan, M.; Wang, P.; Shi, Q.; Xu, H.; Dong, J. One-pot synthesis of SIB@ZIF-8 with enhanced anti-corrosion properties and excellent lubrication properties. *Tribol. Int.* **2020**, *151*, 106491. [CrossRef]
8. Acharya, B.; Pardue, T.N.; Avva, K.S.; Krim, J. In situ, real time studies of thermal reaction film formation temperatures for iron and 304SS surfaces immersed in 5% tricresyl phosphate in base oil. *Tribol. Int.* **2018**, *126*, 106–115. [CrossRef]
9. Mistry, K.K.; Morina, A.; Erdemir, A.; Neville, A. Tribological Performance of EP Lubricants with Phosphorus-Based Additives. *Tribol. Trans.* **2013**, *56*, 645–651. [CrossRef]
10. Sato, K.; Watanabe, S.; Sasaki, S. High Friction Mechanism of ZDDP Tribofilm Based on in situ AFM Observation of Nano-Friction and Adhesion Properties. *Tribol. Lett.* **2022**, *70*, 94. [CrossRef]
11. Wang, J.; Wang, J.; Li, C.; Zhao, G.; Wang, X. A study of 2,5-dimercapto-1,3,4-thiadiazole derivatives as multifunctional additives in water-based hydraulic fluid. *Ind. Lubr. Tribol.* **2014**, *66*, 402–410. [CrossRef]
12. Guegan, J.; Southby, M.; Spikes, H. Friction Modifier Additives, Synergies and Antagonisms. *Tribol. Lett.* **2019**, *67*, 83. [CrossRef]
13. Liu, S.; Yu, T.; Lich, L.V.; Yin, S.; Bui, T.Q. Size and surface effects on mechanical behavior of thin nanoplates incorporating microstructures using isogeometric analysis. *Comput. Struct.* **2019**, *212*, 173–187. [CrossRef]
14. Baig, N. Two-dimensional nanomaterials: A critical review of recent progress, properties, applications, and future directions. *Compos. Part A Appl. Sci. Manuf.* **2023**, *165*, 107362. [CrossRef]
15. Zhao, J.; Gao, T.; Li, Y.; He, Y.; Shi, Y. Two-dimensional (2D) graphene nanosheets as advanced lubricant additives: A critical review and prospect. *Mater. Today Commun.* **2021**, *29*, 102755. [CrossRef]
16. Zhao, Y.; Li, Y.; Li, J.; Xiao, Y.; Mu, W.; Wang, Z.; Song, L.; Yu, J. Flame Retardancy of Epoxy Resins Modified with Few-Layer Black Phosphorus. *Polymers* **2023**, *15*, 1655. [CrossRef]
17. Zang, C.; Yang, M.; Liu, E.; Qian, Q.; Zhao, J.; Zhen, J.; Zhang, R.; Jia, Z.; Han, W. Synthesis, characterization and tribological behaviors of hexagonal boron nitride/copper nanocomposites as lubricant additives. *Tribol. Int.* **2022**, *165*, 107312. [CrossRef]
18. Yang, J.; Xu, X.; Liu, L. Plasma-assisted friction control of 2D MoS_2 made by atomic layer deposition. *Nanotechnology* **2020**, *31*, 395711. [CrossRef]

19. Yi, S.; Guo, Y.; Li, J.; Zhang, Y.; Zhou, A.; Luo, J. Two-dimensional molybdenum carbide (MXene) as an efficient nanoadditive for achieving superlubricity under ultrahigh pressure. *Friction* **2023**, *11*, 369–382. [CrossRef]
20. Wang, H.; Wang, Y.; Liu, Y.; Zhao, J.; Li, J.; Wang, Q.; Luo, J. Tribological behavior of layered double hydroxides with various chemical compositions and morphologies as grease additives. *Friction* **2021**, *9*, 952–962. [CrossRef]
21. Kang, J.; Sangwan, V.K.; Wood, J.D.; Hersam, M.C. Solution-based processing of monodisperse two-dimensional nanomaterials. *Acc. Chem. Res.* **2017**, *50*, 943–951. [CrossRef] [PubMed]
22. Ling, Z.; Wang, Z.; Zhang, M.; Yu, C.; Wang, G.; Dong, Y.; Liu, S.; Wang, Y.; Qiu, J. Sustainable synthesis and assembly of biomass-derived B/N co-doped carbon nanosheets with ultrahigh aspect ratio for high-performance supercapacitors. *Adv. Funct. Mater.* **2016**, *26*, 111–119. [CrossRef]
23. He, Q.; Zeng, Z.; Yin, Z.; Li, H.; Wu, S.; Huang, X.; Zhang, H. Fabrication of flexible MoS_2 thin-film transistor arrays for practical gas-sensing applications. *Small* **2012**, *8*, 2994–2999. [CrossRef] [PubMed]
24. Zhao, Y.; Xie, Y.; Liu, Z.; Wang, X.; Chai, Y.; Yan, F. Two-dimensional material membranes: An emerging platform for controllable mass transport applications. *Small* **2014**, *10*, 4521–4542. [CrossRef] [PubMed]
25. Luo, Y.; Tang, L.; Khan, U.; Yu, Q.; Cheng, H.-M.; Zou, X.; Liu, B. Morphology and surface chemistry engineering toward pH-universal catalysts for hydrogen evolution at high current density. *Nat. Commun.* **2019**, *10*, 269. [CrossRef]
26. Lu, P.; Xiang, S.; Xu, S.; Chen, H.; Wang, H.; Wang, J.; Zhang, G. Tribological study on preparation of two-dimensional MoS_2 as grease additive by ultrasonic liquid phase stripping. *J. Mater. Eng.* **2023**, *51*, 160–168.
27. Ren, S.; Yu, Q.; Yu, X.; Rong, P.; Jiang, L.; Jiang, J. Graphene-supported metal single-atom catalysts: A concise review. *Sci. China Mater.* **2020**, *63*, 903–920. [CrossRef]
28. Sun, X.; Liu, Z.; Welsher, K.; Robinson, J.T.; Goodwin, A.; Zaric, S.; Dai, H. Nano-graphene oxide for cellular imaging and drug delivery. *Nano Res.* **2008**, *1*, 203–212. [CrossRef]
29. O'Neill, A.; Khan, U.; Coleman, J.N. Preparation of high concentration dispersions of exfoliated MoS_2 with increased flake size. *Chem. Mater.* **2012**, *24*, 2414–2421. [CrossRef]
30. Chaurasia, A.; Verma, A.; Parashar, A.; Mulik, R.S. Experimental and computational studies to analyze the effect of h-BN nanosheets on mechanical behavior of h-BN/polyethylene nanocomposites. *J. Phys. Chem. C* **2019**, *123*, 20059–20070. [CrossRef]
31. Peng, L.; Zhu, Y.; Li, H.; Yu, G. Chemically integrated inorganicgraphene two-dimensional hybrid materials for flexible energy storage devices. *Small* **2016**, *12*, 6183–6199. [CrossRef] [PubMed]
32. Liu, Y.; Li, J.; Li, J.; Yi, S.; Ge, X.; Zhang, X.; Luo, J. Shear-Induced Interfacial Structural Conversion Triggers Macroscale Su-perlubricity: From Black Phosphorus Nanoflakes to Phosphorus Oxide. *ACS Appl. Mater. Interfaces* **2021**, *13*, 31947–31956. [CrossRef] [PubMed]
33. Hamrock, B.J.; Dowson, D.C. Isothermal Elastohydrodynamic Lubrication of Point Contacts: Part III—Fully Flooded Results. *J. Lubr. Technol.* **1976**, *99*, 264–275. [CrossRef]
34. Zhang, Y.; Li, P.; Ji, L.; Liu, X.; Wan, H.; Chen, L.; Li, H.; Jin, Z. Tribological properties of MoS_2 coating for ultra-long wear-life and low coefficient of friction combined with additive g-C_3N_4 in air. *Friction* **2021**, *9*, 789–801. [CrossRef]
35. Han, Y.; Qiao, D.; Guo, Y.; Feng, D.; Shi, L. Influence of Competitive Adsorption on Lubricating Property of Phosphonate Ionic Liquid Additives in PEG. *Tribol. Lett.* **2016**, *64*, 22. [CrossRef]
36. Lu, Z.; Cao, Z.; Hu, E.; Hu, K.; Hu, X. Preparation and tribological properties of WS_2 and WS_2/TiO_2 nanoparticles. *Tribol. Int.* **2019**, *130*, 308–316. [CrossRef]
37. Rabaso, P.; Dassenoy, F.; Ville, F.; Diaby, M.; Vacher, B.; Le Mogne, T.; Belin, M.; Cavoret, J. An Investigation on the Reduced Ability of IF-MoS_2 Nanoparticles to Reduce Friction and Wear in the Presence of Dispersants. *Tribol. Lett.* **2014**, *55*, 503–516. [CrossRef]
38. Geng, S.; Yang, W.; Liu, Y.; Yu, Y. Engineering sulfur vacancies in basal plane of MoS_2 for enhanced hydrogen evolution reaction. *J. Catal.* **2020**, *391*, 91–97. [CrossRef]
39. Hu, H.; He, Y.; Wang, Q.; Tao, L. In-situ research on formation mechanisms of transfer films of a Polyimide-MoS_2 composite in vacuum. *Tribol. Int.* **2023**, *180*, 108211. [CrossRef]
40. Wei, X.; Li, W.; Fan, X.; Zhu, M. MoS_2-functionalized attapulgite hybrid toward high-performance thickener of lubricating grease. *Tribol. Int.* **2023**, *179*, 108135. [CrossRef]

lubricants

MDPI

Article

Effects of Sliding Speed on Wear Behavior of High-Velocity Oxygen Fuel-Sprayed FeCrMoNiCuBSiC Metallic Glass Coatings

Lei Qiao [1,2,*], Botao Zhou [1], Ruifeng Li [1], Taotao Li [1], Yue Zhao [1], Xiaoqiang Zhang [1] and Chul-Hee Lee [3]

1 School of Materials Science and Engineering, Jiangsu University of Science and Technology, Zhenjiang 212100, China
2 China Academy of Machinery Ningbo Academy of Intelligent Machine Tool Co., Ltd., Ningbo 315700, China
3 School of Mechanical Engineering, Inha University, Incheon 22100, Republic of Korea
* Correspondence: qiaolei0306@just.edu.cn

Abstract: A FeCrMoNiCuBSiC metallic glass coating was designed and then deposited by the high-velocity oxygen fuel (HVOF) spraying technique. X-ray diffraction, a scanning electron microscope, and a microhardness tester were applied to characterize the phase, microstructure, and mechanical properties of the coating. The amorphous phase was the main phase in the coating, and crystal phases were almost undetectable in the XRD results. The coating had a dense structure (the porosity was 1.47 ± 0.32%) and high Vickers microhardness (848 ± 22 $HV_{0.3}$). The wear behavior of the coatings sliding against WC-Co was studied with a pin-on-disc wear test system and was compared with that of 316L stainless steel. The coating improved the wear resistance of the steel by around 7–9 times at different sliding speeds. As the sliding speed was increased, the wear loss rate of the steel obviously increased, yet the loss rate of the coating decreased first and then increased. This happened because the contact flash temperature induced by friction increases with the sliding speed, which results in oxidative behavior and crystallization events in the coating. The dominating wear mechanism of the coating is fatigue wear combined with oxidative wear.

Keywords: metallic glass coating; HVOF spray; sliding speed; crystallization; wear mechanism

1. Introduction

Recently, much effort has been focused on fabricating metallic glasses because of their notable properties, such as soft magnetism [1], superconductivity [2], low thermal conductivity [3], ultrahigh hardness and strength [4,5], satisfactory corrosion [6–8], and wear resistance [9,10]. Metallic glasses are new metallic materials that combine the characteristics of glass, metal, and liquid and exhibit prospects for many engineering applications. Unfortunately, however, they require a high cooling rate, so the metallic glasses size is seriously restricted, which limits their application [11,12]. In an effort to overcome this problem, spray coating has been used to deposit metallic glasses with unrestricted size in two dimensions. These metallic glass coatings are an efficient way to improve surface performance.

Owing to their distinct properties along with low cost, Fe-based metallic glass coatings have attracted substantial interest. Thermal deposition techniques like arc spraying [13,14], air plasma spraying [15,16], detonation spraying [17,18], cold spraying [19,20], and high-velocity oxygen fuel spraying [21–23] have attracted attention for preparing Fe-based metallic glass coatings. For example, Zhang et al. [24] developed a group of Fe-Si-B-Nb-Mo-Cr metallic glass coatings by arc spraying and their results revealed that the thermal stability of these coatings increased with the increase in Mo and Cr. Fe-based amorphous/nanocrystalline coatings were prepared by Kumar et al. [25] using a plasma spraying technique, and they found that the plasma power played a significant role in determining the morphology, phase, and wear resistance of the coating. Cui et al. [26] deposited Fe-based metallic glass coatings using a detonation spraying technique, it was reported that the corrosion behavior and mechanism of the coating were changed by the oxygen–fuel ratio. Ziemian et al. [27] fabricated Fe-based amorphous coatings with better corrosion and wear properties through the optimization of the cold spray parameters. Sadeghi et al. [28] prepared Fe-based metallic glass coatings by HVAF and HVOF spraying techniques, and the effect of spray techniques was discussed.

In a variety of thermal deposition techniques, HVOF spraying is prominent for its high flame velocity (~1500 m/s) and low flame temperature (~3000 °C) [29,30]. Due to these characteristics, the powder is easily accelerated and deposited on a substrate with high kinetic energy and cooling rate in the procedure of HVOF spraying. This leads to a dense structure and slight oxidation of the coatings. Therefore, the HVOF spraying technique should be capable of preparing coatings with good properties and non-crystalline structures.

Wear happens in many different engineering components. Preparing protective coatings on the surface of engineering parts is a productive way to enhance their wear resistance. In this study, an FeCrMoNiCuBSiC amorphous coating was designed and deposited on a 316L stainless-steel substrate using the HVOF spraying technique. The microstructure, phase, and microhardness of the coating were evaluated exhaustively. The effect of sliding speeds (0.06, 0.09, and 0.12 m/s) on the wear behavior and mechanism of the coatings was also studied and compared with that of the 316L stainless-steel substrate.

2. Materials and Methods

2.1. Specimen Preparation

A rectangular 316L stainless-steel plate ($400 \times 200 \times 10$ mm^3, provided by Baosteel Co., Ltd., Shanghai, China) was used as a substrate. The substrate was sand blasted with white alumina grit before spraying to guarantee good adhesion. Gas-atomized Fe-based powder (20.0 Cr, 15.0 Mo, 5.0 Ni, 2.5 Cu, 4.5 B, 1.875 Si, and 0.625 C, in wt.%) was prepared as a feedstock. For the spraying experiments, a JP-8000 HVOF spraying system (Praxair, Danbury, CT, USA) was used. The carrier gas was nitrogen, and the spraying distance was 380 mm. The samples for tests were sprayed with 0.43 L/min of kerosene flow, 10.86 L/min of carrier gas flow, 869.75 L/min of oxygen flow, 280 mm/s of gun transverse velocity, and 8 g/min of powder feeding rate.

2.2. Specimen Characterization

The specimens for characterization were cut into cubes with sides of 10 mm using a wire-cut electrical discharge machine. Afterward, the cross-section and surface of the sprayed coatings were ground with abrasive papers, polished with alumina sand to give a uniform finish, and then washed with acetone and alcohol before characterization. Phase identification of the coating was executed with the aid of an X-ray diffractometer (XRD; Bruker D8 Advance) by Cu Kα irradiation (λ = 1.5406 Å). The step size and scanning speed were fixed at 0.02° and 2°/min, respectively.

A scanning electron microscope (SEM; Hitachi-3400N, Hitachi, Tokyo, Japan) and an energy-dispersive X-ray spectroscope (EDX, Hitachi, Tokyo, Japan) were used for the analysis of microstructures and elements of the coating. According to the grayscale threshold method, the coating porosity was measured at a magnification of 500× by DT-2000 image analysis software. A Vickers microhardness tester (HXD-1000 TM/LCD, Shanghai Optical Instrument Co., Ltd., Shanghai, China) was utilized to gauge the cross-sectional microhardness of the coatings. The holding time and load were 15 s and 300 g, respectively. In order to guarantee the credibility of the results, each hardness value was averaged from 10 measurements.

2.3. Wear Tests

A sliding wear test under dry friction was implemented using a commercial wear testing machine (MPW 110, Neoplus Inc., Daejeon, Korea) in the air at an ambient temperature. Figure 1 shows the sketch map of a pin-on-disc model used in this study. The pin, with a diameter of 5 mm, was made of tungsten carbide-cobalt (94 WC, 6 Co, in wt.%), which exhibits extremely high hardness (1534 ± 19 HV). The disc was cut from the steel plate, with dimensions of $\varphi30 \times 10$ mm^3 (Figure 1b). To facilitate the observation of the wear mechanism, the coating surface of the specimens should be polished to a mirror finish and cleaned using ultrasound.

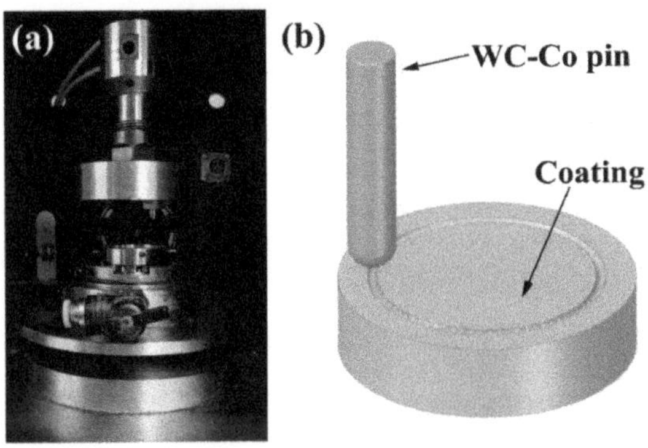

Figure 1. (**a**) Pin-on-disc wear testing machine; (**b**) illustration of a WC-Co pin rotating on the surface of the coating.

The wear tests were conducted by rotating the pin against the disc at a radius of 11.5 mm under an applied load of 20 N. The rotation speeds were set to 50, 75, and 100 r/min, corresponding to sliding speeds of 0.06, 0.09, and 0.12 m/s, respectively. Throughout the wear tests, interruptions were made every hour to measure the mass loss of the specimens. At the end of the 4 h testing period, the wear scar profiles and corresponding wear volumes were measured by a surface testing machine (Mitutoyo SV-3100, Mitutoyo, Kanagawa, Japan). Five wear scar profiles were investigated on each surface to ensure good repeatability. The wear volumes (mm^3) of the coatings were estimated by this equation: $V = 2\pi rS$, where r is the track radius of the wear test and S is the cross-sectional area of the wear scar profile. Each experimental condition was conducted at least twice to ensure the reliability of the results. Worn morphologies were then depicted by SEM equipped with EDX to understand their wear mechanism.

3. Results and Discussion

3.1. Characterization of the Coating

The XRD curves of the powder and deposited coating are presented in Figure 2. As shown in the XRD pattern of the powder, some crystals could be found in the feedstock powder, including α-Fe(Cr), Cr_2B, and (Cr, Fe)$_7C_3$. However, only a single broad peak in the 2θ range of 40–50° could be found from the XRD spectrum of the coating, which implies that the coating is almost entirely amorphous. This phenomenon could be ascribed to the higher cooling rate of the HVOF spraying process than in the gas-atomizing technique [31].

In addition, no obvious oxidation phenomenon could be found within the resolution of the XRD test, which is beneficial for the properties of the coating.

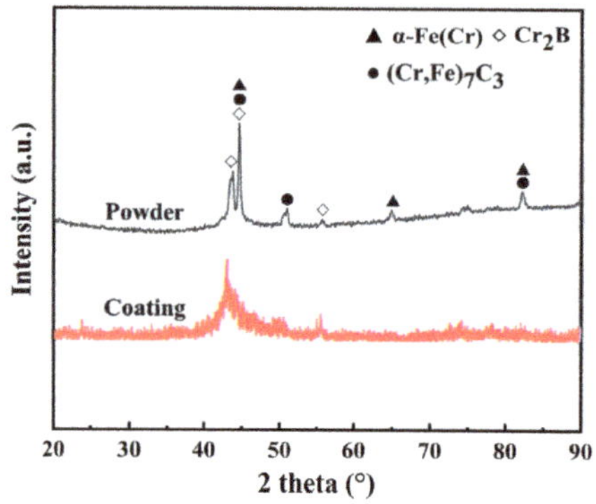

Figure 2. XRD curves of the powder and deposited coating.

Cross-sectional micrographs of the coating are shown in Figure 3. Figure 3a declared that the coating adheres well to the substrate. The coating contains pores and micro-cracks that are generated in association with the release of gases (such as O_2, CO_2, or residual moisture adsorbed on the powder surface) and volume shrinkage during the cooling process of droplets [32]. The calculated porosity of the coating was 1.47 ± 0.32%.

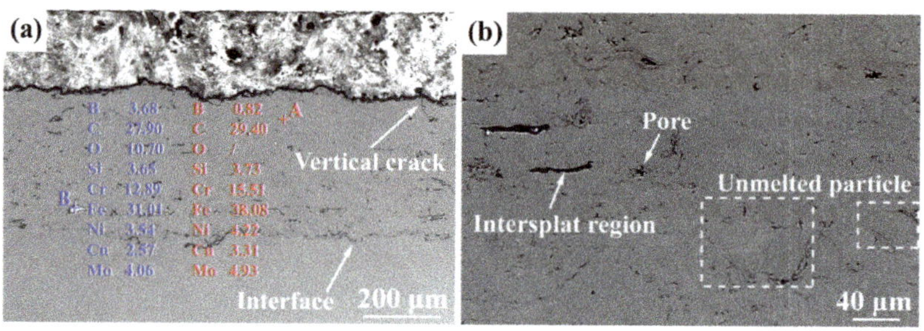

Figure 3. SEM morphologies of cross-section of the Fe-based metallic glass coating: (**a**) low magnification; (**b**) high magnification.

The EDX results shown in Figure 3a reveal that the bright silver area in the coating (point B) contains oxygen. That is to say, oxide inclusions exist in the amorphous matrix. The presence of oxides could be explained by external oxides or the unavoidable particle oxidation in the atmosphere during thermal spraying [33]. The enlarged image in Figure 3b shows the existence of flattened inter-splats, which are caused by the welding of particles [34]. Partially melted and unmelted particles in the coating increase the formation of defects such as pores and inter-splats.

3.2. Microhardness of the Coating

Hardness is a valued indicator in quantifying many material properties, including the estimated endurance limit, strength, and wear performance [35]. Hence, the Vickers microhardness of the Fe-based metallic glass coating was measured in different zones. Figure 4 presents the microhardness values at various distances from the coating/substrate interface. Benefitting from the compact structure, the Fe-based coating exhibits a homogenous microhardness of approximately 848 ± 22 $HV_{0.3}$, which is considerably higher than that of the substrate. Moreover, it is noted that the closer the steel is to the coating/substrate interface, the higher the microhardness is. A theory has been previously postulated for this phenomenon [36]. In the process of sandblasting and spraying, the impact of sand or molten powder generates residual compressive stress on the surface of the substrate, thus generating the effect of work hardening on the substrate.

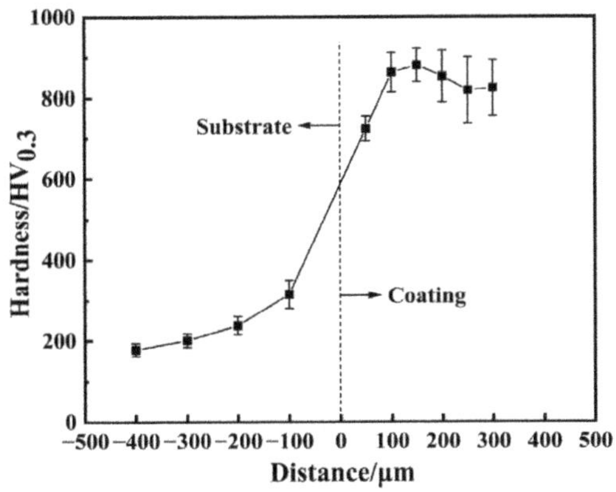

Figure 4. Microhardness–distance profile of the Fe-based metallic glass coating.

3.3. *Dry Sliding Wear Behavior*

The tribological properties and mechanisms of a material are subject to various factors, such as temperature, sliding speed, and applied pressure. In this work, wear tests were conducted at sliding speeds of 0.06, 0.09, and 0.12 m/s to investigate the influence of sliding speed on the tribological performance of the coating. The variation curves of the friction coefficient for the coating and steel with sliding times at different sliding speeds are demonstrated in Figure 5. The average friction coefficients of the coating at sliding speeds of 0.06, 0.09, and 0.12 m/s are 0.688, 0.710, and 0.701, respectively, while those of the steel are 0.699, 0.691, and 0.692, respectively. Figure 5a shows that the curves of the coatings all contain a running-in stage and a steady-state stage. In the running-in stage, the friction coefficient of the coatings increased rapidly and reached a steady state after about 150 s at sliding speeds of 0.06 and 0.09 m/s; for a sliding speed of 0.12 m/s, the steady state was reached after approximately 600 s (as shown in the inset of Figure 5a), with fluctuations within a small range thereafter. This is because, in the early stage of wear, significant microscopic unevenness exists between the coating and the WC-Co pin. As sliding progresses, the protrusions are worn down, leading to an increase in the contact area and friction coefficient. After the running-in stage, the wear process stabilizes [37]. Interestingly, the running-in stage is hard to identify in the friction coefficient curves of the steel, as shown in Figure 5b. In other words, its wear behavior reached a steady state quickly. This happened because its structure is more uniform than that of the coating. In addition, it is implied that the friction coefficient of the steel remained nearly constant in the steady-state stage, but the friction coefficient of the coating still fluctuated within a small range. Except for the inhomogeneity of their phase and microstructure, a drop and rise in the friction coefficient of the coating are also related to the formation and destruction of the transfer layers on the contact surface [38].

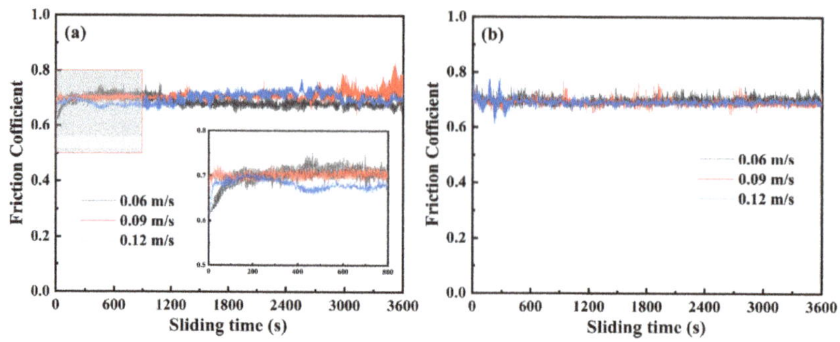

Figure 5. COF vs. sliding time of the Fe-based coating (**a**) and 316L stainless steel (**b**).

Figure 6 reveals the wear tracks of the Fe-based metallic glass coating and the reference steel. All the specimens exhibit continuous and homogenous wear tracks, and the wear tracks of the steel are much wider and deeper than that of the metallic glass coating. This indicates that the 316L stainless steel exhibits poor wear resistance compared to the coating under the same conditions, which is related to the higher hardness of the coating. In addition, in Figure 6a–c, it is obvious that the wear scar width of the coating reduces as the sliding speed accelerates to 0.09 m/s and increases when the sliding speed keeps increasing to 0.12 m/s. However, for the steel, the wear scar width keeps increasing with the sliding speeds (as seen in Figure 6d–f).

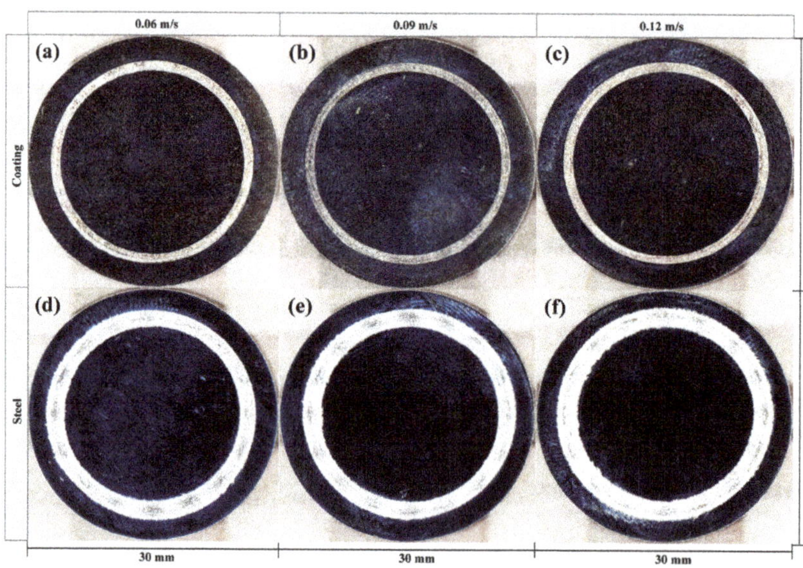

Figure 6. Macroscopic images of the Fe-based coating (**a–c**) and 316L stainless steel (**d–f**) after 4 h wear test at different sliding speeds: (**a**,**d**) 0.06 m/s; (**b**,**e**) 0.09 m/s; (**c**,**f**) 0.12 m/s.

The wear scar profiles of the coating and steel at different sliding speeds were measured, as shown in Figure 7. The influence of sliding speeds on the coating is significantly lighter in comparison to 316L stainless steel, particularly for the width of the wear scar. As shown in the wear scar profiles, it is clear that for the steel, the depth and width of the wear scar increase distinctly as the sliding speed increases. When the sliding speed is 0.06 m/s, the wear volume loss of the steel was calculated to be 10.103 mm^3. As the sliding speed increased to 0.09 and 0.12 m/s, the wear volume losses of the steel increased to 13.652 mm^3 and 17.791 mm^3, respectively. However, the effect of sliding speeds on the wear scar profiles of the coating does not follow this trend. In Figure 7a, the wear volume loss of the coating decreases from 1.956 mm^3 to 1.425 mm^3 as the sliding speed accelerates from 0.06 m/s to 0.09 m/s and then increases to 2.060 mm^3 as the sliding speed accelerates to 0.12 m/s. The enhanced wear resistance of the coating at 0.09 m/s can primarily be attributed to the oxidation process and the increased hardness resulting from the nano-crystallization of the amorphous coating.

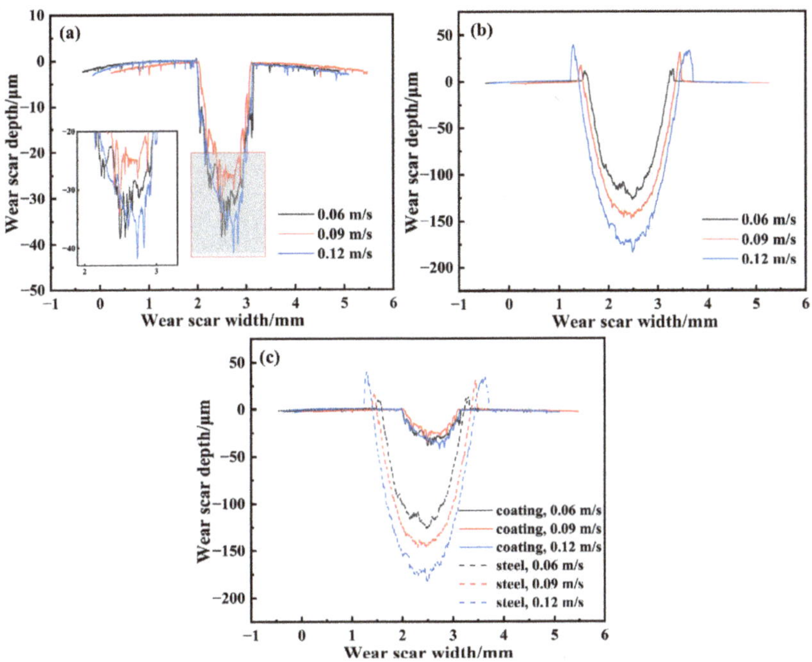

Figure 7. Surface profiles across the wear scars of the Fe-based coating (**a**), 316L stainless steel (**b**), and their comparison (**c**).

In order to compare the wear volume losses between the coating and steel intuitively, their respective wear scar profiles are presented together in Figure 7c. It is evident that the coating exhibits significantly less volume loss than the steel under the same conditions. After 4 h of wear, the volume loss of the steel is about 7~9 times greater than that of the coating at the same sliding speed. Furthermore, unlike the coating, a noticeable material accumulation phenomenon was observed on the worn surface of 316L stainless steel, as shown in Figure 7b. This phenomenon occurred because the steel experienced relatively severe wear, generating more wear debris that accumulated at the edges of the wear scar. The distinct wear scar profiles of the coating and the steel indicate differences in their wear mechanisms.

The wear mass loss rates of the coating and steel were compared under different sliding speeds, as presented in Figure 8. Evidently, it can be found the wear mass loss rates of 316L stainless steel were much greater than that of the coating (Figure 8a). The mass loss rates of the coating and steel under different sliding speeds after being worn for 4 h are illustrated in Figure 8b. As the sliding speed increased, the mass loss rate of the steel obviously increased. However, the mass loss rate of the coating decreased first and then increased. This result is consistent with the analysis of the wear scars.

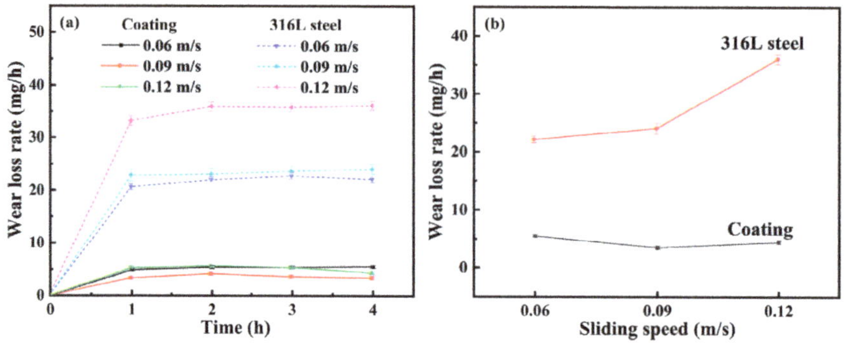

Figure 8. Mass loss rates of the Fe-based coating and 316L stainless steel during the wear test (**a**) and after 4 h wear test (**b**).

3.4. Worn Morphologies

Figure 9 presents the worn morphologies of the coatings under various sliding speeds. All the coatings exhibited similar worn morphologies. In the wear tracks, it is clear that delamination is present on the worn surfaces of the coatings (Figure 9a,c,e). In the worn morphology of each coating, faults resulting from detachment can be observed at different depths and positions. Among all the coatings, the coating worn at 0.09 m/s exhibited the weakest delamination. At higher magnification (Figure 9b,d,f), noticeable cracks can be observed on the worn surface of the coating, which are typical characteristics of fatigue wear.

Oxidation is another phenomenon that can be observed on the worn surfaces of these coatings. The EDX results of points A and B in Figure 9c are shown in Table 1. Compared to the original content of the coating, the alloying element (Fe, Mo, and Cr) content of the worn surface was reduced, while the oxide content of the worn surface (point B) was much higher than that of the original surface (point A). This implies the occurrence of oxidative wear in the coating during the wear process.

Table 1. EDX results of the worn surface of the Fe-based coating (at.%).

Element	C	O	Si	Mo	Cr	Fe	Co	Ni	Cu
Point A	11.64	0.40	5.63	5.82	18.86	46.44	0.42	6.03	4.76
Point B	14.91	24.68	3.78	3.73	12.62	32.74	0.32	3.62	3.59

The worn morphologies of 316L steel were radically different from those of the coatings. The worn morphologies of the steel were consistent across different sliding speeds, so the worn surface morphology at a sliding speed of 0.09 m/s was selected as a typical example. As exhibited in Figure 10, the wear scar width of the steel at 0.09 m/s was 2166 μm, approximately twice that of the coating (1116 μm) at the same sliding speed. The worn morphology of the steel is more severe compared to that of the coating. As shown in Figure 10a, as a result of the comparatively low hardness and serious wear of the steel, a large amount of wear debris was generated and accumulated at the edge of the wear scar [39]. This is consistent with the results from the wear scar profiles. The cracks, smashed regions, and plate-like structures in Figure 10b are regular characteristics of adhesive wear. This mechanism is very common for Fe-based materials/stainless steel pairs [40].

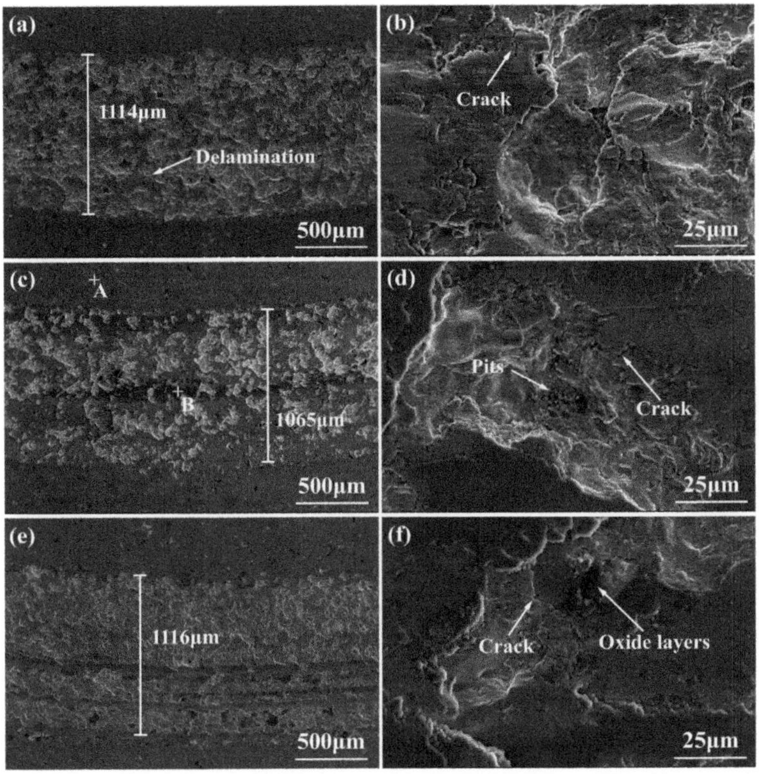

Figure 9. SEM micrographs of the wear scars of the Fe-based coatings at sliding speeds of 0.06 m/s (**a**,**b**), 0.09 m/s (**c**,**d**), and 0.12 m/s (**e**,**f**), (**a**,**c**,**e**): low magnification; (**b**,**d**,**f**): high magnification).

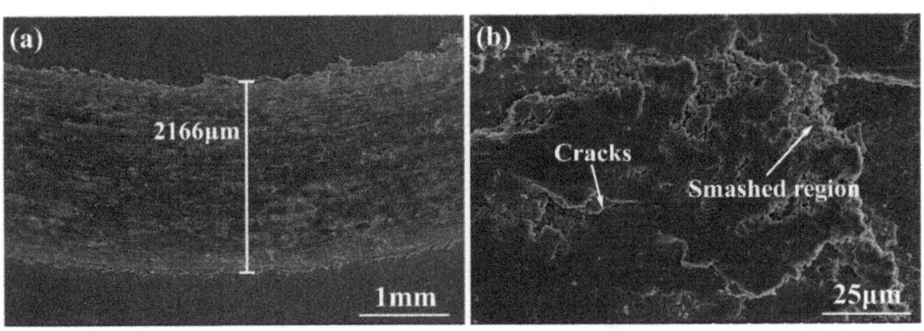

Figure 10. SEM micrographs of 316L stainless steel at sliding speed of 0.09 m/s: (**a**) low magnification; (**b**) high magnification.

Starting off, the WC-Co pin and the steel are in direct contact in the form of two-body wear. As the wear process progresses, the generated wear debris is preferentially transferred from the steel to the surface of WC-Co because of the lower hardness of steel and the strong adhesion of Co with Ni and Cr in 316L stainless steel [41]. Hence, with the addition of wear debris, the wear mechanism was changed to three-body wear. In this process, the wear particles wedge in the steel surface and form abrasive grooves parallel to the sliding direction of the WC-Co pin (Figure 10a). In summary, the wear mechanism of 316L stainless steel is a combination of abrasion, adhesion, and serious plastic deformation.

3.5. Wear Mechanism of the Coating

The wear mechanisms of the coating and steel are quite different. The corresponding wear mechanism of the coating is illustrated in Figure 11. As previously mentioned, due to the high flame velocity of the HVOF spraying technique, the powder can be accelerated to high speed, so the coatings usually have a dense structure and do not easily crack or peel. In addition, the deposited Fe-based metallic glass coating in this study has a high hardness, which also leads to its strong resistance to local deformation and inhibits the cracking and detachment of the coating. However, as thermally sprayed coatings, pores, and cracks can undoubtedly be found in the Fe-based coating. Therefore, these defects will propagate in preference under the effect of pressure and shear force, thus leading to larger cracks. As the wear process progresses, the cracks further expand. Eventually, these cracks spread over the entire coating surface and resulted in delamination and flaked lamellae. The pits on the wear scars also imply the existence of delamination wear on the coating surface. This type of wear mechanism is considered fatigue wear.

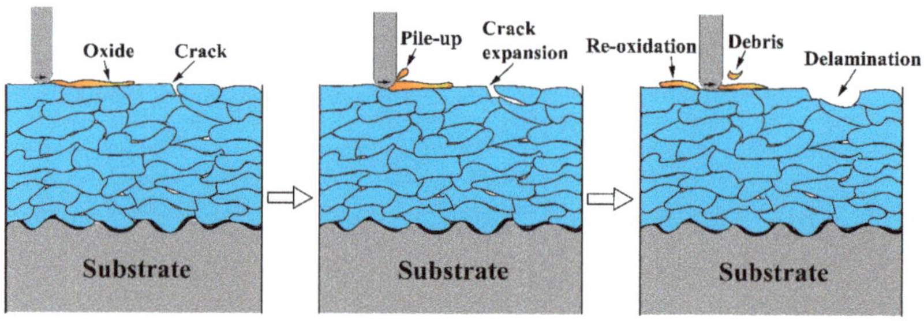

Figure 11. Illustrations of the wear mechanism of the coating.

Oxidative wear is another wear mechanism of the Fe-based metallic glass coating. Under the effect of frictional heat, the coating is prone to oxidation, forming oxide layers (Figure 9f). The previous experimental results (Figures 7 and 8) show that the sliding speed has an obvious influence on the wear loss rate of the coating. This is because, in the process of frictional sliding, the high temperature caused by friction will result in the oxidation and crystallization of the coating. Liu's model can be used to estimate the flash temperature T_f (°C) between the friction pair and the coating [42]:

$$T_f = \frac{\mu v \sqrt{\pi F_N P_y}}{4(k_1 + k_2)}$$

where μ is the friction coefficient, v is the sliding speed, F_N is the normal load, P_y is the nanoindentation hardness of the coating, and k_1 and k_2 are the thermal conductivities of the coating and friction pair, respectively. It can be seen that the flash temperature is proportional to the sliding speed. For Fe-based amorphous coatings, due to their laminar microstructure and porosity, the thermal conductivity is generally poor, leading to higher flash temperatures during the wear process. The flash temperature can typically reach several hundred or even over a thousand degrees Celsius [43,44], which exceeds the crystallization temperature of Fe-based amorphous coatings, thereby causing surface oxidation and crystallization of the coating.

To be more specific, upon increasing the sliding speed from 0.06 m/s to 0.09 m/s, the contact flash temperature rises, resulting in the formation of thicker and more oxide layers on the wear scar surface. As shown in Figure 9a,c,e, there are many disconnected oxide layers distributed on the worn surface of the coating. When the oxide layers increase to a certain thickness, it will be easy to be broken under the action of shear force. The oxide layers will successively undergo the course of formation, break, and peeling in the process of frictional sliding. Compared to the sliding speed of 0.06 m/s, there is more oxidized debris on the coating surface at the sliding speed of 0.09 m/s, which reduces metal-to-metal contact. Although the hardness of the oxidized debris is higher, its presence generally reduces the wear rate. On the other hand, when the sliding speed reaches 0.09 m/s, the higher flash temperature results in more serious crystallization in the coating, leading to the formation of more nano-scale crystalline phases. It is generally accepted that a proper amount and size of crystals in an amorphous matrix can improve the hardness [45]. Thus, the coating worn at 0.09 m/s shows a slight decrease in the wear rate compared to that at 0.06 m/s.

Compared with the sliding speed of 0.09 m/s, the wear loss rate of the coating increased when increasing the sliding speed to 0.12 m/s. This happened because the shear force improves even though severe oxidation and crystallization are generated at 0.12 m/s. The higher shear force leads to more severe delamination of the coating. In addition, it is generally accepted that excessive crystallization behavior may decrease the hardness of amorphous/crystalline composites. This is because when the grains of the second phase in composite materials are larger, the composite materials usually exhibit lower hardness, resulting in poorer wear resistance. In summary, the wear behavior of the Fe-based metallic glass coating is affected by the joint action of crack propagation, oxidation, and crystallization. The wear mechanism of the coating is fatigue wear combined with oxidative wear.

4. Conclusions

In this study, a detailed investigation was conducted on the phases, microstructures, and tribological properties of FeCrMoNiCuBSiC coating prepared by an HVOF spraying technique. The effect of the sliding speeds on the tribological performance of the coating was investigated and compared with that of 316L stainless steel. The following major conclusions were drawn from this study:

- The amorphous phase was the major phase of the coating. The thickness of the coating was around 370 μm. The coating exhibited a low porosity of 1.47 ± 0.32% and a high hardness of 848 ± 22 $HV_{0.3}$.
- Under dry sliding wear conditions, the coating exhibited much better resistance properties than that of the stainless steel. Compared with the steel, the wear volume loss rate of the coating at the same sliding speed was 7–9 times lower.
- As a consequence of the oxidation and crystallization behavior in the coating during the wear process, the wear resistance of the coating increased first and then decreased with the increase in sliding speed. The main wear mechanism of the coating was fatigue wear combined with oxidative wear.

Author Contributions: Conceptualization, L.Q. and C.-H.L.; methodology, L.Q.; validation, L.Q., R.L. and C.-H.L.; data curation, B.Z.; writing—original draft, L.Q. and B.Z.; writing—review and editing, T.L.; visualization, Y.Z.; supervision, C.-H.L. and X.Z.; funding acquisition, R.L. All authors have read and agreed to the published version of the manuscript.

Funding: This research was funded by the National Natural Science Foundation of China, grant number 52075228.

Data Availability Statement: The data generated during the current study are available from the author upon reasonable request.

Conflicts of Interest: Author Lei Qiao was employed by China Academy of Machinery Ningbo Academy of Intelligent Machine Tool Co., Ltd., the remaining authors declare no conflicts of interest.

References

1. Zhang, X.; Xiao, Z.; Zhou, X.; Liu, G.; Li, C.; Huang, Q.; Axinte, E.; Zhou, G.; Zhao, D.; Wang, Y. Dependence of structures and performances of soft-magnetic FeCo-based metal glasses on cryogenic treatments. *J. Mater. Res. Technol.* **2024**, *33*, 3492–3500. [CrossRef]
2. Baggioli, M.; Setty, C.; Zaccone, A. Effective theory of superconductivity in strongly coupled amorphous materials. *Phys. Rev. B* **2020**, *101*, 214502. [CrossRef]
3. Yao, H.; Wang, L.; Zhou, Z.; Wang, B.; Tan, Z.; He, D.; Xue, Y. Thermal transport property correlated with microstructural evolution of Fe-based amorphous alloy. *Acta Mater.* **2020**, *200*, 793–802. [CrossRef]
4. Zhong, J.; Hou, B.; Zhang, W.; Zhang, S.; Zhao, Y.; Zhao, C.; Li, W. High Hardness, Excellent Hydrophobicity, and Favorable Corrosion Resistance of Plasma- Sprayed FeCrMoSi Amorphous Coatings on 304 Stainless Steel. *Molecules* **2023**, *28*, 6718. [CrossRef]
5. Jiang, Q.; Zhang, P.; Tan, J.; Tian, Y.; Ma, S.; Wu, D. Influence of the microstructure on mechanical properties of SLM additive manufacturing Fe-based bulk metallic glasses. *J. Alloys Compd.* **2022**, *894*, 162525. [CrossRef]
6. Deng, L.; Gebert, A.; Zhang, L.; Chen, H.; Gu, D.; Kühn, U.; Zimmermann, M.; Kosiba, K.; Pauly, S. Mechanical performance and corrosion behaviour of Zr-based bulk metallic glass produced by selective laser melting. *Mater. Design* **2020**, *189*, 108532. [CrossRef]
7. Nayak, S.; Kumar, A.; Laha, T. Fe-based metallic glass coatings by thermal spraying: A focused review on corrosion properties and related degradation mechanisms. *Int. Mater. Rev.* **2023**, *68*, 404–485. [CrossRef]
8. Wei, X.; Jin, J.; Jiang, Z.; Liang, D.; Shen, J. FeCrMoWCBY metallic glass with high corrosion resistance in molten lead–bismuth eutectic alloy. *Corros. Sci.* **2021**, *190*, 109688. [CrossRef]
9. Aditya, A.; Wu, H.F.; Arora, H.; Mukherjee, S. Amorphous metallic alloys: Pathways for enhanced wear and corrosion resistance. *JOM* **2017**, *69*, 2150–2155. [CrossRef]
10. Zhao, P.; Shi, Z.; Wang, X.; Li, Y.; Cao, Z.; Zhao, M.; Liang, J. A Review of the Laser Cladding of Metal-Based Alloys, Ceramic-Reinforced Composites, Amorphous Alloys, and High-Entropy Alloys on Aluminum Alloys. *Lubricants* **2023**, *11*, 482. [CrossRef]
11. Wang, Q.; Han, P.; Yin, S.; Niu, W.J.; Zhai, L.; Li, X.; Mao, X.; Han, Y. Current research status on cold sprayed amorphous alloy coatings: A review. *Coatings* **2021**, *11*, 206. [CrossRef]
12. Feng, Z.; Geng, H.; Zhuang, Y.; Li, P. Progress, applications, and challenges of amorphous alloys: A critical review. *Inorganics* **2024**, *12*, 232. [CrossRef]
13. Pan, P.; Zhou, W.; Zhao, Y.; Wang, Y.; Zhang, N. Hot corrosion behavior of an arc sprayed Fe-based amorphous coating in a simulated biomass firing environment. *Corros. Sci.* **2022**, *194*, 109938. [CrossRef]
14. Pan, T.J.; Leng, Y.; Liu, S.; Jiang, J.; Shen, J.; Xiang, J.H. Wear and Corrosion Resistance Performance of Iron-Based Metallic Amorphous Coating Produced by Twin Wire Arc Spraying. *J. Mater. Eng. Perform.* **2023**, *32*, 7556–7566. [CrossRef]
15. Faridi, M.A.; Nayak, S.K.; Prasad, D.; Mustafa, S.E.; Babu, D.A.; Murthy, J.K.N.; Laha, T. Evolution of Wear Mechanism Under Different Sliding Conditions of an Fe-Based Amorphous Coating Synthesized by Plasma Spraying. *J. Therm. Spray Technol.* **2024**, *33*, 2593–2610. [CrossRef]
16. Kim, K.W.; Ham, G.S.; Cho, G.S.; Kim, C.P.; Park, S.C.; Lee, K.A. Microstructures and corrosion properties of novel $Fe_{46.8}$-$Mo_{30.6}$-$Cr_{16.6}$-$C_{4.3}$-$B_{1.7}$ metallic glass coatings manufactured by vacuum plasma spray process. *Intermetallics* **2021**, *130*, 107061. [CrossRef]
17. Zhai, H.; Ou, M.; Cui, S.; Li, W.; Zhang, X.; Cheng, B.; He, D.; Li, X.; Cai, A. Characterizations the deposition behavior and mechanical properties of detonation sprayed Fe-based amorphous coatings. *J. Mater. Res. Technol.* **2022**, *18*, 2506–2518. [CrossRef]
18. Cui, S.; Zhai, H.M.; Li, W.S.; Tong, W.; Li, X.S.; Cai, A.H.; Fan, X.J.; Li, X.Q.; Xiong, D.S. Superhydrophobic Fe-based amorphous coating fabricated by detonation spraying with excellent anti-corrosion and self-cleaning properties. *Rare Met.* **2023**, *42*, 629–644. [CrossRef]
19. Su, J.; Kang, J.; Yue, W.; Ma, G.; Fu, Z.; Zhu, L.; She, D.; Wang, H.; Wang, C. Comparison of tribological behavior of Fe-based metallic glass coatings fabricated by cold spraying and high velocity air fuel spraying. *J. Non-Cryst. Solids* **2019**, *522*, 119582. [CrossRef]
20. Wang, Q.; Han, P.; Niu, W.; Li, N.; Mao, X. High-speed impact behavior of cold-sprayed Fe-based amorphous particle on Mg alloy substrate. *Intermetallics* **2023**, *163*, 108070. [CrossRef]
21. Li, C.; Wang, H.; Ding, J.; Wang, S.; Li, J.; Kou, S. Effects of heat treatment on HVOF-sprayed Fe-based amorphous coatings. *Surf. Eng.* **2021**, *37*, 590–598. [CrossRef]
22. Nayak, S.K.; Kumar, A.; Pathak, A.; Banerjee, A.; Laha, T. Multi-scale mechanical properties of Fe-based amorphous/nanocrystalline composite coating synthesized by HVOF spraying. *J. Alloys Compd.* **2020**, *825*, 154120. [CrossRef]
23. Qiao, L.; Wu, Y.; Hong, S.; Cheng, J.; Zhu, S. Influence of annealing on microstructure and cavitation erosion resistance of iron-based metallic glass coatings synthesized by HVOF thermal spraying. *Intermetallics* **2023**, *161*, 107970. [CrossRef]

24. Zhang, B.; Cheng, J.; Liang, X. Effects of Cr and Mo additions on formation and mechanical properties of Arc-sprayed FeBSiNb-based glassy coatings. *J. Non-Cryst. Solids* **2018**, *499*, 245–251. [CrossRef]
25. Kumar, A.; Kumar, R.; Bijalwan, P.; Dutta, M.; Banerjee, A.; Laha, T. Fe-based amorphous/nanocrystalline composite coating by plasma spraying: Effect of heat input on morphology, phase evolution and mechanical properties. *J. Alloys Compd.* **2019**, *771*, 827–837. [CrossRef]
26. Cui, S.; Zhai, H.; Li, W.; Fan, X.; Li, X.; Ning, W.; Xiong, D. Microstructure and corrosion resistance of Fe-based amorphous coating prepared by detonation spray. *Surf. Coat. Technol.* **2020**, *399*, 126096. [CrossRef]
27. Ziemian, C.W.; Wright, W.J.; Cipoletti, D.E. Influence of impact conditions on feedstock deposition behavior of cold-sprayed Fe-based metallic glass. *J. Therm. Spray Techn.* **2018**, *27*, 843–856. [CrossRef]
28. Sadeghi, E.; Joshi, S. Chlorine-induced high-temperature corrosion and erosion-corrosion of HVAF and HVOF-sprayed amorphous Fe-based coatings. *Surf. Coat. Technol.* **2019**, *371*, 20–35. [CrossRef]
29. Mardali, M.; SalimiJazi, H.R.; Karimzadeh, F.; Luthringer, B.; Blawert, C.; Labbaf, S. Comparative study on microstructure and corrosion behavior of nanostructured hydroxyapatite coatings deposited by high velocity oxygen fuel and flame spraying on AZ61 magnesium based substrates. *Appl. Surf. Sci.* **2019**, *465*, 614–624. [CrossRef]
30. Zhao, X.; Li, C.; Li, S.; Han, X.; Jiang, H. Mechanism study on the influence of combustion models and spray gun geometric parameters on high-velocity oxygen-fuel (HVOF) thermal spraying. *J. Manuf. Process.* **2023**, *98*, 173–185. [CrossRef]
31. Chau, J.L.H.; Alfred, I.; Pan, T.; Yang, C.C. Preparation of gas-atomized Fe-based alloy powders and HVOF sprayed coatings. *Adv. Mater. Res.* **2017**, *6*, 343–348.
32. Qiao, L.; Wu, Y.; Duan, J.; Gao, W.; Hong, S. Corrosion behavior of arc-sprayed pore-sealed Zn and Al coatings in seawater containing sulfate-reducing bacteria (SRB). *J. Therm. Spray Techn.* **2021**, *30*, 1557–1565. [CrossRef]
33. Lou, M.; White, D.R.; Banerji, A.; Alpas, A.T. Dry and lubricated friction behaviour of thermal spray low carbon steel coatings: Effect of oxidational wear. *Wear* **2019**, *432*, 102921. [CrossRef]
34. Alidokht, S.A.; Gao, Y.; Castilho, B.D.; Sharifi, N.; Harfouche, M.; Stoyanov, P.; Makowiec, M.; Moreau, C.; Chromik, R.R. Microstructure and mechanical properties of Tribaloy coatings deposited by high-velocity oxygen fuel. *J. Mater. Sci.* **2022**, *57*, 20056–20068. [CrossRef]
35. Liang, D.; Ma, J.; Cai, Y.; Liu, X.; Xie, S.; Wei, X.; Xu, G.; Shen, J. Characterization and elevated-temperature tribological performance of AC–HVAF-sprayed Fe-based amorphous coating. *Surf. Coat. Technol.* **2020**, *387*, 125535. [CrossRef]
36. Cheng, J.; Wu, Y.; Hong, S.; Cheng, J.; Qiao, L.; Wang, Y.; Zhu, S. Spray parameters optimization, microstructure and corrosion behavior of high-velocity oxygen-fuel sprayed non-equiatomic CuAlNiTiSi medium-entropy alloy coatings. *Intermetallics* **2022**, *142*, 107442. [CrossRef]
37. Venema, J.; Matthews, D.T.A.; Hazrati, J.; Wörmann, J.; Boogaard, A.H. Friction and wear mechanisms during hot stamping of AlSi coated press hardening steel. *Wear* **2017**, *380*, 137–145. [CrossRef]
38. Cho, M.H.; Cho, K.H.; Kim, S.J.; Kim, D.H.; Jang, H. The role of transfer layers on friction characteristics in the sliding interface between friction materials against gray iron brake disks. *Tribol. Lett.* **2005**, *20*, 101–108. [CrossRef]
39. Nie, M.H.; Jiang, P.F.; Li, X.R.; Zhu, D.D.; Zhang, Z.H. Microstructure and tribological properties of laser directed energy deposited 316-NiTi heterogeneous bionic sandwich structure coatings. *J. Mater. Res. Technol.* **2024**, *29*, 5090–5106. [CrossRef]
40. Li, C.X.; Bell, T. Sliding wear properties of active screen plasma nitrided 316 austenitic stainless steel. *Wear* **2004**, *256*, 1144–1152. [CrossRef]
41. Sun, F.; Li, Z.; Jiang, D.; Chen, B. Adhering wear mechanism of cemented carbide cutter in the intervallic cutting of stainless steel. *Wear* **1998**, *214*, 79–82.
42. Meghwal, A.; Pinches, S.; King, H.J.; Schulz, C.; Stanford, N.; Hall, C.; Berndt, C.C.; Ang, A.S.M. Fe-based amorphous coating for high-temperature wear, marine and low pH environments. *Materialia* **2022**, *25*, 101549. [CrossRef]
43. Li, X.; Zhai, H.; Li, W.; Cui, S.; Ning, W.; Qiu, X. Dry sliding wear behaviors of Fe-based amorphous metallic coating synthesized by d-gun spray. *J. Non-Cryst. Solids* **2020**, *537*, 120018. [CrossRef]
44. Al-Abboodi, H.; Fan, H.; Mhmood, I.A.; Al-Bahrani, M. The dry sliding wear rate of a Fe-based amorphous coating prepared on mild steel by HVOF thermal spraying. *J. Mater. Res. Technol.* **2022**, *18*, 1682–1691. [CrossRef]
45. Koga, G.Y.; Ferreira, T.; Guo, Y.; Coimbrão, D.D.; Jorge Jr, A.M.; Kiminami, C.S.; Bolfarini, C.; Botta, W.J. Challenges in optimizing the resistance to corrosion and wear of amorphous Fe-Cr-Nb-B alloy containing crystalline phases. *J. Non-Cryst. Solids* **2021**, *555*, 120537. [CrossRef]

lubricants

MDPI

Article

Study on the Effect of CeO_2 on the Performance of WC + Ni60 Laser Cladding Coating

Jingquan Wu [1,2], Jianwen Zhang [1], Dianlong Chen [1], Jiang Huang [3], Wenqing Shi [4], Fenju An [1] and Xianglin Wu [1,2,*]

1 School of Mechanical Engineering, Guangdong Ocean University, Zhanjiang 524088, China; wujq@gdou.edu.cn (J.W.); jianwencheung@foxmail.com (J.Z.); cdl8940@gdou.edu.cn (D.C.); afj1022@126.com (F.A.)
2 Guangdong Engineering Technology Research Center of Ocean Equipment and Manufacturing, Zhanjiang 524088, China
3 School of Electronics and Information Engineering, Guangdong Ocean University, Zhanjiang 524088, China; huangjiang@gdou.edu.cn
4 School of Materials and Science and Engineering, Guangdong Ocean University, Yangjiang 529500, China; swqafj@163.com
* Correspondence: wuxiangl@gdou.edu.cn

Abstract: The aim of this investigation was to examine how CeO_2 powder influences the performance of WC + Ni60 composite powder. Various cladding layers of WC + Ni60, incorporating differing mass fractions of CeO_2, were created on the surface of Q235 steel utilizing laser cladding technology. To analyze the microscopic structure of the resulting cladding layer, scanning electron microscopy was employed. Additionally, the abrasion and corrosion resistance properties were assessed through experimentation with a pin-and-disc friction and wear tester and an electrochemical workstation, respectively. The results of the study showed that when the mass fraction of CeO_2 was 1%, the grain on the surface of the coating was refined, the carbide formation was reduced, and the uniformity of the cladding layer was the best. In terms of corrosion resistance, the coating with 1% CeO_2 had a self-corrosion potential of 0.07 V and a self-corrosion current density of 1.82×10^{-5} A·cm^{-2}, showing the best corrosion resistance, and the coating self-corrosion potential was higher than that of the coating and substrate without CeO_2. In terms of abrasion resistance, coatings with 1% CeO_2 had a lower coefficient of friction (0.47) and a smaller wear rate 0.034 mm^3, and the wear amount was only 23.5% of that of coatings without CeO_2, resulting in the best wear resistance. In conclusion, coatings containing 1% CeO_2 exhibit the minimal coefficient of friction and the lowest wear rates, while simultaneously providing optimal corrosion resistance.

Keywords: laser cladding; CeO_2; wear; corrosion resistance

1. Introduction

Bare starling harvesters have been working in seawater and shoal environments for a long time; the working conditions are harsh, and the sand casting device, made of Q235, is prone to wear and is subject to corrosion from seawater, which leads to the failure of the parts and the need for the frequent replacement of parts. This not only reduces the working efficiency, but also increases the cost of use. Therefore, improving the abrasion and corrosion resistance of Q235 steel landing gear and extending its service life has become essential.

Laser cladding technology represents a sophisticated approach to surface modification, employing a high-energy-density laser beam to simultaneously melt both the substrate and the coating material [1]. This process facilitates rapid solidification, resulting in a surface coating characterized by minimal dilution and a strong metallurgical bond with the underlying material. Such attributes significantly enhance the surface properties of the substrate. By integrating innovative materials and optimizing laser parameters, researchers have continued to expand the applicability of laser cladding across various industries, thereby advancing the field of surface engineering and contributing to the development of more resilient components. Wu, T. et al. [2] used laser cladding technology to prepare two types of coatings on 60Si2Mn spring steel. They examined a cladding configuration that consists of a singular Fe/WC layer, alongside another variant featuring a Ni60 transition layer incorporated with the Fe/WC cladding. The study found that the Ni60 transition layer reduced porosity and cracks, promoting columnar crystal growth. Both coatings consisted of dendritic and eutectic structures, with snowflake-like equiaxed crystals at the top and middle of the Fe/WC coating, and columnar crystals at the bottom. The diffusion of Ni elements decreased microhardness and increased the friction coefficient.

Husen Yang et al. [3] created a composite coating consisting of 60% Ni60 and 60% tungsten carbide (WC) on a base of 316 L stainless steel. They used a technique called laser-directed energy deposition to apply the coating. The aim of their work was to understand how different levels of laser power would change the way the coating's microscopic structure formed and its overall mechanical qualities. They also looked at how the coating's chemical composition, its internal structure, its hardness, and its ability to resist wear and friction were all connected. The research revealed that the amount of energy delivered by the laser, as measured by its power, played a key role in how much energy was absorbed by the material. This energy level determined how much of the Ni60 phase around the WC particles would melt and how well the reinforcing particles would bond with the substrate or the existing coating. As the laser power was increased from 800 W to 1400 W, the coating became more compact, and its surface hardness increased at first before it started to decrease. At the same time, the friction that the coating experienced decreased, with the least wear occurring at a laser power setting of 1200 W. These results emphasize that the laser power is a crucial factor in adjusting the microstructure and enhancing the properties of the coatings produced through laser deposition.

In a study by Xu et al. [4], two nickel-based composite coatings were synthesized on TC4 titanium alloy substrates via laser cladding techniques. These coatings were formulated as Ni60-Ti-Cu-xB4C and Ni60-Ti-Cu-B4C-xCeO_2. Notably, the inclusion of 8 wt.% B4C in the coating composition resulted in a maximum hardness of 1078 HV, which is 3.37 times greater than the hardness of the TC4 base material. Furthermore, the friction coefficient was reduced by 24.7%, and the wear rate was a mere 2.7% of that observed for the TC4 substrate. Similarly, when the CeO_2 content was increased to 3 wt.%, the coating hardness increased to an average of 1105 HV, which is 3.45 times higher than the TC4 substrate's hardness. This enhancement in hardness was accompanied by a further decrease in the friction coefficient of 33.7% and a wear rate that was just 1.8% of the TC4 material's rate. These findings highlight the potential of laser cladding to produce coatings with superior mechanical properties for titanium alloys.

In a collaborative effort led by Xue, K. et al. [5], a Ni25 alloy was utilized to create a transition layer on the compromised surface of H13 tool steel through the application of laser cladding. Following this, an Fe104 alloy was applied to form a reinforcing layer. The researchers conducted a comprehensive analysis to examine how the Ni25 transition layer influenced the phase composition, the internal structure, the hardness at a microscopic level, and the wear characteristics of the Fe104 layer. The results showed that the Ni25 transition layer could improve the comprehensive performance of the Fe104 layer, but could also lead to a decline in some mechanical properties.

In research conducted by Yang et al. [6], the investigators incorporated differing concentrations of ceria dioxide (CeO_2) into the composition of an $AlCoCrFeNi_{2.1}$ alloy powder. Subsequently, they applied this alloyed powder to fabricate coatings on the H13 steel substrate via the laser cladding method. The investigation revealed that the inclusion of 3% CeO_2 within the alloy yielded multiple advantageous outcomes. Initially, this addition was found to significantly decrease the prevalence of cracks, pores, and inclusions, which are detrimental imperfections capable of compromising the structural integrity of coatings. Furthermore, the presence of CeO_2 stimulated the growth of a finer grain structure within the coatings, a characteristic that contributes to the material's increased strength and robustness. Moreover, the integration of CeO_2 resulted in a more consistent microstructure and a higher hardness for the surface coatings. These enhancements not only augmented the coating's resistance to wear but also substantially improved its resistance to corrosion, thereby rendering the composite coating more resilient and appropriate for environments demanding robustness against both wear and corrosive agents.

Based on existing research findings, various aspects have been explored regarding the combination of Fe/WC overlay layers and Ni60 transition layers, Ni60/60% WC composite coatings, and nickel-based composite coatings containing B4C and CeO_2. Among these, WC has the effect of improving hardness [7], while Ni60 can impact the bonding performance between the coating and substrate [8–10]. CeO_2 has also been shown to have a significant influence on the coating hardness and friction coefficient in previous studies. However, there is relatively limited research on the simultaneous combination of CeO_2, WC, and Ni60 in overlay layers. Conducting research on CeO_2 + WC + Ni60 overlay layers, by combining these three components, holds the promise of enhancing the coating's mechanical attributes, including its hardness and resistance to frictional wear, thus augmenting its performance in engineering contexts.

2. Experimental Materials, Methods, and Equipment

2.1. Materials

This experiment chooses Q235 steel as the base material, which has a general hardness between HB 100 and 150; at a relatively low hardness, it is easy to carry out a variety of mechanical processing techniques. The wear resistance is general; in some cases that require high wear resistance, it may be necessary to take measures such as surface treatment to improve its wear resistance. Due to its good mechanical properties and economy, it is widely used in construction and manufacturing industries. To improve the wear and corrosion resistance of Q235 steel, a surface coating consisting of Ni60, WC, and CeO_2 is employed, as detailed in Table 1. The Ni60 alloy, known for its superior properties, serves as the primary coating material, thereby significantly enhancing the material's overall performance and extending its service life.

Table 1. The proportions of Ni60, WC and CeO_2 (mass fraction, %).

Ni60	WC	CeO_2
75	25	0
74	25	1
73	25	2

According to the SEM results, the quality fraction of the elements in Ni60 is shown in Table 2, and their microscopic morphology is mostly spherical, as shown in Figure 1a. This morphology helps the coating to form a uniform coating layer on the substrate surface, thus enhancing the adhesion and wear resistance of the coating. The size distribution of powder also has important effects on the formation of the coating and its properties. The Ni60

(Tsuengyue Metal Materials Co., Santai, Sichuan, China) powder is shown in Figure 1b; the maximum particle size is 133.43 μm, the minimum particle size is 45.89 μm, the average particle size is 85.14 μm, and the powder size meets the normal distribution. This property indicates that the particle size distribution of the chosen powder is appropriate for the coating procedure. Such a distribution enhances both the mobility and filling capabilities, ultimately facilitating the more cohesive and uniform application of the coating.

Table 2. Ni60 powder and Q235 chemical composition (mass fraction, %).

Element	C	Cr	B	Mn	Si	Fe	P	S	Ni
Ni60	0.8–1.2	14–16	3–3.5	-	3.5–4.0	14–15	0.02	0.02	Bal
Q235	0.22	-	-	0.3–0.7	0.35	Bal	0.045	0.05	-

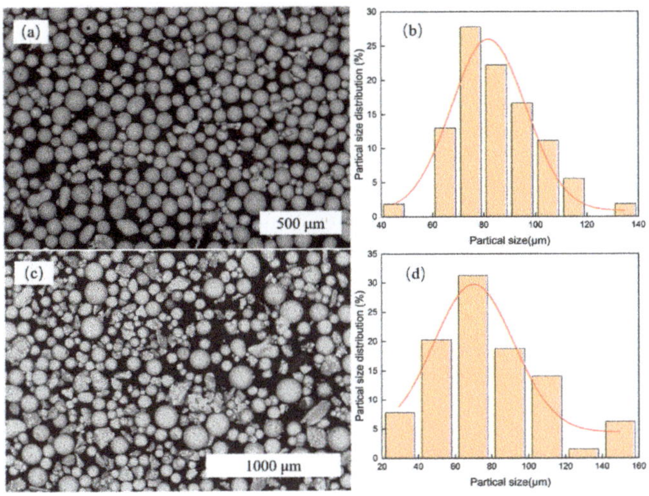

Figure 1. Morphology and particle size distribution of mixed powder. (**a**) Ni60 material morphology, (**b**) Ni60 material size, (**c**) WC material morphology, (**d**) WC material size.

The coating contains WC (Tsuengyue Metal Materials Co., Santai, Sichuan, China), as shown in Figure 1c, which is a hard material that significantly improves the hardness and wear resistance of the coating. The particle size coincidence analysis of WC is shown in Figure 1d; the standard deviation is 15.65, the maximum particle size is 156.96 μm, the minimum particle size is 24.16 μm, the average particle size is 77.86 μm, and the powder size conforms to the normal distribution. This particle size characteristic makes it possible to strengthen the coating, and the strength and toughness of the coating can be effectively enhanced through reasonable particle size selection.

2.2. Equipment and Methods

Laser cladding: In this experimental process, the laser cladding test was carried out by using the XL-F2000T laser (Guangzhou Xingrhenium Laser Technology Co., Ltd., Guangzhou, China) as the experimental facility, with the power parameter set at 1200 W, the scanning rate set at 800 mm/min, and the defocusing amount constant set at 3 mm. CeO_2, WC, and Ni60 powders were selected for the experiments and mixed according to specific ratios, and then pre-packed onto the Q235 steel plate substrate, with the thickness of the powder precisely controlled to 1 mm; then, the multi-pass cladding process was implemented.

Scanning electron microscopy: Following the polishing and treatment of the experimental specimens, their surface morphology was analyzed and recorded using a Czech MIRA3 scanning electron microscope (SEM) (Brno, Czech Republic). This investigation focused on understanding the interactions between the matrix and the reinforcing phases while also characterizing the microstructural attributes of the coating. Additionally, the elemental distribution within the coating was assessed using an energy-dispersive spectrometer (EDS), allowing for both qualitative and quantitative evaluations.

The phase composition of the coatings was assessed through the use of a Shimadzu XRD-6100 X-ray diffractometer (Shimadzu, Tokyo, Japan). This analysis involved a diffraction scanning range that spanned from 10 to 90 degrees, with a meticulous scanning step of 0.04 degrees and a scanning speed of 6° per minute. By employing this method, we ensured a comprehensive examination of the coatings' structural characteristics, thereby facilitating a clearer understanding of their properties and potential applications.

Friction and wear: To investigate the abrasion resistance of the coatings, the cross-sectional abrasion resistance of the coated samples was assessed utilizing the SFT-2M friction and wear testing apparatus manufactured by CKH Technology Development Co. (Zhongkehua Science and Technology Development Co., Ltd., Lanzhou, China). The dimensions of the friction and wear test specimen were 15 mm by 15 mm, with GCr15 steel serving as the material for the friction vice. The parameters for the wear test included a loading force of 60 N, a duration of 30 min, an operational speed of 200 revolutions per minute, a rotational radius of 2.5 mm, and a scanning length of 3 mm.

The characterization of electrochemical workstations: In this study, the CS Studio electrochemical workstation was employed to evaluate the corrosion resistance properties of the tested samples. Given that seawater predominantly consists of a high concentration of sodium chloride (NaCl), along with trace amounts of other salts, the corrosive medium was configured as an aqueous solution of NaCl with a mass fraction of 0.035 to mimic the corrosive seawater environment. This was done to prevent the inclusion of additional contaminants that could introduce variability into the assessment of corrosion factors. The experimental setup included a pre-test electrode system, which utilized a three-electrode electrochemical cell. The coated sample served as the working electrode, while the platinum sheet and the saturated calomel electrode (SCE) were designated as the auxiliary and reference electrodes, respectively. The scanning rate was adjusted to 0.5 mV per second, and the sampling frequency was set at 1 hertz. Polarization curves were derived from the analysis of the experimental data, and the corrosion potentials and currents for each sample were determined from these curves, thereby enabling an evaluation of the materials' corrosion resistance.

3. Results and Analysis

3.1. Coating Morphology

Figure 2 clearly indicates that there are no significant pores or cracks present. This feature is crucial for the performance of the coating, as the presence of pores and cracks often greatly weakens the mechanical properties, corrosion resistance, and other key performance indicators of coatings.

In the fusion-coated coating without CeO_2 addition, a high number of WC particles can be clearly recognized. As shown in Figure 2a, there are 15 white particles. These WC particles are responsible for enhancing the hardness and wear resistance of the coating system [11,12], and contribute to the strengthening of the overall performance of the coating. After the introduction of CeO_2, the microstructure of the coatings underwent a significant evolution. Specifically, the CeO_2-added coating exhibits a clearly recognizable white bright band in the area of the fusion mark. At the same time, the number of WC particles decreases compared to the coating without CeO_2. The number of white particles decreases to 11 in Figure 2b, and further decreases to 9 in Figure 2c. This phenomenon is most likely attributed to the substantial interference and remodeling of the solidification process and the microstructure formation mechanism of the coatings with CeO_2. During the solidification process, CeO_2 may affect the temperature distribution, elemental diffusion behavior and crystal growth kinetics of the molten pool, which may alter the dissolution of WC particles and ultimately lead to a reduction in the number of WC particles [13], which is a special microstructural feature; this microstructural transformation may be further associated with a change in the macroscopic properties of the coating [14], which is worthy of in-depth investigation and analysis.

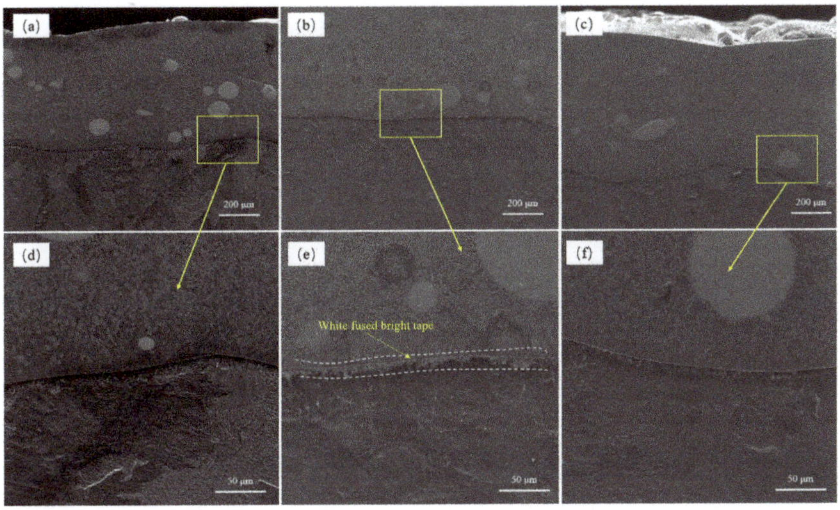

Figure 2. Interface morphology with different amounts of CeO_2. (**a**) Topography of the cross-section without CeO_2 and (**b**) 1% CeO_2. (**c**) Cross-sectional morphology of 2% CeO_2. (**d**) Enlarged view at the weld line without CeO_2 addition. (**e**) Enlarged view at the 1% CeO_2 weld line. (**f**) Enlarged view at the 2% CeO_2 weld line.

When CeO_2 is added, the width of the white bright band at the fusion mark increases, in which case the diffusion and fusion between the cladding material and the substrate are more significant, which positively affects the bond strength. During this process, Ce elements may react with Fe, Cr, Ni and other elements in Ni60 alloys and form intermetallic compounds. The formation and distribution of these intermetallic compounds at the fusion marks can effectively strengthen the bonding strength between the cladding layer and the substrate [11]. At the same time, the introduction of CeO_2 may lead to the formation of more rare-earth compound phases. These rare-earth compound phases are enriched or diffusely distributed at the fusion marks, and their presence has a significant impact on the diffusion and fusion mechanism between the cladding material and the substrate [12], thus changing the bonding characteristics and overall performance of the cladding layer and the substrate at the microstructural level.

Figure 2b,e correspond to a coating with 1% CeO_2. In both images, the bright white band at the weld mark is the widest and most distinct. This indicates that the microstructure of the coating has undergone a special microstructural change during the welding process with the addition of CeO_2. This variation may be related to the effects of CeO_2 on the solidification kinetics, crystallization behavior, and elemental diffusion of the coating.

Figure 2c,f show the topography of the 2% CeO_2 coating and an enlarged view near the weld scar. As is evident, the weld line appears as white and wavy. This unique morphology may be due to a combination of specific thermal stresses, hydrodynamic effects, and chemical reactions during the welding process at this CeO_2 addition. Further study of the formation mechanism of this wavy weld line is of great significance for understanding the effect of CeO_2 on coating properties and for optimizing the preparation process of coatings.

3.2. Phase Analysis

Through the analysis of XRD patterns (Figure 3), it was found that the effect of CeO_2 additives in the laser cladding Ni60 + WC powder system was determined by precise X-ray diffraction (XRD) analysis, which performed a detailed phase characterization of the coatings without and with different proportions of CeO_2. The experiment showed that the presence of r-(Fe, Ni) solid solution, carbide $M_{23}C_6$, and $FeNi_3$ phases was mainly detected in the reference coating without CeO_2 [13]. However, when CeO_2 was introduced into the cladding system as an additive, it not only maintained the basal phase mentioned above, but also significantly induced the formation of a new phase, CeNi, which directly confirmed the chemical activity of CeO_2 during the cladding process and its effect on the phase composition [14].

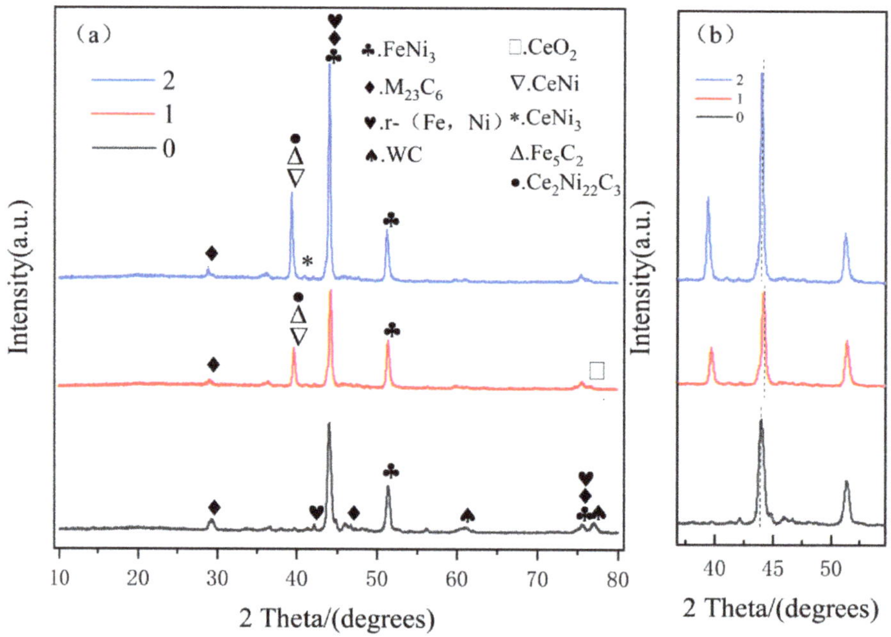

Figure 3. XRD diffraction spectrum analysis. (**a**) Diffraction patterns of fused cladding layers with different contents of CeO_2. (**b**) The main diffraction peak magnification shows a different amount of CeO_2.

By further analyzing the XRD pattern, it can be observed that in Figure 3b, with the addition of CeO_2, there is a subtle but identifiable shift in the position of specific diffraction peaks. This is typically attributed to changes in lattice parameters, possibly due to the formation of new phases or the influence of CeO_2 on the existing phase structure. The intensity of the 1% CeO_2 diffraction peak decreased, indicating that the grain in the coating was refined [15]. The intensity of the diffraction peak at 2% CeO_2 is significantly enhanced compared to the coating without CeO_2, with a significant increase in the carbide concentration. CeO_2, as a rare-earth oxide with excellent catalytic performance, may promote the chemical reaction between WC powder and the Ni60 matrix through multiple paths, including, but not limited to, reducing the activation energy of the chemical reaction, so as to overcome the energy barrier, so that the reaction that does not occur easily at high temperatures can be carried out smoothly, thereby promoting the formation and accumulation of carbide phases. The addition of CeO_2 also significantly improves the fluidity and nucleation dynamics of the melt pool [16]. Its unique physical properties, such as reducing the surface tension and critical nucleation resistance of liquid metal, effectively promote the uniform mixing and full reaction of various chemical components in the molten pool, and create more favorable conditions for the formation of carbides. At the same time, CeO_2 may also act as a nucleating agent [17], promoting grain refinement and increasing the number of grain boundaries and phase boundaries; these interfaces serve as preferred sites for carbide nucleation and growth, further improving the efficiency and quantity of carbide formation.

As the CeO_2 content increased to 2%, the chemical reaction path and equilibrium state in the molten pool changed, resulting in more WC powder participating in the carbide formation reaction. In addition, CeO_2 may undergo complex chemical reactions with other molten pool elements to form new compounds or interphases, which further promote the stabilization and growth of the carbide phase, which is manifested as a significant increase in the intensity of the corresponding diffraction peaks [18].

3.3. Coated SEM and Energy Spectrum Analysis

In the laser cladding Ni60 + WC composite powder system, the effect of CeO_2 on the microstructure and chemical composition of the cladding layer was systematically studied by introducing CeO_2 as an additive and increasing its content from 1% to 2%; the following detailed observation and analysis results were obtained.

Figure 4a demonstrates the results of the energy spectral analysis of Spectrum A in the microscopic region (Table 3), at the middle position of the fused cladding layer without added CeO_2, where the mass percentage of W element is as high as 26.43%, and the atomic ratio of carbon (C) is 10.88%. These data combinations show the aggregation of WC phases, possibly due to the lack of an efficient refinement or dispersion mechanism. The analysis of Ni in Spectrum B revealed a mass percentage of 58.59%, indicating that Ni is the main component in this region.

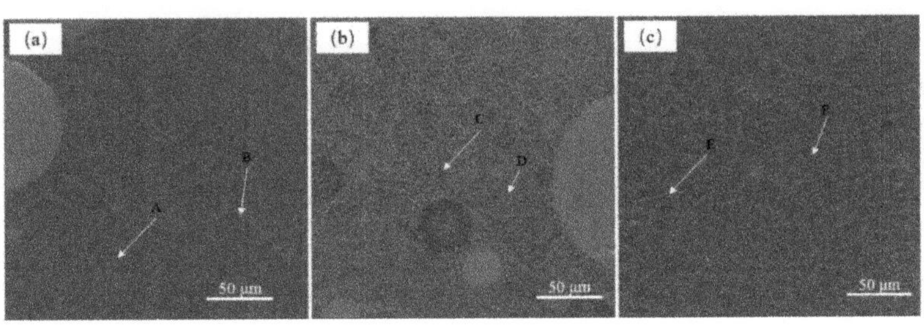

Figure 4. Coating EDS scan points with different CeO_2 contents. (**a**) EDS scanning points without added CeO_2. (**b**) EDS scanning points with 1% CeO_2 content. (**c**) EDS scanning points with 2% CeO_2 content.

Table 3. Distribution of the main elements of EDS scan points.

Element		B	C	Si	Ce	Cr	Fe	Ni	W
Spectrum A	Weight %	0.54	2.12	0.41	-	10.78	18.07	41.11	26.43
	Atomic %	3.09	10.88	0.89	-	12.8	19.98	43.24	8.88
Spectrum B	Weight %	1.74	1.11	0.77	0.74	5.29	24.93	58.59	6.83
	Atomic %	8.61	4.94	1.46	0.28	5.45	23.88	53.39	1.99
Spectrum C	Weight %	1.33	3.27	0.64	0.55	14.22	15.79	34.12	30.09
	Atomic %	7.16	15.78	1.32	0.23	15.87	16.41	33.73	9.5
Spectrum D	Weight %	3.05	1.82	0.25	0.47	11.63	21.4	47	14.37
	Atomic %	14.62	7.84	0.46	0.17	11.57	19.84	41.44	4.05
Spectrum E	Weight %	0.39	2.26	0.4	0.55	14.12	11.41	30.5	40.37
	Atomic %	2.45	12.94	0.98	0.27	18.63	14.02	35.64	15.07
Spectrum F	Weight %	0.87	1.56	0.44	0.49	11.85	19.78	43.76	21.26
	Atomic %	4.8	7.75	0.93	0.21	13.64	21.18	44.58	6.91

Figure 4b and Table 3 shows that the atomic ratios of Cr, Fe, and Ni are 15.87%, 16.41%, and 33.73%, respectively, while the atomic ratios of carbon (C) increase to 15.78%, respectively. Notably, the atomic ratios of these elements are close to the stoichiometric ratios (23:6) of the $M_{23}C_6$ (M stands for Ni, Fe, Cr) complex carbides, thus strongly supporting the conclusion that dendritic tissues are identified as $M_{23}C_6$ complex carbides [19]. The mass percentage of Ni and the mass percentage of Fe in Spectrum D are 47%, and the mass percentage of Fe is 21.4%; the supersaturated solid solution with C, W, Cr and other elements is solidly dissolved [20,21].

In addition, the discovery of boron (B) in Spectrum D with a mass percentage of 3.05% and an atomic ratio of 14.62% confirms the presence of borides, which may be due to impurities in the original powder or the formation of by-products during the reaction.

The changes in the microstructure when the CeO_2 content is increased to 2% are shown in Figure 4c. The energy spectrum analysis of Spectrum E showed that the mass ratio of W increased significantly to 40.37%, while the atomic ratio reached 15.07% and the atomic ratio of carbon (C) was 12.94%. At this time, the atomic ratio of W to C approached 1:1, and the aggregation of the WC phase appeared, which may be related to the effect of CeO_2 on the growth kinetics of the WC phase at a high concentration. The analysis of Spectrum F for Ni showed that its mass percentage was 43.76%, and when adding 2% CeO_2, r-(Ni) supersaturated solid solution became the dominant phase state in this region, which may be related to the stabilizing effect of CeO_2 on the Ni matrix phase and its ability to promote the formation of solid solutions.

Figure 5a shows a scanning electron microscope (SEM) image of the cladding layer without CeO_2, and after a careful microstructural analysis, small holes in the cladding layer can be observed. The formation of these holes is most likely due to the failure of the internally generated CO_2 gas to escape in time during the rapid cooling of the melt pool, leaving these defects in the solidified coating [22]. Further observation shows that the coating is mainly composed of irregularly shaped blocks and secondary dendrites, and these structural characteristics indicate that the distribution of the strengthening phase in the coating is uneven, which may adversely affect the overall performance of the coating.

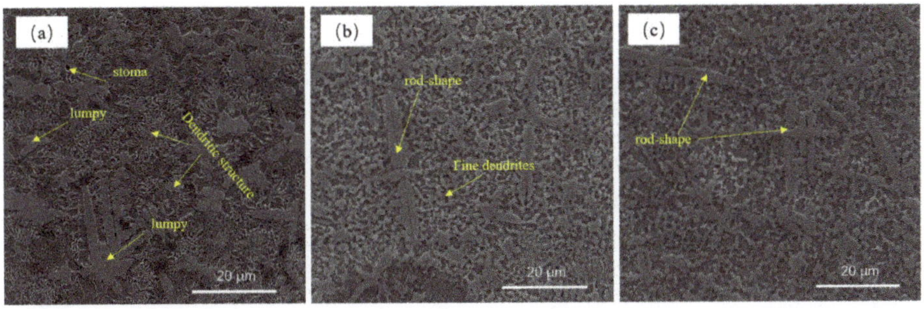

Figure 5. Microstructure with different CeO_2 contents. (**a**) 2000-fold microstructure without added CeO_2. (**b**) 2000-fold microstructure containing 1% CeO_2. (**c**) 2000-fold microstructure containing 2% CeO_2.

When 1% CeO_2 is added to the feedstock, the microstructure of the cladding layer changes significantly, as shown in Figure 5b. At this time, the cladding layer is mainly composed of reticulated and needle-like structures, the grain size is significantly reduced, which can make the microstructure size more uniform, and the refinement effect is significant; the grain size can be determined by the XRD diagram shown in Figure 3a [23]. The reason for this change is that trace amounts of active Ce ions play a key role in the melt pool. They are easily adsorbed onto the surface of the crystal nucleus, forming a barrier layer, which effectively hinders the further growth of the grain and allows the tissue to be refined [24]. At the same time, the addition of Ce ions also reduces the supercooling phenomenon of the components in the solidification process, reduces the degree of segregation between the components, and weakens the directionality of dendrite growth, which makes the structure of the cladding layer more uniform.

Figure 5c shows the SEM image of the coating when 2% CeO_2 is added to the composite feedstock. At this time, the cladding layer is mainly composed of equiaxed crystals and rods, and the number of small blocks increases. This phenomenon indicates that with the increase in the amount of CeO_2 added, the ability of rare earth elements to promote the fluidity of liquid metals gradually emerges. The addition of rare-earth elements can reduce the melting point and surface tension of liquid metal, thereby improving the fluidity of liquid metal. This increased fluidity facilitates the chemical reaction, which facilitates the generation of more compounds [25]. However, it is also important to note that excess REEs may lead to more lumps in the coating, which may have an impact on the mechanical properties and corrosion resistance of the coating. Therefore, when preparing the cladding layer, it is necessary to reasonably control the addition of rare-earth elements to obtain the best coating performance.

3.4. Friction and Wear Analysis

Figure 6a–c show the data for the coefficient of friction of the coating. Figure 6a shows that the coefficient of friction of the cladding coating surface fluctuates within the range of 0.50 to 0.53 without the addition of CeO_2.

When 1% CeO_2 is added, the coefficient of friction stabilizes at around 0.47. With the addition of 2% CeO_2, the coefficient of friction fluctuates between 0.52 and 0.67. Through a comparative analysis of the coefficient of friction under different addition ratios, it can be clearly seen that the coefficient of friction of the cladding surface with 1% CeO_2 is the smallest. Figure 6b shows the friction and wear topography. A closer comparison shows that the mixed-powder coating with 1% CeO_2 performs best in terms of wear and wears the least. Figure 6c shows a wear of 0.146 mm^3 on the cladding coating surface without the addition of CeO_2. When 1% CeO_2 is added, the amount of wear is significantly reduced to 0.034 mm^3, which is 0.112 mm^3 less than the amount of coating wear without CeO_2. With the addition of 2% CeO_2, the amount of wear is 0.039 mm^3. It can be seen that adding the right amount of CeO_2 can effectively reduce the amount of wear on the cladding coating surface, and that the best results are achieved at a 1% addition ratio. This phenomenon may be due to the addition of CeO_2, which alters the microstructure and mechanical properties of the coating [26], thereby improving the wear and friction resistance of the coating. Further research could explore the mechanism of the influence of CeO_2 on coating performance and the mechanism of coating performance change under different addition ratios, so as to provide a theoretical basis and technical support for optimizing the coating design and improving the coating performance.

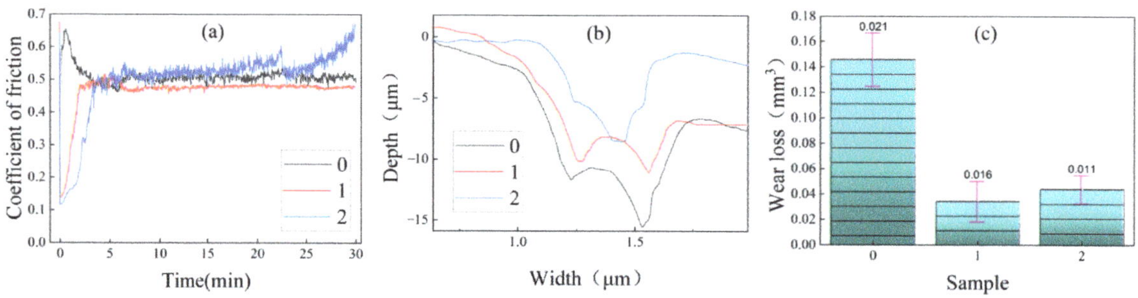

Figure 6. Friction and wear. (**a**) Coefficient of friction of each coating surface. (**b**) Wear profile. (**c**) Amount of wear.

Without the addition of CeO_2, as shown in Figure 7a, a furrow topography with a width of 760 μm is presented. This furrow topography is one of the typical features of abrasive wear. Abrasive wear is usually caused by hard particles or hard protrusions cutting and scratching the surface of the material during friction. In this case, it is the detached particles produced by the friction pair itself during the friction process that act as an abrasive. From a microscopic point of view, these abrasives are constantly cutting the surface of the material under the action of relative motion, resulting in a pronounced furrow. At the same time, the electron microscope image with a local magnification of 2000 times, as shown in Figure 7d, shows the presence of friction debris, which further confirms that it is mainly abrasive wear. These chips may be formed by the continuous shedding of the material under the cutting action of the abrasive. The degree of abrasive wear is usually closely related to factors such as the hardness, shape, and size of the abrasive, as well as the relative movement speed of the friction pair.

The abrasion topography for when 1% CeO_2 is added is shown in Figure 7b. The width of the abrasion mark is approximately 620 μm, which is reduced compared to that without the addition of CeO_2. Very little debris can be seen at this point, and there are obvious sticking marks and strains on the worn surface, indicating that it may be mainly adhesive wear. Adhesive wear is caused by the instantaneous high temperature generated at the contact point under the action of high pressure and the temperature generated when the two contact surfaces move relative to each other, which softens or melts the surface of the material, resulting in the adhesion of the material on the contact surface. In the subsequent relative motion, the adhesive point is sheared, resulting in adhesive wear. This is because fine grains can increase the strength and hardness of the material, but they can also change the surface properties of the material. After grain refinement, the surface of the material is more uniform and the contact area increases, which makes it easier to form adhesive points during friction. Reducing the coefficient of surface friction in CeO_2 also has an effect on the adhesive wear. A decrease in the coefficient of surface friction means that during relative motion, the frictional force decreases. A lower coefficient of friction may result in a smoother relative motion between the contact surfaces, increasing the likelihood of adhesion [27]. When there is less friction between two surfaces, they are more likely to come into close contact, forming sticking points.

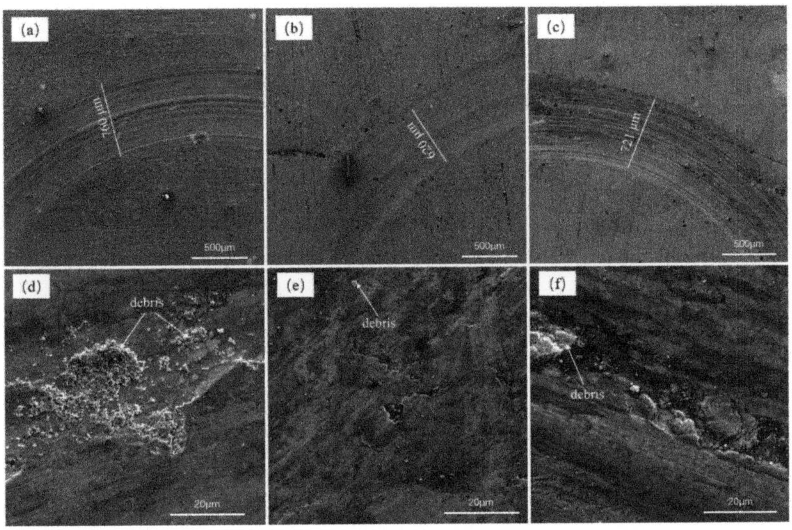

Figure 7. Scanning electron microscope (SEM) image after wear. (**a**) Post-wear morphology of fused cladding layer with CeO_2 mass fraction 0. (**b**) Post-wear morphology of fused cladding layer with CeO_2 mass fraction 1. (**c**) Fused cladding layer with CeO_2 mass fraction of 2 after abrasion. (**d**) Localized magnification of the furrows after abrasion of the fused cladding layer with a CeO_2 mass fraction of 0 by a factor of 2000. (**e**) Localized magnification of the furrows after the abrasion of the fused cladding layer with a CeO_2 mass fraction of 1 by a factor of 2000. (**f**) Localized magnification of the furrows after the abrasion of the fused cladding layer with a CeO_2 mass fraction of 2 by a factor of 2000.

The topography after wear for when 2% CeO_2 is added is shown in Figure 7c, with a width of about 721 μm in the wear scar groove. A partial enlargement of Figure 7f shows streamlined marks, the appearance of debris, and strain on the wear surface, suggesting possible mixed wear. There are multiple wear mechanisms operating at the same time during friction. In this case, it may be a combination of abrasive wear, adhesive wear, and other wear mechanisms [28–30]. The appearance of chips can be a sign of abrasive wear, while the phenomenon of strain is related to adhesive wear. The case of mixed wear is complex and requires a combination of factors.

3.5. Corrosion Resistance Analysis

The polarization curves in Figure 8 show that the corrosion resistance of the cladding layer is better than that of the base material (Q235). Combined with the values in Table 4, it is clear that the self-corrosion potential of the substrate is −0.78 V, the self-corrosion current density is 5.81×10^{-4} A·cm^{-2}, the self-corrosion potential of the mixed-powder cladding coating without CeO_2 is −0.67 V, and the self-corrosion current density is 2.87×10^{-5} A·cm^{-2}; the current density is significantly lower because the high temperature causes the WC and Ni60 powders to melt and mix rapidly. This is followed by rapid cooling and solidification during the laser cladding process. This rapid thermal cycling process contributes to the formation of a dense coating structure, reducing defects such as porosity and cracks [31,32]. Porosity and cracks are often channels for the intrusion of corrosive media, and reducing them can effectively prevent the penetration of corrosive media, thereby improving the corrosion resistance of coatings [33]. The dense structure also improves the physical barrier of the coating, making it more difficult for corrosive media to reach the base material, further enhancing the corrosion resistance. Ni60 is a nickel-based alloy powder containing a high content of nickel, chromium and other elements. Nickel and chromium are able to form dense oxide films, such as nickel–chromium oxides, which have good stability and corrosion resistance.

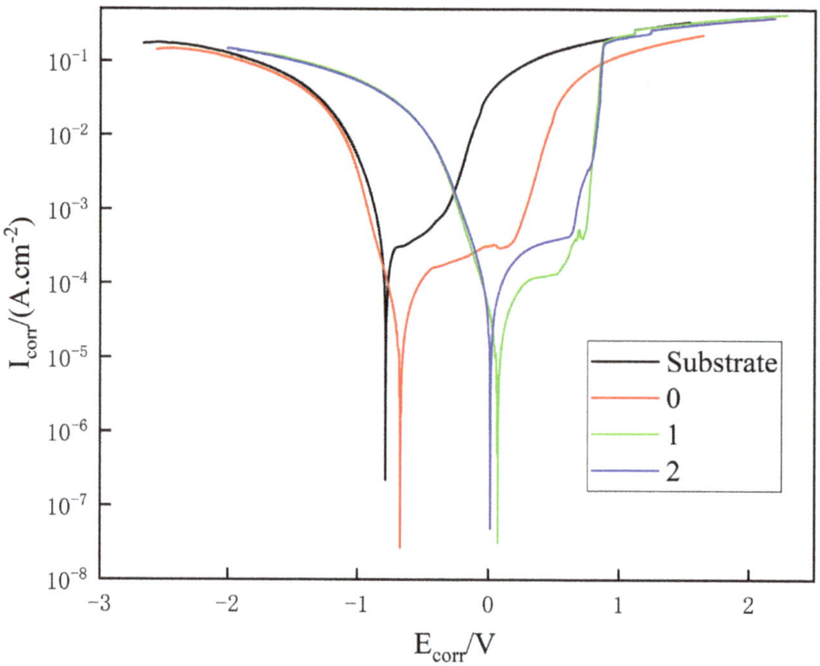

Figure 8. Polarization curves of cladding layers with different mass fractions of CeO_2.

Table 4. Corrosion test results of samples in 3.5% NaCl solution.

Sample	Base Metal	0	1	2
E_{corr}/V	−0.78	−0.67	0.07	0.01
$I_{corr}/(A{\cdot}cm^{-2})$	5.81×10^{-4}	2.87×10^{-5}	1.82×10^{-5}	8.01×10^{-5}

The coating with 1% CeO_2 + WC + Ni60 mixed-powder laser fusion has a self-corrosion potential of 0.07 V and a self-corrosion current density of 1.82×10^{-5} $A{\cdot}cm^{-2}$, which is greater than the coating without added CeO_2; the current density is less than that of the coating without added CeO_2, which refines the grains, and the grain refinement makes the size of each grain become smaller, which reduces the microcells. The electrode potential difference reduces the corrosion current and improves the corrosion resistance of the material [34]. CeO_2 promotes the alloying elements in Ni60 in the molten pool to improve the strength and hardness of the coating through mechanisms such as solid solution strengthening and precipitation strengthening, and enhances the corrosion resistance of the coating. Chromium improves the oxidation and pitting resistance of the coating, and nickel improves the reduction and stress corrosion resistance of the coating [35]. The generation of a high percentage of solid solution improves the corrosion resistance.

The self-corrosion potential of the coating with 1% CeO_2 + WC + Ni60 mixed-powder laser cladding is 0.07 V, the self-corrosion current density is 1.82×10^{-5} A·cm^{-2}, the self-corrosion potential is greater than that of the coating without CeO_2, and the current density is less than that of the coating without CeO_2, which refines the grains; the number of grain boundaries can be increased by the refined grains. Grain boundaries are often high in energy and chemical activity, which can hinder the propagation of corrosion [34]. When corrosion occurs at grain boundaries, the path of corrosion becomes more tortuous due to the increase in the number of grain boundaries, which prolongs the time needed for the corrosive medium to reach the substrate and improves the corrosion resistance of the coating. CeO_2 promotes the alloying elements in Ni60 in the molten pool to improve the strength and hardness of the coating through mechanisms such as solution strengthening [12] and precipitation strengthening, and enhances the corrosion resistance of the coating. Chromium can improve the oxidation and pitting resistance of coatings, and nickel can improve the reducing and stress corrosion resistance of coatings [35]. A large proportion of solid solution is generated, which improves the corrosion resistance.

The self-corrosion potential of the coating with 2% CeO_2 mixed-powder laser fusion is 0.01 V, and the self-corrosion current density is 8.01×10^{-5} A·cm^{-2} greater than that of the 1% CeO_2 fusion coating layer, whose self-corrosion density is 1.82×10^{-5} A·cm^{-2}. It can be seen in Figure 5 that the overall number of carbides in the 1% CeO_2 fusion layer is less than that in the 2% CeO_2 fusion coating layer, which is due to the greater number of carbides generated; the large number of carbides generated increases the possibility of intergranular corrosion [36] and carbide generation, which increases the possibility of intergranular corrosion [36]. The intergranular region tends to have high chemical activity due to its different chemical composition and intragranular structure. When carbides are present in large quantities at grain boundaries, the chemical stability at grain boundaries decreases and they are more susceptible to attack by corrosive media. Corrosive media preferentially erode at grain boundaries, leading to the occurrence of intergranular corrosion, which reduces the overall corrosion resistance of the coating. Elements such as nickel and chromium in Ni60 often form a dense oxide film, such as nickel–chromium oxide, which plays a vital role in improving the corrosion resistance of the coating. However, when chromium is combined with carbon, it can lead to the uneven distribution of chromium, which in turn adversely affects the corrosion resistance of the coating. On the one hand, chromium plays a key role in the formation of oxide films. The dense oxide film can effectively block the intrusion of corrosive media and provide good protection for the coating [37]. When chromium is evenly distributed, the oxide film is formed more stably and continuously, and its protective effect can be better exerted.

On the other hand, when chromium is combined with carbon, carbides are formed [38]. The presence of these carbides may disrupt the microstructure of the coating, resulting in a decrease in the density of the coating [12]. At the same time, due to the formation of carbides, part of the chromium is consumed, so that the distribution of chromium in the coating becomes uneven. At the same time, due to the formation of carbides, part of the chromium is consumed, so that the distribution of chromium in the coating becomes uneven. In areas with a low chromium content, the formation of an oxide film is inhibited, which reduces the corrosion resistance of the coating [39].

4. Conclusions

1. The coating with 1% CeO_2 has a lower coefficient of friction (0.47) and a lower amount of wear (0.034), which is 0.112 mm^3 less than the coating without CeO_2, showing the best wear resistance. A decrease in the coefficient of friction means that during the friction process, the friction between the coating surfaces decreases.

2. The coating with 1% CeO_2 has a self-corrosion potential of 0.07 V and a self-corrosion current density of 1.82×10^{-5} $A \cdot cm^{-2}$, showing the best corrosion resistance. The self-corrosion current density is significantly reduced, indicating that the corrosion resistance of the coating has been enhanced. In summary, coatings with 1% CeO_2 exhibit significant advantages in terms of grain refinement, abrasion resistance, and corrosion resistance. This provides an important reference for the development of high-performance coating materials.

Author Contributions: Conceptualization, W.S. and J.H.; methodology, J.W.; software, X.W.; validation, J.H., X.W., D.C. and W.S.; formal analysis, X.W.; investigation, X.W and J.Z.; resources, W.S. and J.H.; data curation, J.W.; writing—original draft preparation, X.W.; writing—review and editing, X.W., D.C. and J.H.; visualization, J.W.; supervision, F.A. and J.Z.; project administration, W.S.; funding acquisition, J.W. and J.Z. All authors have read and agreed to the published version of the manuscript.

Funding: Supported by the National Natural Science Foundation Project (No. 62073089), the Zhanjiang Science and Technology Plan Project (No. 2021A05171), and the Laser Processing Team Project of Guangdong Ocean University (No. CCTD201823).

Institutional Review Board Statement: Not applicable.

Informed Consent Statement: Not applicable.

Data Availability Statement: Data are contained within the article.

Conflicts of Interest: The authors declare no conflicts of interest.

References

1. John, M.; Kuruveri, U.B.; Menezes, P.L. Laser Cladding-Based Surface Modification of Carbon Steel and High-Alloy Steel for Extreme Condition Applications. *Coatings* **2022**, *12*, 1444. [CrossRef]
2. Wu, T.; Shi, W.; Xie, L.; Gong, M.; Huang, J.; Xie, Y.; He, K. Study on the effect of Ni60 transition coating on microstructure and mechanical properties of Fe/WC composite coating by laser cladding. *Opt. Laser Technol.* **2023**, *163*, 109387. [CrossRef]
3. Yang, H.; Li, W.; Liu, Y.; Li, F.; Yi, J.; Eckert, J. The Microstructure and Properties of Ni60/60% WC Wear-Resistant Coatings Prepared by Laser-Directed Energy Deposition. *Micromachines* **2024**, *15*, 1071. [CrossRef] [PubMed]
4. Xu, S.; Han, K.; Wang, H.; Xi, Y.; Wang, L.; Dong, X. The Influence of Adding B_4C and CeO_2 on the Mechanical Properties of Laser Cladding Nickel-Based Coatings on the Surface of TC4 Titanium Alloy. *Materials* **2024**, *17*, 3823. [CrossRef]
5. Xue, K.N.; Lu, H.F.; Luo, K.Y.; Cui, C.Y.; Yao, J.H.; Xing, F.; Lu, J.Z. Effects of Ni25 transitional layer on microstructural evolution and wear property of laser clad composite coating on H13 tool steel. *Surf. Coat. Technol.* **2020**, *402*, 126488. [CrossRef]
6. Yang, C.; Jing, C.; Fu, T.; Lin, T.; Guo, W.; Liu, N. Effect of CeO_2 on the microstructure and properties of $AlCoCrFeNi_{2.1}$ laser cladding coatings. *J. Alloys Compd.* **2024**, *976*, 172948. [CrossRef]
7. Xue, S.L.; Li, J.H.; Yao, F.P.; Yang, Y.H.; Li, X.X. Preparation and Structural Properties of Gradient Composite Coatings for Laser Cladding. *Lasers Eng. (Old City Publ.)* **2023**, *54*, 343.
8. Li, D.; Cui, X.; Yuan, C.; Zhang, D.; Jin, G.; Zheng, W.; Yang, Y. Effect of Ni modified graphene on microstructure and properties of Ni60 composite coatings prepared by laser cladding. *Opt. Laser Technol.* **2021**, *136*, 106756. [CrossRef]
9. Yang, L.; Li, B.; Yao, J.; Li, Z. Effects of diamond size on the deposition characteristic and tribological behavior of diamond/Ni60 composite coating prepared by supersonic laser deposition. *Diam. Relat. Mater.* **2015**, *58*, 139–148. [CrossRef]
10. Chen, S.; Feng, A.; Chen, C.; Song, X. Research on the Process of Laser Cladding Ni60 Coating on High-Nickel Cast Iron Surfaces. *Processes* **2024**, *12*, 647. [CrossRef]
11. Luo, J.; Gao, J.J.; Gou, S.W.; Li, Y.L.; Lin, H.X.; Wu, X.R.; Qi, S.Y. Study on microstructure and mechanical properties of Ni60+WC/Ni35/AISI1040 functional surface gradient structure of remanufacturing chute plate for the mining scraper by a low cost high power CO_2 laser cladding technique. *Mater. Res. Express* **2020**, *7*, 086521. [CrossRef]
12. Lian, G.; Gao, W.; Chen, C.; Huang, X.; Feng, M. Review on hard particle reinforced laser cladding high-entropy alloy coatings. *J. Mater. Res. Technol.* **2024**, *33*, 1366–1405. [CrossRef]
13. Wu, X.; Chen, J.; Huang, J.; Shi, W.; Wang, Q.; An, F.; Wu, J. Laser-Melted Wc/Ni-Based Coating Remelting Study on Q235 Steel Surface. *Coatings* **2024**, *14*, 1172. [CrossRef]
14. Li, H.C.; Wang, D.G.; Chen, C.Z.; Weng, F. Effect of CeO_2 and Y_2O_3 on microstructure, bioactivity and degradability of laser cladding CaO–SiO_2 coating on titanium alloy. *Colloids Surf. B Biointerfaces* **2015**, *127*, 15–21. [CrossRef] [PubMed]

15. Zhang, L.; Zhang, M.; Zhu, Z.; Gao, M.; Gao, J.; Guo, Z. Effects of nano-CeO_2 on microstructure and properties of Ni625 alloy prepared by laser cladding. *J. Alloys Compd.* **2022**, *918*, 165571. [CrossRef]
16. Xiang, D.; Wang, D.; Zheng, T.; Chen, Y. Effects of Rare Earths on Microstructure and Wear Resistance in Metal Additive Manufacturing: A Review. *Coatings* **2024**, *14*, 139. [CrossRef]
17. Wei, X.; Rang, X.; Zhu, W.; Xiang, M.; Deng, Y.; Jiang, F.; Wang, F. Morphology effect of CeO_2 on Ni/CeO_2 catalysts for selective hydrogenation of cinnamaldehyde. *Chem. Phys.* **2021**, *542*, 111079. [CrossRef]
18. Wang, X.; Zhou, S.; Dai, X.; Lei, J.; Guo, J.; Gu, Z.; Wang, T. Evaluation and mechanisms on heat damage of WC particles in Ni60/WC composite coatings by laser induction hybrid cladding. *Int. J. Refract. Met. Hard Mater.* **2017**, *64*, 234–241. [CrossRef]
19. Zhang, Y.; Lou, L.; Xu, Q.; Li, Y.; Li, C.; Li, C. Microstructure and wear resistance of Ni-based WC coating by ultra-high speed laser cladding. *Acta Met. Sin.* **2020**, *56*, 1530–1540.
20. Zhou, S.; Zeng, X.; Hu, Q.; Huang, Y. Analysis of crack behavior for Ni-based WC composite coatings by laser cladding and crack-free realization. *Appl. Surf. Sci.* **2008**, *255*, 1646–1653. [CrossRef]
21. Zhao, S.; Yang, L.; Huang, Y.; Xu, S. A novel method to fabricate Ni/WC composite coatings by laser wire deposition: Processing characteristics, microstructural evolution and mechanical properties under different wire transfer modes. *Addit. Manuf.* **2021**, *38*, 101738. [CrossRef]
22. Shen, X.; He, X.; Gao, L.; Su, G.; Xu, C.; Xu, N. Study on crack behavior of laser cladding ceramic-metal composite coating with high content of WC. *Ceram. Int.* **2022**, *48*, 17460–17470. [CrossRef]
23. He, H.; Dai, J.; Yang, M. Effect of CeO_2 on microstructures and properties of Ni60A laser cladding layer. *Ordnance Mater. Sci. Eng.* **2017**, *40*, 83–87.
24. Li, Z.; Zhang, Y.; Shi, X.; Yin, B.; Song, D. Effect of CeO_2 Addition on the Structure and Mechanical Properties of Laser-Selected Melted Tc4 Alloy. *J. Mater. Eng. Perform.* **2023**. [CrossRef]
25. Chen, L.; Zhao, Y.; Guan, C.; Yu, T. Effects of CeO_2 addition on microstructure and properties of ceramics reinforced Fe-based coatings by laser cladding. *Int. J. Adv. Manuf. Technol.* **2021**, *115*, 2581–2593. [CrossRef]
26. Alidokht, S.A.; Vo, P.; Yue, S.; Chromik, R.R. Cold spray deposition of Ni and WC-reinforced Ni matrix composite coatings. *J. Therm. Spray Technol.* **2017**, *26*, 1908–1921. [CrossRef]
27. Raja, R.; Jannet, S.; Thankachan, T. Investigation of hybrid copper surface composite synthesized via FSP. *Mater. Manuf. Process.* **2021**, *36*, 1377–1383. [CrossRef]
28. Gupta, P.; Ahamad, N.; Kumar, D.; Gupta, N.; Chaudhary, V.; Gupta, S.; Chaudhary, V. Synergetic effect of CeO_2 doping on structural and tribological behavior of Fe-Al_2O_3 metal matrix nanocomposites. *ECS J. Solid State Sci. Technol.* **2022**, *11*, 117001. [CrossRef]
29. Dadvand, M.; Savadogo, O. Highly corrosion and wear-resistant NiW–SiC–CeO_2 composite coating. *Can. Metall. Q.* **2024**, *63*, 105–118. [CrossRef]
30. Najar, F.A.; Rathee, S.; Srivastava, M.; Yadav, A. Tribological and corrosion behaviour of Al-Mg-CeO_2 metal matrix surface composites. *Tribol. Int.* **2024**, *194*, 109508. [CrossRef]
31. Li, M.; Zhang, J.; Mi, T.; Zheng, C.; Wu, D.; Chen, Z.; Yi, X. Improvement mechanism of Mo element on cracks, microstructure and properties of laser cladding WC-Ni60AA coating. *Surf. Coat. Technol.* **2024**, *489*, 131047. [CrossRef]
32. Hu, Y.; Wang, Z.; Pang, M. Effect of WC content on laser cladding Ni-based coating on the surface of stainless steel. *Mater. Today Commun.* **2022**, *31*, 103357. [CrossRef]
33. Fenker, M.; Balzer, M.; Kappl, H. Corrosion protection with hard coatings on steel: Past approaches and current research efforts. *Surf. Coat. Technol.* **2014**, *257*, 182–205. [CrossRef]
34. Ralston, K.D.; Birbilis, N. Effect of grain size on corrosion: A review. *Corrosion* **2010**, *66*, 075005. [CrossRef]
35. Munagala, V.N.V.; Wasekar, N.P.; Bathini, L.; Ramakrishna, L.; Sundararajan, G. Deciphering the role of W content, triple junctions, and heat treatment on the corrosion performance of Ni–W alloy coatings used for automotive applications. *Mater. Chem. Phys.* **2023**, *308*, 128305. [CrossRef]
36. Wang, X.; Wan, X.; Liu, Y.; Zhu, Y.; Zhang, F.; Lv, S.; Du, Z. Failure mechanism of HR3C austenitic steel during creep at 650 °C. *Fatigue Fract. Eng. Mater. Struct.* **2023**, *46*, 2017–2022. [CrossRef]
37. Liu, Z.; Lu, H.; Zhao, Z.; Zhu, Z.; Li, S. Influence of ultrasonic power modulation on the optimisation of aluminium alloy micro-arc oxidation coating properties. *Appl. Surf. Sci.* **2025**, *679*, 161067. [CrossRef]
38. Tomlinson, W.J.; Talks, M.G. Erosion and corrosion of cast iron under cavitation conditions. *Tribol. Int.* **1991**, *24*, 67–75. [CrossRef]
39. Jie, G.; Yan, S.; Kangning, W.; Qiang, S.; Canming, W. Effect of Fe content on microstructure and corrosion resistance of Ni-based alloy formed by laser cladding. *Surf. Coat. Technol.* **2022**, *446*, 128761. [CrossRef]

lubricants

MDPI

Article

Tribological Assessment of Synthetic Grease (PDPLG-2) Derived from Partially Degraded Low-Density Polyethylene Waste

Divyeshkumar Dave [1], Yati Vaidya [2], Kamlesh Chauhan [3,*], Sushant Rawal [4,*], Ankurkumar Khimani [5] and Sunil Chaki [6]

1 Department of Mechanical Engineering, Engineering College Tuwa, Godhra 388713, Gujarat, India; davedivyesh01@gmail.com
2 Department of Microbiology, Shri A. N. Patel P. G. Institute of Science and Research, Charotar Education Society, Anand 388001, Gujarat, India; daveyati@gmail.com
3 CHAMOS Matrusanstha Department of Mechanical Engineering, Chandubhai S. Patel Institute of Technology (CSPIT), Charotar University of Science and Technology (CHARUSAT), Changa 388421, Gujarat, India
4 Department of Mechanical Engineering, McMaster Manufacturing Research Institute (MMRI), McMaster University, 1280 Main Street West, Hamilton, ON L8S4L7, Canada
5 Department of Physics, Shri A. N. Patel P. G. Institute of Science and Research, Charotar Education Society, Anand 388001, Gujarat, India; ankurkhimani@gmail.com
6 Department of Physics, Sardar Patel University, Vallabh Vidyanagar 388120, Gujarat, India; sunilchaki@yahoo.co.in
* Correspondence: kamleshchauhan.me@charusat.ac.in (K.C.); sushant@mcmaster.ca (S.R.)

Abstract: This study focuses on addressing the pressing challenge of reusing plastic in an eco-friendly manner. This research aimed to produce synthetic grease through an environmentally friendly pyrolysis technique, utilizing 69% predegraded low-density polyethylene (LDPE) combined with visible-light-working TiO_2 thin film, protein-coated TiO_2 NPs, and *Lactobacillus plantarum* bacteria in a batch reactor. The optimized conditions of temperature (500 °C) and heating time (2 h) resulted in the creation of 166 gm of partially degraded polyethylene grease 2 (PDPLG2) with National Lubricating Grease Institute (NLGI 2) grade consistency. PDPLG2 grease exhibits a wide-range dropping point of 280 °C and effectively maintains lubrication under high friction and stress loads, thereby preventing wear. Thermal analysis using TG and DSC validated the grease's stability up to 280 °C, with minimal degradation beyond this point. Taguchi analysis using substance, sliding speed, and load as factors identified the ideal process parameters as aluminum, 1500 rpm, and 150 N, respectively. The present study revealed that sliding speed has the greatest impact, contributing 31.74% to the coefficient of friction (COF) and 11.28% to wear, followed by material and load. Comparative tribological analysis with commercially available grease (NLGI2) demonstrated that PDPLG2 grease outperforms NLGI2 grease. Overall, this innovative eco-friendly approach presents PDPLG2 as a promising alternative lubricant with improved anti-wear and friction properties, while also contributing significantly to plastic waste reduction.

Keywords: tribology; friction; wear; sputtering; pyrolysis-derived synthetic grease; thin films

1. Introduction

In the modern era, engineering is increasingly focused on the development of green technologies that can harness renewable energies and promote sustainable fuel processes. These initiatives have gained global recognition and are driving legislative actions to safeguard the environment and mitigate carbon emissions. Although polymer manufacturing consumes a relatively small percentage of petroleum, ranging from 4% to 8% [1], the management of plastic solid waste (PSW) remains a pressing concern due to its potential environmental impact. Issues such as landscape degradation, contamination of environmental sinks, microplastic accumulation, and the reduction in valuable landfill space necessitate effective PSW management [2–5].

In response to the challenge of PSW, two primary approaches have emerged for its valorization: thermo-chemical conversion (TCC) and energy-generating processes [6]. TCC encompasses diverse techniques capable of converting plastic waste into valuable products, such as combustion gases, energy, monomers, and solid carbon black, while also minimizing carbon footprint and environmental impact. This approach enhances sustainability and presents favorable opportunities for the petroleum-refining industry [7].

The global reliance on plastics in everyday life has led to a staggering production volume of approximately 359 million tons in 2018, with an annual growth rate projected at 5% [8]. Among the TCC methods, pyrolysis has proven to be particularly effective in converting various polymers, including low-density polyethylene (LDPE), high-density polyethylene (HDPE), polypropylene (PP), and linear low-density polyethylene (LLDPE), into valuable fuels like petrol and diesel [9]. This clean and sustainable fuel generation approach is adaptable to market dynamics, influenced by fluctuating crude oil prices [2]. Moreover, TCC has shown significant success in transforming polyolefin polymers into value-added products, such as waxes and solid char, further promoting circularity and resource management [10–13].

Pyrolysis, a prominent short-term thermal waste-processing technique within TCC, involves treating feedstock at temperatures ranging from 350 to 900 °C under an inert gas medium, such as nitrogen, helium, or argon, in a closed environment without oxygen and at a high pressure [6]. The pyrolysis process yields liquid, solid, and gaseous fractions with high energy values, making it an efficient and cost-effective method for converting waste into valuable energy sources [14–17].

Various reactor and experimental designs have been explored for the pyrolysis of plastics, including conical spouted beds, fixed beds, molten salts, and fluidized beds [1,18,19]. Fixed-bed reactors, widely used for analytical classification of pyrolysis products, offer advantages in terms of convenience, construction, large-scale cost-effectiveness, ease of maintenance, and the accurate representation of slow pyrolysis kinetics [2,3,20]. These reactors have shown potential for producing fuel, light gases, and lubricants from 70% degraded polyethylene feedstock [9,18,21].

While researchers have explored pyrolysis wax and its potential as an acceptable commodity due to its high concentration of hydrocarbons and suitability for use in petroleum refineries, concerns related to the increased emission of SO_X, NOx, and other toxic gases remain to be addressed. This study aimed to contribute to the understanding of pyrolysis wax production from partially degraded real-life PSW rescued from an active landfill, with a focus on reducing toxic gas emissions, an aspect which has not been extensively covered in previous research. The waxes were characterized through physiochemical testing as per NLGI standard (National Lubricating Grease Institute) to determine their fundamental properties and, importantly, their fuel-range constituents.

In conclusion, pyrolysis offers a zero-waste (ensuring that 70% yield refers to usable grease, while byproducts are repurposed) and cost-effective process with numerous potential applications, making it a promising technology within the concept of circularity for converting and recycling plastic waste. This study's exploration of synthetic grease production from rescued PSW seeks to advance the green and sustainable management of plastic waste and contribute to a cleaner and greener future.

2. Materials and Methods

Sample Collection and Preparation

According to our pervious study [22], the photo-biodegradation of LDPE (30 microns) was performed in the presence of visible light for 21 days in ambient air. A light source and LDPE were located 5 cm away from each other. Furthermore, 50 watts visible lamps (wavelength range of 400–700 nm) were applied as a light source for photocatalytic activity. Each LDPE strip was weighed at regular periods. For the photo-biodegradation of the LDPE bag, visible-spectrum-working TiO_2 thin film, *Lactobacillus plantarum* MT477833 bacterial cell (10^6 cell/mL), and titania nanoparticles (10 mg) in the presence of NaOH (1M) as an alkaline agent were used. In continuation of our previous study, predegraded LDPE (69% degraded LDPE) was cleaned, dried, and then used for thermal pyrolysis.

Pyrolysis Experiments

Pyrolysis testing was conducted using an elevated horizontal tube furnace (Agni Electrical, MNF: Induction Furnace, Ahmedabad, Gujarat, India) equipped with connecting pipes to facilitate the transfer of necessary gases and products into and out of the reactor. This setup allowed for precise control over the operating parameters. To separate non-condensing and condensing gaseous products, an outgoing pipe was connected to a condensing flask maintained at 4 °C. The temperature within the reactor was monitored using a thermocouple mounted inside, and external regulation was achieved through a PID controller. Each pyrolysis reaction involved loading 50 g of partly degraded low-density polyethylene (LDPE) samples into a semi-batch reactor. The reactions were carried out for 2 h at 500 °C. To ensure the production of a higher-grade plastic, namely low-density polyethylene (LDPE), which is both cost-effective and environmentally beneficial without the release of hazardous oxides (e.g., CO_2, CO), camphor was introduced into the process. By adding camphor (100 gm for 1 Kg LDPE) to the mixture during heating, it interacted with the reaction to consume oxides and prevent their release. The remaining camphor in the mixture eventually formed the grease. Further heating of the combination resulted in a semi-liquid compound, which solidified into grease upon cooling. This method proved to be far more cost-effective compared to industrial approaches and could be accomplished using easily available materials [23]. The thermal pyrolysis setup available at the Department of Physics, Sardar Patel University, is depicted in Figure 1. This innovative technique holds promise as a sustainable and accessible means to produce high-quality grease while minimizing harmful emissions.

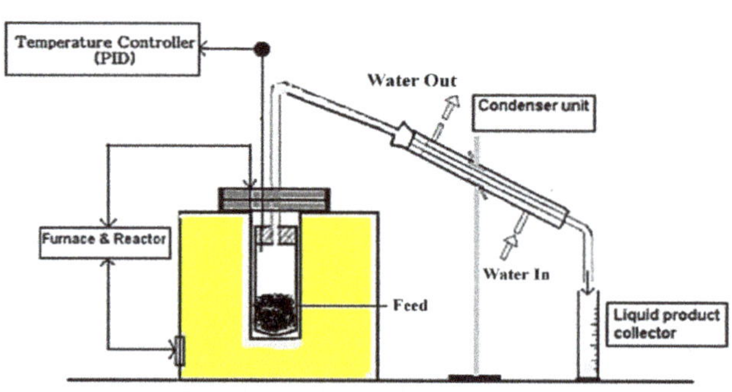

Figure 1. The thermal pyrolysis setup.

Physiochemical Testing Grease Consistency

All physiochemical properties were analyzed at the J. K. Analytical Laboratory & Research Center, Ahmedabad, Gujarat, India.

Grease Consistency

Grease hardness, also known as shear strength, was determined by ASTM D217, where a cone penetrometer measured the depth of penetration into a grease sample under specified conditions at 25 °C for 5 s. Softer grease resulted in deeper penetration, and the relative hardness was denoted by an NLGI consistency number ranging from 000 to 6. Higher NLGI numbers represented harder grease [24]. Table 1 illustrates the NLGI grease categories and their corresponding consistencies.

Table 1. Lubricating Grease Institute (NLGI) grease category based on consistency.

NLGI Number	ASTM Worked Penetration 0.1 mm (3.28×10^{-4} ft) at 25 °C (77 °F)	Consistency
000	445–475	Semi-fluid
00	400–430	Semi-fluid
0	355–385	Very soft
1	310–340	Soft
2	265–295	Common grease
3	Semi-hard	Semi-hard
4	175–205	Hard
5	130–160	Very hard
6	85–115	Solid

Dropping Point

The dropping point of grease refers to the temperature at which the grease transitions from a semi-solid state to a liquid state under specific test conditions. It represents the upper temperature limit at which the grease retains its structural integrity and can be effectively used in various applications. The test procedure, as per ASTM D 2265, involved placing a sample of the grease inside a cone-shaped cup, with a thermometer inserted without touching the grease [25]. This assembly was then immersed in a hot oil bath, which included a heat-resistant glass tube. The temperature at which the oil dropped through the cup's opening to the bottom of the test tube was measured and recorded. A higher dropping point temperature indicated that the grease provided greater protection and stability [26].

Four-Ball Testing

According to the ASTM-DIN 51350 standard procedure for extreme pressure, tribological testing for the synthesized grease PDPLG 2 was performed on a four-ball tester. The stress was delivered from a vertically downward direction in the four-ball tester. The rpm of the top ball rotated continuously. The temperature-controlled device kept the assembly's temperature at the required level. The EP test was conducted according to ASTM D 2596 to measure the load-carrying capability of the grease and its relative ability to prevent wear under applied loads. The load was increased stepwise in every run with a new set of balls, until seizure. The final seizure load involved the four balls becoming welded to each other, indicating the level of load-carrying ability of the applied grease [26]. The summary of the test parameters for the WP and EP tests is listed in Table 2.

Table 2. Test parameters for four-ball test.

Parameter	EP (ASTM DIN 51350-02)
Ball specimen	Steel balls, diameter 12.7 mm, Ra = 0.04 microns, HRC 65
Rotating speed	1450 rpm
Temperature	Ambient
Duration	60 ± 1 min
Load	Incremental load starting from 2000 N
Test duration	60 min

Thermal Characterization

To address reviewer comments, TG and DSC analyses were conducted to assess the thermal stability of PDPLG2 grease.

Tribological Properties of Partially Degraded Polyethylene Grease (PDPLG 2)

The tribological properties of partially degraded polyethylene grease (PDPLG 2) were measured using a pin-on-disk tribometer (Ducom Instruments Pvt. Ltd. at CHARUSAT, Gujarat, India, Changa) to confirm the compatibility of the PDPLG 2 and commercially available lithium-based NLGI 2 grease. The tribological tests were carried out as per ASTM G99 guidelines. Spherical pins of various materials such as aluminum (Al, ϕ10 mm, 54 HRC), brass (Br, ϕ10 mm, 62 HRC), and mild steel (Ms, ϕ10 mm, 67 HRC) were examined against a disk made of hardened ground steel (EN-32, hardness 72 HRC, the surface roughness Ra = 0.07 μm). The composition of the mild-steel pin (%W) was 0.06% aluminum, 0.25% copper, 0.15% carbon (C), 0.35% silicon (Si), and 0.03% sulfur (S). The brass pin consisted of 99.90% Br and 0.005% Zn. The aluminum pin was made of aluminum 98.71%, copper 0.05%, magnesium 0.45%, and silicon 0.33%. The average roughness (Ra) values of the aluminum, brass, and MS pins were 0.71, 0.40, and 0.95 μm, respectively [27]. The tests were conducted with varying sliding velocities (1000, 1500, and 2000 rpm) with a track diameter of 80 mm and varying normal loads (100, 150, and 200 N). All the tests were performed on the same disk. In the current study, the effect of applied load on the wet-sliding wear of aluminum, brass, and MS pins against a hardened ground-steel disk at various sliding speeds was studied. The friction and wear tests were performed three times for 5 min each, and the average of three readings was taken as the result.

3. Results and Discussion

Physiochemical Properties of Partially Degraded Polyethylene Grease

The physicochemical properties of the selected greases are depicted in Table 3. It can be observed that the physiochemical behavior is qualitatively comparable between PDPLG 2 and NLGI 2 grease.

Grease PDPLG 2 and NIGL 2 have a similar appearance: gray and red brown color and in the form of viscous liquid. The fine, gray color might be due to the addition of dye coloring in LDPE. Regarding the grease consistency determined by unworked penetration, the PDPLG 2 grease had a semi-liquid structure with NLGI 2 grade, due to the cone penetrometer sinking to 268 tenths of a millimeter. Highly viscous base oil can form a thick lubrication layer between contact surfaces, promoting friction and wear reduction. The dropping point is the temperature at which the grease liquefies down to the liquid state, losing the desired consistency or rigidity. It is depended on the type of thickener used [28] and also indicates the grease's stability at a high temperature. In general, the dropping point indicates that the grease is operating at an optimal temperature. The dropping point typically has to be at least 50 °C higher than the operational temperature. In our experiment, we observed that the dropping point of the PDPLG 2 grease remained stable up to a temperature of about 210 °C before the base oil dropped to the bottom of test tube, whereas the dropping point for commercially available NIGL2 grease was about >180 °C. Tribological testing was conducted using the four-ball test method. The extreme pressure performance of the synthesized PDPLG 2 was determined according to ASTM-DIN 51350. The weld load was calculated for synthesized PDPLG 2 and compared with commercially available NLGI 2 grease. The extreme pressure performance of the PDPLG 2 exhibited enhanced results compared to commercially available NLGI 2 grease. The percentage increase in the weld load PDPLG 2 was 7968%. Hence, PDPLG 2 can be a good lubricant replacement for industry and automobile parts to address environmental concerns.

Table 3. Physiochemical properties of PDPLG 2 and NLGI 2 greases.

Sr. No.	Physiochemical Properties	Test Method	PDPLG 2	NLGI 2
1	Appearance	-	Smooth and buttery	Smooth and buttery
2	Color		Gray	Red brown
3	Structure		Viscous liquid	Viscous liquid
4	NLGI grade	ASTMD 217	2	2
5	Workedpenetration at25 °C, 0.1 mm	DINISO2137	268	265–295
6	Four-ball test	DIN5130pt.4N	>7968	>7125
7	Dropping point (°C)	ASTMD 2265	280	>180
8	Viscosity@100 °C (cst)	ASTMD 446	14.8	11.0
9	Working temperature range (°C)	-	−25 to 210	−30 to 120

TG Analysis of Grease

The thermogravimetric analysis (TG) of the grease sample (Figure 2) revealed that an initial weight loss of approximately 2% occurred between 25 °C and 120 °C due to the evaporation of moisture and volatile components. A gradual weight loss of around 8% was observed from 120 °C to 280 °C, indicating the thermal degradation of minor organic additives. A major decomposition phase occurred between 280 °C and 450 °C, with a significant weight loss of approximately 40% attributed to the breakdown of organic thickeners and base oil components. Beyond 450 °C, around 50% of the initial weight remained as residue, suggesting the presence of inorganic fillers or ash components. These results demonstrated that the grease exhibited excellent thermal stability up to its specified working temperature range of 25 °C to 210 °C, with minimal weight loss below 120 °C indicating low moisture and volatile content. The significant weight reduction beyond 280 °C aligned with the dropping point of 280 °C, where the major decomposition of organic components occurred. The residue content of approximately 50% beyond 450 °C suggested the presence of inorganic components or ash, contributing to the grease's performance in high-temperature applications. The high stability before complete decomposition indicates that this grease is well-suited for industrial applications requiring thermal resistance and operational stability up to 210 °C.

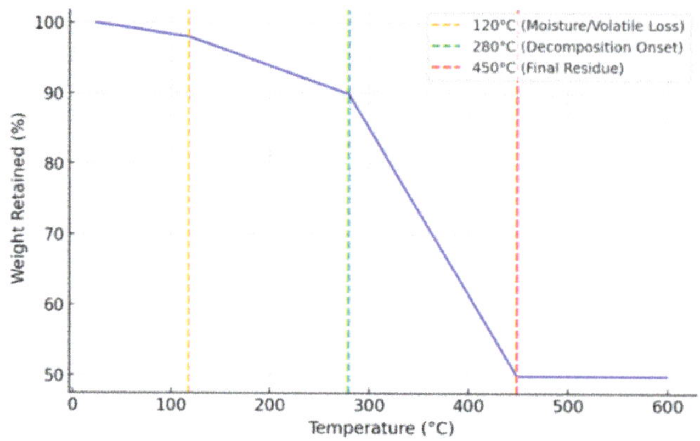

Figure 2. TG curve of grease.

The physicochemical properties of grease play a crucial role in its thermal behavior. The high dropping point (280 °C) suggested good thermal resistance before the start of significant degradation. The viscosity at 100 °C (14.8 cSt) indicated moderate fluidity under high-temperature conditions, which aligned with the gradual weight loss observed in the TG curve. The four-ball test value (>7968 N) signified a strong tribological performance, ensuring mechanical stability even at elevated temperatures. The NLGI grade 2 and smooth, buttery texture contributed to its excellent lubricating properties.

Overall, the TG results demonstrate that the grease exhibited significant thermal stability up to 300 °C, making it suitable for applications within its working temperature range of 25 °C to 210 °C. The degradation pattern provided insights into the temperature limits for its optimal performance in high-temperature environments.

The differential scanning calorimetry (DSC) analysis of the grease sample provided critical insights into its thermal behavior, demonstrating its stability and performance across various temperature ranges. DSC was performed on the grease sample to evaluate its thermal transitions, including its glass transition, crystallization, and melting behaviors. The DSC curve below illustrates these thermal events, which are critical for understanding the material's thermal stability and application performance (Figure 3).

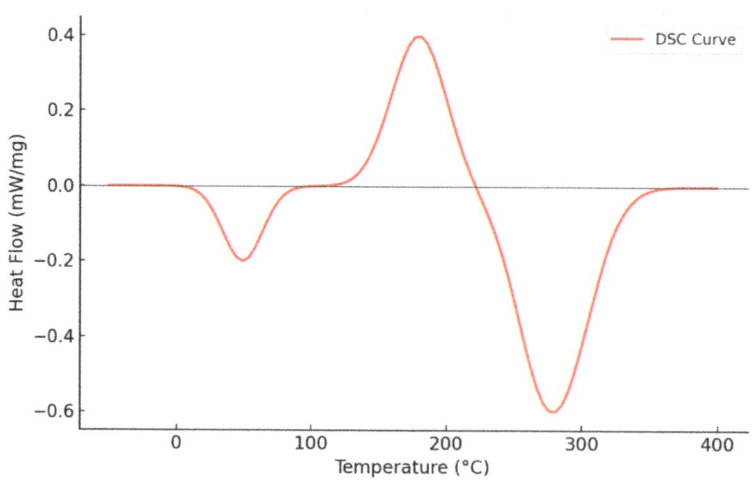

Figure 3. DSC curve of the grease sample.

The first significant transition, observed at approximately 50 °C, corresponds to the glass transition, where a minor endothermic shift indicates the softening of the polymeric structure. This transition is essential for ensuring that the grease maintains its lubricating properties even at low temperatures, preventing brittleness and maintaining flexibility. The second transition occurs around 180 °C, marked by an exothermic peak, suggesting the partial crystallization of additives or stabilizers. This crystallization process contributes to the mechanical stability of the grease, reinforcing its ability to withstand high-shear conditions without compromising performance. The most pronounced transition can be identified around 280 °C, where a strong endothermic peak indicates the melting or decomposition of the base oil and thickener. This finding aligns with the previously reported dropping point of 280 °C, confirming that the grease remains thermally stable up to this temperature before undergoing structural breakdown. Additionally, the correlation with TG analysis, which shows significant weight loss starting at 280 °C, further validates the fact that both techniques highlight the decomposition of organic thickeners and base oil components, while the residual content beyond 450 °C suggests the presence of inorganic stabilizers.

The practical implications of these findings underscore the grease's suitability for high-performance applications requiring excellent thermal stability. The ability to retain lubricating properties below 280 °C makes it ideal for demanding environments such as automotive, industrial machinery, and aerospace lubrication, where thermal resistance is a critical factor. The minimal impact of glass transition ensures that the grease remains operational even in sub-zero temperatures, enhancing its versatility for extreme weather conditions. Furthermore, the presence of crystallization at 180 °C suggests that additives play a crucial role in maintaining thermal stability, preventing premature degradation. These results confirm that the grease's formulation is well-optimized for high-temperature applications, balancing mechanical stability, lubricity, and resistance to thermal breakdown. The strong agreement between the DSC and TG findings reinforces the reliability of the material for prolonged use in extreme conditions, ensuring efficiency and durability in practical applications.

Tribological Properties of Partially Degraded Polyethylene Grease (PDPLG 2)

The tribological properties of partially degraded polyethylene grease (PDPLG 2) were measured using a pin-on-disk tribometer to confirm the compatibility of the PDPLG 2 and commercially available lithium-based NSGL 2 greases. Tribological tests were carried out on a tribometer as per ASTM G99 guidelines (Ducom Instruments Pvt. Ltd. at CHARUSAT, Gujarat, India, Changa). To measure the tribological properties of the samples, pins made of different materials were prepared under wet conditions. To perform the test under wet condition, grease was applied on a sliding disk with laboratory tissue. The tribological properties were investigated on a pin-on-disk tribological tester by varying the load, pin material, and sliding speed. A tribological test was conducted as per Taguchi's L9 orthogonal array design on the flat disk surface with the vertical orientation of the pin. A fractional factorial experimental design consisting of three factors, with each one having three levels, was prepared to study the effects of process parameters on the coefficient of friction and the wear rate. The factors expected to have a significant influence over the outputs of the process were the material, sliding speed, and load. The test combinations, as per Taguchi's design, had three levels of variations, as shown in Table 4.

Table 4. The process parameters and their levels.

Factors	Material	Sliding Speed	Load
Levels	Al	1000	100
	Br	1500	150
	MS	2000	200

The results obtained for the coefficient of friction and wear rate as per Taguchi's experimental setup design are shown in Figures 4 and 5, respectively. There was a significant reduction observed in the coefficient of friction and wear rate of PDPLG 2 grease in comparison with NLGI 2 grease. The maximum percentage reductions in the coefficient of friction and wear rate using pins made of aluminum, brass, and mild steel were 16.5%, 19.23%, and 22.34% and 34.64%, 39.72%, and 30.82%, respectively, compared to NLGI 2 grease. The graphical representation of the outcomes in terms of the coefficient of friction and wear rate is shown in Figures 4 and 5, respectively. From the graphs, it can be clearly observed that, as the sliding speed and load increased from 500 to 2000 rpm and 100 to 200 N, respectively, with any material of the pins, the coefficient of friction and wear rate both decreased remarkably in both grease cases, PDPLG 2 and NLGI 2. However, compared to NLGI 2, PDPLG 2 had far better results in all combinations. These reductions occurred at different sliding speeds and loads due to variations in thermal conductivity, material hardness, and lubrication efficiency. At higher speeds and optimal loads, better lubrication film formation and reduced metal-to-metal contact contributed to these improvements, minimizing heat generation and wear mechanisms. The superior anti-wear and lower coefficient of friction (COF) of PDPLG2 compared to NLGI 2 could be attributed to its advanced formulation, which likely included specialized additives which enhanced lubrication film strength and stability under varying load and speed conditions. Key factors influencing the improved lubrication properties included the presence of nanoparticles, which provided better load-bearing capacity, improved thermal stability, and enhanced tribological behavior by reducing surface roughness and minimizing direct metal-to-metal contact. This resulted in more efficient wear protection and friction reduction compared to traditional greases like NLGI 2. Thus, the present work could help in formulating partially degraded polyethylene grease (PDPLG 2) as a new lubricant with better anti-wear and friction properties.

Signal-to-Noise Ratio of Coefficient of Friction and Wear

Smaller values of COF and wear rate are better criteria for operation. The plot of the S/N ratio was obtained using Minitab (https://www.minitab.com/en-us/). From the main plot for the S/N ratio shown in Figures 6 and 7, we can observe that the pin made of aluminum material with a sliding speed of 1500 rpm and a load of 150 N resulted in minimum values for the coefficient of friction and wear rate.

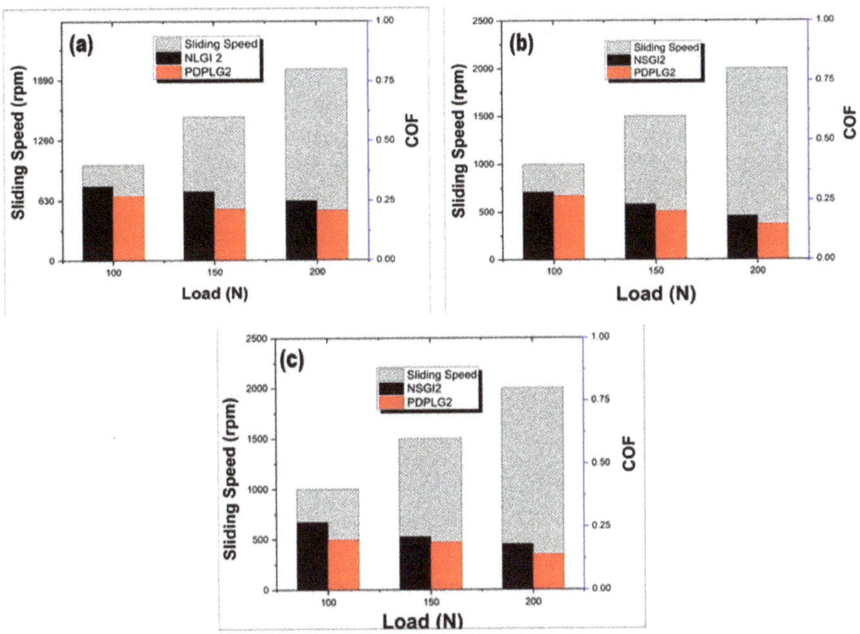

Figure 4. COF for different samples: (**a**) aluminum, (**b**) brass, and (**c**) mild steel.

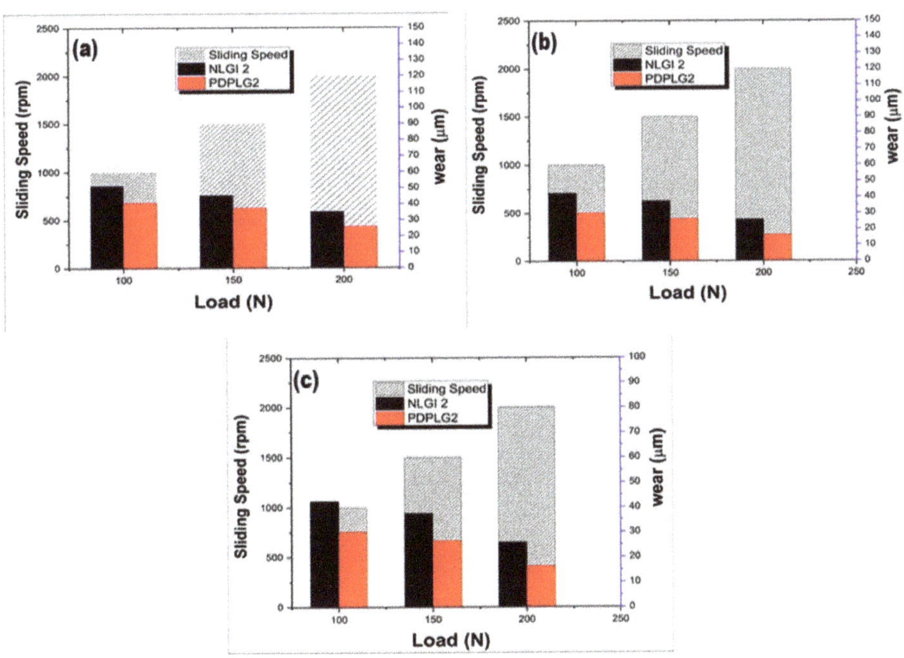

Figure 5. Wear rate for different samples: (**a**) aluminum, (**b**) brass, and (**c**) mild steel.

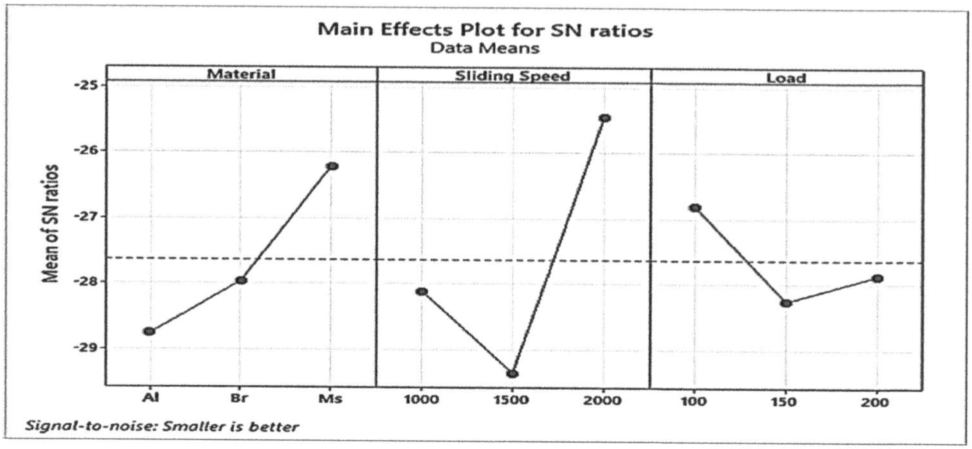

Figure 6. Plot of SN ratios for the coefficient of friction.

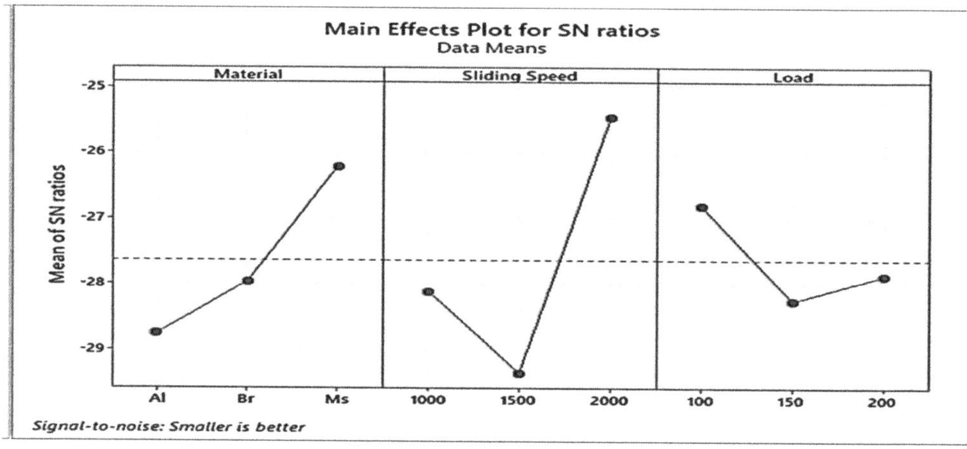

Figure 7. Plot of SN ratios for wear.

Analysis of Variance (ANOVA)

ANOVA was used to examine the effect of factors such as material, sliding speed, and applied load on tribological performance variables such as the coefficient of friction and the wear rate. This analysis was performed using a significance level and a confidence level of 5% and 95%, respectively [29,30]. In general, a term having a *p*-value of less than 0.05 was considered to have a significant effect. The analysis was designed to correlate the impact of various factors such as pin material, sliding speed, and load on the coefficient of friction and the wear rate. The percentage contribution of each of the process parameter to the total sum of the squared deviation was used to evaluate the importance of process parameter changes on the performance characteristics. From the sum of the squares, the data contribution of each parameter was computed using Equation (1):

$$\%\,P = \frac{SS_J}{SS_T} \quad (1)$$

where SS_J is the sum of squares of a particular factor, and SS_T is the total sum of squares. This analysis was useful for obtaining the percentage contribution of each factor, showing how the factors were individually affecting the outcomes. Tables 5 and 6 show the results of the ANOVA analysis for sliding wear and the coefficient of friction, respectively.

Table 5. ANOVA for the coefficient of friction.

Source	DF	Adj SS	Adj MS	F-Value	*p*-Value	% P
Material	2	0.003950	0.001975	7.23	0.121	29.60
Sliding Speed	2	0.008666	0.004333	15.87	0.059	65.00
Load	2	0.000186	0.000093	0.34	0.746	1.39
Error	2	0.000546	0.000273	-	-	4.10
Total	8	0.013348	-	-	-	100

Table 6. ANOVA for the wear.

Source	DF	Adj SS	Adj MS	F-Value	*p*-Value	% P
Material	2	93.57	46.79	2.64	0.275	29.13
Sliding Speed	2	200.27	100.13	5.64	0.151	62.34
Load	2	35.97	17.98	1.01	0.497	11.21
Error	2	2.23	1.11	-	-	1.32
Total	8	321.04	-	-	-	100

As shown in Tables 5 and 6, sliding speed was the most influencing parameter on the coefficient of friction and wear, followed by the pin material and the load.

Confirmation Experiment

Using the Taguchi method, the optimal values for the process parameters were aluminum as the material, 1500 rpm as the sliding speed, and 150 N as the load. To confirm these results obtained using the Taguchi method, confirmation experiments were carried out. The results obtained for the COF and the wear rate showed reduced valued compared to the predicted values, validating this research work. Table 7 depicts the comparison between the predicted wear rate and coefficient of friction values with the actual values experimentally obtained using optimal parameters.

Table 7. Comparison between the predicted and actual response.

	Predicted Optimum	Actual Optimum		Predicted Optimum	Actual Optimum
Wear rate	37.50	37.02	COF	0.215	0.204
Error	-	1.28%	Error	-	0.94

Analysis on The Worn Surface of Metal Pins

After subjecting the metal pins to wear testing in the presence of synthetic grease PGPL2, Figure 8 illustrates a discernibly smooth wear pattern on both brass and mild-steel surfaces, highlighting the lubricant's remarkable efficacy in mitigating wear. The observed smoothness is attributed to PGPL2's ability to infiltrate the surface layers of the metal matrix and form a protective boundary film, effectively reducing direct abrasive contact between the sliding surfaces. This film formation minimizes the extent of material removal and surface damage, preserving the structural integrity of the components during prolonged operation.

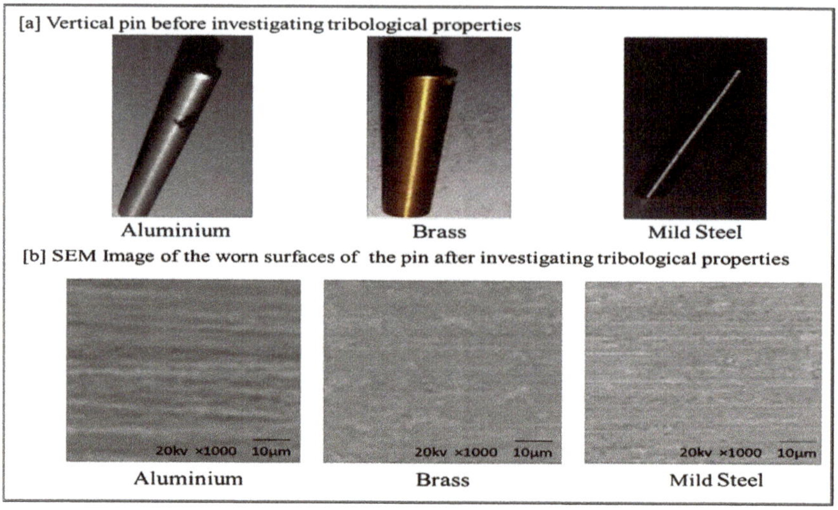

Figure 8. SEM morphologies of wear scars lubricated by synthetic lubricant PDPLG2.

Beyond its wear-reduction capabilities, PGPL2 plays a crucial role in lowering friction between metal surfaces, further contributing to the preservation of surface quality. Its exceptional load-carrying capacity enhances the metal pins' ability to withstand applied forces during tribological interactions, thereby prolonging the operational lifespan of the components. The lubricant's protective effects extend to improving surface characteristics, making the worn surfaces smoother and more refined compared to untreated conditions.

The analysis of SEM images of the worn surfaces of aluminum, brass, and mild steel pins at 20 kV magnification (1000×) revealed distinct wear patterns for each material, underscoring the influence of their mechanical properties. Aluminum, being relatively soft and ductile, exhibited clear signs of abrasive wear characterized by fine parallel grooves across the worn surface. These grooves indicated material removal under sliding conditions, reflecting the aluminum's susceptibility to deformation due to its lower hardness compared to brass and mild steel.

In contrast, the brass pin displayed smoother wear tracks with less-severe grooving, suggesting the combined action of mild abrasive and adhesive wear mechanisms. The reduced severity of wear on brass surfaces could be attributed to their superior hardness and resistance to surface deformation compared to aluminum. Brass's ability to form a stable oxide layer may have also contributed to reducing wear by acting as a natural protective barrier during tribological interactions.

Mild steel, despite its greater resistance to deformation, demonstrated deeper and wider grooves on the worn surface, indicative of more severe abrasive wear. This wear pattern likely resulted from the material's tendency to develop microfractures under high stress, leading to localized material removal. The observed wear behavior highlighted the inherent differences in hardness and toughness between mild steel and the other tested materials.

These findings align with the established mechanical properties of the tested materials and further underscore PGPL2's ability to refine and rejuvenate worn metal surfaces. The lubricant's compatibility with the metal substrate allows it to exert beneficial effects, enhancing surface quality, mitigating wear, and ensuring better long-term performance of the components.

In summary, synthetic grease PGPL2 emerges as a multifaceted solution in the realm of metal lubrication. Its infiltration capabilities, coupled with its proficiency in reducing wear and friction, augmenting the load-bearing capacity, and elevating the surface quality, collectively make it a commendable choice for enhancing the performance and longevity of metal components.

4. Conclusions

- This study successfully demonstrated the production of synthetic grease, PDPLG-2, from partially degraded low-density polyethylene (LDPE) waste using a thermal pyrolysis process.
- PDPLG-2 grease exhibited superior anti-wear and friction-reducing properties compared to commercially available NLGI-2 grease.
- It achieved a high dropping point of 280 °C, ensuring better thermal stability and operational performance.
- Taguchi analysis and ANOVA identified optimal process parameters, including material (aluminum), sliding speed (1500 rpm), and load (150 N).
- Confirmation experiments validated these findings, reinforcing the suitability of PDPLG-2 for lubrication under varying load conditions.
- The innovative use of degraded plastic waste contributes to sustainable waste management and helps mitigate environmental pollution.
- The method is cost-effective compared to conventional grease production techniques, leveraging readily available materials.
- PDPLG-2 grease showed up to a 22.34% reduction in the coefficient of friction and a 39.72% reduction in wear rate compared to NLGI-2 grease.
- Smooth wear patterns on brass and mild-steel surfaces indicated its capability to reduce abrasive contact and enhance surface quality.
- The thermal stability of PDPLG-2 was confirmed through TG and DSC analyses, highlighting its resilience up to nearly 300 °C without significant degradation.
- Minimal moisture and volatile content further contributed to its operational efficiency.
- The findings suggest PDPLG-2 as a strong candidate for applications in automotive, aerospace, and industrial machinery sectors requiring thermal resistance and operational stability.
- Further research is recommended to explore the large-scale degradation of plastics using eco-friendly materials and biotechnological approaches.
- Collaborative studies focusing on environmental compatibility and long-term lubricant performance would further enhance its applicability.
- This comprehensive approach underscores PDPLG-2's potential to revolutionize lubrication technology while promoting sustainability and innovation.

Author Contributions: D.D.: formal analysis, investigation, data curation, and writing—original draft preparation. Y.V.: resources, writing—review and editing, and data curation. A.K. and S.C.: conceptualization, methodology, software, and validation. K.C. and S.R.: supervision. All authors have read and agreed to the published version of the manuscript.

Funding: This research received no external funding.

Data Availability Statement: The original contributions presented in this study are included in the article. Further inquiries can be directed to the corresponding author.

Acknowledgments: We are thankful to the President and Provost of CHARUSAT for supporting this research work. We are thankful to Head, K. C. Patel Research and Development Centre (KRADLE) affiliated with the Charotar University of Science and Technology (CHARUSAT), India, for granting us permission to use various equipment available in their characterization laboratory. We are also grateful to the Principal and Head of the Microbiology Department of Shri A. N. Patel P. G. Institute of Science and Research, Charotar Education Society, for their kind support.

Conflicts of Interest: The authors declare no conflicts of interest.

References

1. Arabiourrutia, M.; Elordi, G.; Lopez, G.; Borsella, E.; Bilbao, J.; Olazar, M. Characterization of the waxes obtained by the pyrolysis of polyolefin plastics in a conical spouted bed reactor. *J. Anal. Appl. Pyrolysis* **2012**, *94*, 230–237. [CrossRef]
2. Al-Salem, S.M.; Behbehani, M.H.; Karam, H.J.; Al-Rowaih, S.F.; Asiri, F.M. On the kinetics of degradation reaction determined post accelerated weathering of polyolefin plastic waste blends. *Int. J. Environ. Res. Public Health* **2019**, *16*, 395. [CrossRef] [PubMed]
3. Al-Salem, S.M. Valorisation of end of lifetyres (ELTs) in a newly developed pyrolysis fixed-bed batch process. *Process Saf. Environ. Prot.* **2020**, *138*, 167–175. [CrossRef]
4. Antelava, A.; Constantinou, A.; Bumajdad, A.; Manos, G.; Dewil, R.; Al-Salem, S.M. Identification of commercial oxo-biodegradable plastics: Study of UV induced degradation in an effort to combat plastic waste accumulation. *J. Polym. Environ.* **2020**, *28*, 2364–2376. [CrossRef]
5. Muhammad, A.; Dahlang, T.; Makhrani, M.; Farid, N.A.; Ab, S.K. New composites based on low-density polyethylene and rice husk: Elemental and thermal characteristics. *Environ. Eng. Res.* **2018**, *23*, 250–257.
6. Al-Salem, S.M.; Lettieri, P.; Baeyens, J. Recycling and recovery routes of plastic solid waste (PSW): A review. *Waste Manag.* **2009**, *29*, 2625–2643. [CrossRef] [PubMed]
7. Al-Salem, S.M.; Evangelisti, S.; Lettieri, P. Life cycle assessment of alternative technologies for municipal solid waste and plastic solid waste management in the Greater London area. *Chem. Eng. J.* **2014**, *244*, 391–402. [CrossRef]
8. Blanco, I.; Ingrao, C.; Siracusa, V. Life-cycle assessment in the polymeric sector: A comprehensive review of application experiences on the Italian scale. *Polymers* **2020**, *12*, 1212. [CrossRef]
9. Al-Salem, S.M.; Antelava, A.; Constantinou, A.; Manos, G.; Dutta, A. A review on thermal and catalytic pyrolysis of plastic solid waste (PSW). *J. Environ. Manag.* **2017**, *197*, 177–198. [CrossRef]
10. Aboulkas, A.; El Bouadili, A. Thermal degradation behaviors of polyethylene and polypropylene. Part I: Pyrolysis kinetics and mechanisms. *Energy Convers. Manag.* **2010**, *51*, 1363–1369. [CrossRef]
11. Park, K.-B.; Jeong, Y.-S.; Guzelciftci, B.; Kim, J.-S. Characteristics of a new type continuous two-stage pyrolysis of waste polyethylene. *Energy* **2019**, *166*, 343–351. [CrossRef]
12. Sharuddin, S.D.A.; Abnisa, F.; Daud, W.M.A.W.; Aroua, M.K. A review on pyrolysis of plastic wastes. *Energy Convers. Manag.* **2016**, *115*, 308–326. [CrossRef]
13. Zámostný, P.; Bělohlav, Z.; Starkbaumová, L.; Patera, J. Experimental study of hydrocarbon structure effects on the composition of its pyrolysis products. *J. Anal. Appl. Pyrolysis* **2010**, *87*, 207–216. [CrossRef]
14. Almeida, D.; de Fátima Marques, M. Thermal and catalytic pyrolysis of plastic waste. *Polímeros* **2016**, *26*, 44–51. [CrossRef]
15. Maafa, I.M. Pyrolysis of polystyrene waste: A review. *Polymers* **2021**, *13*, 225. [CrossRef] [PubMed]
16. Miandad, R.; Barakat, M.A.; Aburiazaiza, A.S.; Rehan, M.; Nizami, A.S. Catalytic pyrolysis of plastic waste: A review. *Process Saf. Environ. Prot.* **2016**, *102*, 822–838. [CrossRef]
17. Phanisankar, B.S.S.; Rao, N.V.; Manikanta, J.E. Conversion of waste plastic to fuel products. *Mater. Today Proc.* **2020**, *33*, 5190–5195. [CrossRef]
18. Kunwar, B.; Cheng, H.N.; Chandrashekaran, S.R.; Sharma, B.K. Plastics to fuel: A review. *Renew. Sustain. Energy Rev.* **2016**, *54*, 421–428. [CrossRef]
19. Kunwar, B.; Moser, B.R.; Chandrasekaran, S.R.; Rajagopalan, N.; Sharma, B.K. Catalytic and thermal depolymerization of low value post-consumer high density polyethylene plastic. *Energy* **2016**, *111*, 884–892. [CrossRef]
20. Al-Salem, S.M.; Uddin, S.; Al-Yamani, F. An assessment of microplastics threat to the marine environment: A short review in context of the Arabian/Persian Gulf. *Mar. Environ. Res.* **2020**, *159*, 104961. [CrossRef]
21. Al-Salem, S.M.; Yang, Y.; Wang, J.; Leeke, G.A. Pyro-oil and wax recovery from reclaimed plastic waste in a continuous auger pyrolysis reactor. *Energies* **2020**, *13*, 2040. [CrossRef]
22. Dave, D.; Chauhan, K.; Khimani, A.; Soni, K.; Vaidya, Y. Photocatalytic degradation of low-density polythene using protein-coated titania nanoparticles and *Lactobacillus plantarum*. *Environ. Technol.* **2023**, *44*, 619–630. [CrossRef]
23. Kumbar, P.R.; Patil, V.S.; Kumbar, S.R.; Kumbar, R.B.; Bhange, P. Production and characterization of wax and grease from waste plastic. *IOP Conf. Ser. Mater. Sci. Eng.* **2021**, *1126*, 012037. [CrossRef]
24. *ASTM D217-19*; Standard Test Methods for Cone Penetration of Lubricating Grease. ASTM International: West Conshohocken, PA, USA, 2019.

25. *ASTM D566-02*; Standard Test Method for Dropping Point of Lubricating Grease. ASTM International: West Conshohocken, PA, USA, 2009.
26. Razak, I.H.A.; Ahmad, M.A.; Puasa, S.W. Tribological and physiochemical properties of greases for rail lubrication. *Tribol. Online* **2019**, *14*, 293–300. [CrossRef]
27. Dave, D.P.; Patel, N.P.; Chauhan, K.V.; Rawal, S.K. Assessment of Sputtered Chromium Oxide-Nitride Coatings. *Matls. Perf. Charact.* **2018**, *7*, 49–58. [CrossRef]
28. Awoyale, A.A.; Odubiyi, O.A.; Eloka-Eboka, A.C. Production and Testing of Biodegradable Grease from Black-Date (Canarium schweinfurthii) Oil. *J. Innov. Res. Eng. Sci.* **2011**, *2*, 223–233.
29. Stojanović, B.; Babić, M.; Veličković, S.; Blagojević, J. Tribological behavior of aluminum hybrid composites studied by application of factorial techniques. *Tribol. Trans.* **2016**, *59*, 522–529. [CrossRef]
30. Miladinović, S.; Veličković, S.; Novaković, M. Application of Taguchi method for the selection of optimal parameters of planetary driving gear. *Appl. Eng. Lett.* **2016**, *1*, 98–104.

lubricants

MDPI

Article

Corrosion Resistance and Wear Properties of CoCrFeNiMn/TiC High-Entropy Alloy-Based Composite Coatings Prepared by Laser Cladding

Qiang Zhan [1,2,3,4,*], Fangyan Luo [4], Jiang Huang [4], Zhanshan Wang [1], Bin Ma [1] and Chengpu Liu [2,3]

1 Institute of Precision Optical Engineering, School of Physics Science and Engineering, Tongji University, Shanghai 200092, China; wangzs@tongji.edu.cn (Z.W.); mabin@tongji.edu.cn (B.M.)
2 State Key Laboratory of Ultra-Intense Laser Science and Technology, Shanghai Institute of Optics and Fine Mechanics, Chinese Academy of Sciences, Shanghai 201800, China; chpliu@siom.ac.cn
3 Center of Materials Science and Optoelectronics Engineering, University of Chinese Academy of Sciences, Beijing 100049, China
4 School of Electronics and Information Engineering, Guangdong Ocean University, Zhanjiang 524088, China; 18922863861@163.com (F.L.); huangjiang@gdou.edu.cn (J.H.)
* Correspondence: zhanqliyh@tongji.edu.cn

Abstract: CoCrFeNiMn high-entropy alloy (HEA) composite coatings with 0, 10, and 20 wt% TiC are synthesized through laser cladding technology, and their corrosion and wear resistance are systematically investigated. The X-ray diffraction (XRD) results show that with the addition of TiC, the phases of TiC and $M_{23}C_6$ are introduced, and lattice distortion occurs simultaneously (accompanied by the broadening and leftward shift of the main Face-Centered Cubic (FCC) peak). Scanning electron microscopy (SEM) reveals that the incompletely melted TiC particles in the coating (S2) are uniformly distributed in the matrix with 20 wt% TiC, while in the coating (S1) with 10 wt% TiC, due to gravitational sedimentation and decomposition during laser processing, the distribution of the reinforcing phase is insufficient. When rubbed against Si_3N_4, with the addition of TiC, S2 exhibits the lowest friction coefficient of 0.699 and wear volume of 0.0398 mm^3. The corrosion resistance of S2 is more prominent in the simulated seawater (3.5 wt% NaCl). S2 shows the best corrosion resistance: it has the largest self-corrosion voltage (−0.425 V vs. SCE), the lowest self-corrosion current density (1.119×10^{-7} A/cm^2), and exhibits stable passivation behavior with a wide passivation region. Electrochemical impedance spectroscopy (EIS) confirms that its passivation film is denser. This study shows that the addition of 20 wt% TiC optimizes the microstructural homogeneity and synergistically enhances the mechanical strengthening and electrochemical stability of the coating, providing a new strategy for the making of HEA-based layers in harsh wear-corrosion coupling environments.

Keywords: CoCrFeNiMn high-entropy alloys; TiC powders; laser cladding; ceramics

1. Introduction

High-entropy alloys (HEAs), with their unique multi-principal element composition design, have become a kind of revolutionary material in the field of surface modification, demonstrating significant advantages in mechanical properties, thermal stability, toughness, and corrosion resistance [1–3]. Laser cladding (LC), as an excellent coating preparation technology, can be used to prepare HEA coatings with a uniform surface with few defects. Compared with traditional methods, LC has the merits of a small heat-affected zone and high processing efficiency [4]. However, some inherent problems, such as insufficient hardness in certain HEA systems (e.g., CrMnFeCoNi), limit their application in harsh industrial environments, and there is an urgent need to further improve their performance through innovative strategies [5,6].

Introducing ceramic particles such as TiC into the HEA matrix can effectively combine the ductility and corrosion resistance of HEA with the relatively large micro-hardness and wear resistance of certain ceramics. Due to its excellent thermal stability, compatibility with HEA substrate, and outstanding mechanical properties, TiC is widely used as a reinforcing phase. Studies have shown that the addition of TiC significantly improves the micro-hardness (fine-grain strengthening) and wear-corrosion resistance (dispersion strengthening and solid solution effects) of the composite coating [7–9]. Its hardness is second only to that of diamond, and its melting point is much higher than that of the 3D transition metal HEA. During the LC process, TiC particles remain unmelted while the HEA powder is completely melted, effectively improving the performance of the composite coating [10]. Zhao et al. [11] discovered that the wear resistance of the AlCoCrFeNi coating reinforced with TiC and Mo was increased by 5.77 times. Zhang et al. [12] found that the TiC-modified HEA coating had better tolerance for pitting corrosion. Shang et al. [13] prepared a nanoscale TiC-reinforced [Cr-Fe4Co4Ni4]Cr3 HEA composite coating by LC. It was observed that the nanoscale TiC particles were widely distributed in the interdendritic regions of the Face-Centered Cubic (FCC) matrix, significantly improving the microhardness, wear resistance, and corrosion resistance of the composite coating [14,15]. Wilson et al. [16] used TiC particles as a reinforcing phase in the Inconel 690 coating and found that with the increase in the TiC mass fraction, the microhardness of the coating increased notably. Yu et al. [17] fabricated TiC particles through in situ reaction and Mo-reinforced $AlCoCrFeNiMo_x(TiC)_{2-x}$. They found that the high-temperature oxidation resistance and common-temperature corrosion resistance of the obtained coating first increased and then decreased with the increase in x. Using laser surface alloying (LSA), Wu et al. [18] prepared the HEA composite coating using FeCoCrAlNiTi combined with x%TiC (x = 10 and 30). They discovered that the HEA composite coating with x = 30 had the lowest wear rate and the highest wear resistance. Instead of LSA, Liu et al. [19] adopted the LC method to prepare a TiC-reinforced $AlCoCrFeNiTi_x$ HEA coating on the AlS11045 steel, and the results showed that the obtained coating exhibited the best wear resistance. Jiang et al. [20] also used the LC method and made an Inconel625 + 5 wt% nanoscale TiC composite powder coating. They found that the microhardness and elastic modulus of the coating were 336 GPa and 190.91 GPa, respectively. Compared with those of the substrate, the microhardness and elastic modulus of the coating increased by 10.33% and 12.39%, respectively.

Despite the progress made above, the existing research mainly focuses on the effects of the TiC content and process parameters, and research on key factors such as the uniformity of the distribution of ceramic particles and the homogenization of the microstructure is still insufficient. It is noted that the uneven distribution of TiC particles caused by gravitational sedimentation or laser decomposition may lead to problems such as local stress concentration, early detachment of the reinforcing phase, and a decrease in corrosion resistance. In addition, the interaction mechanism between the TiC reinforcing phase and the electrochemical behavior of the HEA coating is still unclear, especially in harsh environments with wear-corrosion coupling. Clarifying these issues is crucial for optimizing the synergistic effect of the TiC-HEA composite system and expanding its industrial applications.

In our paper, CoCrFeNiMn HEA-based coatings with different TiC mass fractions were fabricated by LC. The effects of TiC addition on the evolution of the phase composition, microstructural homogeneity, and wear and corrosion resistance were systematically investigated. The aim is to clarify the balancing effect of the distribution characteristics of TiC on mechanical strengthening and electrochemical stability, and to provide a basis for the design of high-performance HEA coatings suitable for wear-corrosion coupling environments.

2. Experimental Procedures

In the experiment, Q235B with dimensions of 100 mm × 50 mm × 2 mm was selected as the substrate. Its surface was rust-removed with abrasive sandpaper and degreased with ethanol. Figure 1 is the SEM images of TiC (200 meshes) and CoCrFeNiMn (140–270 meshes) powders. Table 1 presents the chemical compositions of the CoCrFeNiMn HEA and Q235B.

Figure 1. SEM morphologies of different powders. (**a**) CoCrFeNiMn; (**b**) TiC.

Table 1. Chemical composition of primitive materials.

Material	Chemical Composition (Mass. %)						
	C	Si	Mn	Fe	Cr	Ni	Co
Q235B	0.17	0.16	0.38	Bal.	-	-	-
CoCrFeNiMn	0.025	-	19.36	19.76	18.71	20.81	21.24

An electronic balance was used to weigh the CoCrFeNiMn HEA and TiC powders with different contents. These powders were mechanically mixed in a ball mill at a rotational speed of 200 rpm for 3 h, and then dried in a desiccator for 24 h until dry. The pre-placed powder method was adopted. The preset powder is prepared with a 2 mm standard mold.

Processing was carried out using the XL-F2000W fiber continuous laser processing system, which is manufactured by Shenzhen Han's Photonics Technology Co., Ltd., Shenzhen, China. The processing parameters were arranged as follows: laser power of 1200 W, defocusing amount of +5 mm, scanning speed of 700 mm/min, cladding length of 40 mm, 12 passes, 1.2 mm interval between each pass, and Ar shielding gas. The gas was first ventilated for 20 min and then kept flowing until the end of the experiment. The schematic diagram of the laser processing system is shown in Figure 2. Table 2 illustrates the naming method of samples.

The three samples were precisely trimmed to a cuboidal shape with dimensions of 10 mm × 10 mm × 2 mm. Subsequently, these trimmed samples were carefully embedded in a mold by means of a cold-mounting solution. A field-emission SEM (FEI, QUANT 250, Eindhoven) equipped with an energy-dispersive spectrometer (EDS, Noran System 7, Thermo Fisher Scientific, Waltham, MA, USA) was used to analyze the microstructure and elemental distribution. An X-ray diffractometer (Model: XRD-6100, Manufacturer: Shimadzu Corporation, Kyoto, Japan) was employed to measure the peak position and angle of the crystal. The diffraction angle ranged from 10° to 100° with a scanning speed of 4°/min. Based on experience and test results, we require an angle ranging from 10° to 90°.

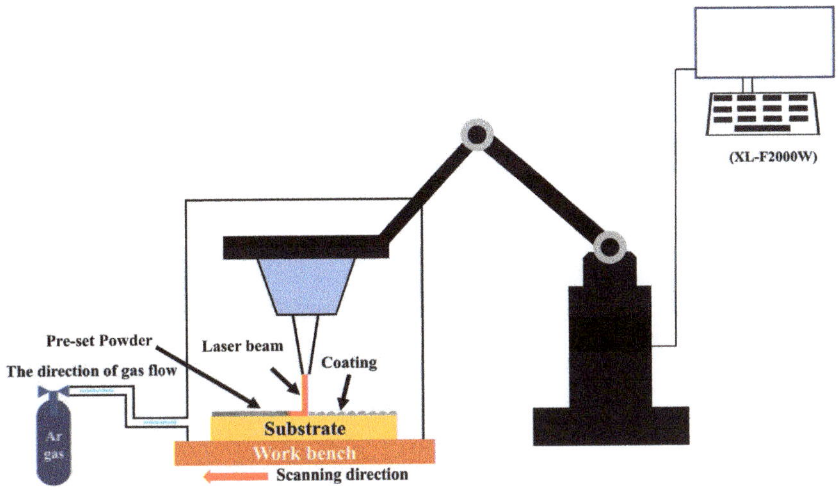

Figure 2. Schematic of the laser cladding process system.

Table 2. Names of coatings with various powder mass fractions.

Name of Coatings	Mass Fractions/(Mass.%)
S0	CoCrFeNiMn
S1	CoCrFeNiMn + 10 wt% TiC
S2	CoCrFeNiMn + 20 wt% TiC

A pin-disc-type friction wear tester, Model SFT—2M, manufactured by Lanzhou Zhongkehua Science and Technology Development Co., Ltd. in Lanzhou, China, and equipped with a displacement sensor, was utilized to determine the coefficient of friction (COF) and wear volume. The indenter is a Si_3N_4 steel ball with a diameter of 5 mm, the experimental load is 30 N, the rotational radius is 2 mm, the test duration is 60 min, and the rotational velocity is 200 rpm.

Electrochemical tests were conducted by the three-electrode method on the samples using an electrochemical workstation (CS350M, Corrtest, Wuhan, China) in a 3.5 wt% NaCl solution to analyze their corrosion resistance.

The open circuit potential (OCP) of the samples was measured for 3600 s. Subsequently, EIS was performed on the samples at a frequency range of 10^{-1}–10^5 Hz. Finally, the potentiodynamic polarization curves were required in the scanning range of −0.5 V–1.5 V (vs. OCP), with a scanning rate set at 0.01 V/s.

3. Results and Analysis

3.1. Metallographic Analysis

Figure 3a shows the phase constitutions of the S0, S1, and S2 coatings. One can see that the pure CoCrFeNiMn HEA without the addition of TiC is composed of a single FCC phase. As we know, the basic phase of CoCrFeNiMn HEA is FCC [21–23]. After adding TiC, the main peak of the coating is composed of the FCC phase, indicating that the addition of TiC does not cause a phase change in the main peak of the CoCrFeNiMn coating. First, one can observe the S1 sample. After adding 10% TiC, no new peaks appear in the coating. The main reason may be that the amount of additional TiC is too small [24]. The denser

TiC moves to the bottom due to the influence of gravity, and thus there is less TiC at the top. In addition, the small amount of TiC may decompose due to the action of the laser, so there is less TiC in the coating, resulting in no obvious change in the XRD pattern. When the amount of additional TiC reaches 20%, the TiC and the $M_{23}C_6$ begin to appear. This indicates that part of the TiC remains unmelted in the coating. Part of the TiC melts and decomposes, and the decomposed C atoms form metal carbides. The $M_{23}C_6$ phase has a relatively high hardness. When it is dispersed in the alloy in the form of fine particles, it can play a role of dispersion strengthening, hindering the movement of dislocations, thus improving the hardness and strength of the alloy.

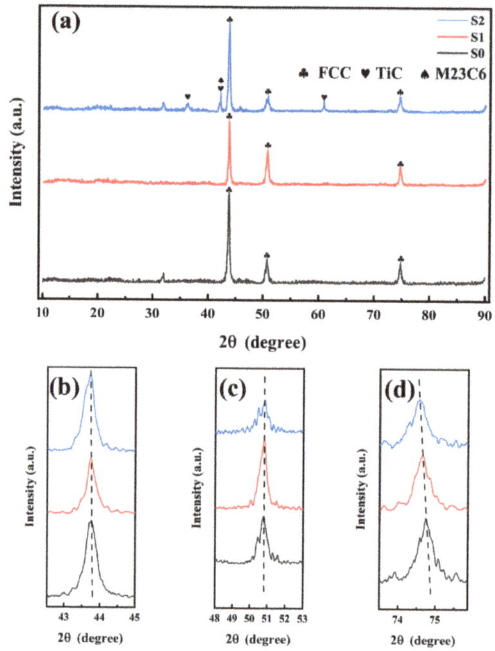

Figure 3. (**a**) XRD pattern of the sample; (**b–d**) enlarged spectrum of the position of the main peak in XRD.

Based on Bragg's law ($n\lambda = 2d\sin\theta$), when there is an increase in the lattice spacing, the main peak position shifts leftward [25,26]. During the LC cooling process, a large amount of residual tensile stress is likely to be generated, which will lead to an increase in the lattice spacing and cause the position of the main peak to shift to the left [27]. In addition, the atomic radius of TiC particles is relatively large, which can easily cause lattice distortion of the matrix [28]. Therefore, as the amount of additional TiC increases, the position of the main peak gradually shifts to the left. The microstress caused by lattice distortion does not have a certain direction and magnitude, which will lead to irregular changes in the interplanar spacing and ultimately cause the main peak to broaden [27]. From the comparison of the main peaks in Figure 3b–d, it is not difficult to find that the peak intensity of the XRD diffraction peak is weakened, and the width is increased, which is direct proof of the existence of microstress in the coating.

3.2. Microscopic Morphology

Figure 4 demonstrates the morphology of the samples, where all morphological transitions in different regions can be observed. The cooling rate of LC is very large.

According to the constitutional supercooling criterion [29,30], the temperature gradient (G) and the solidification rate (R) have an important influence on the crystal growth of the coating. In the micro-molten pool, planar grains are formed, and simultaneously, cellular grains are formed in the middle part of the HEA coating. These laterally grown grains weaken the heat diffusion, thus reducing the G/R ratio. With the consumption of the heat, the surface layer of the molten pool begins to solidify, and the original grains become unstable and are turned into small columnar crystals, which is clearly shown in Figure 4a. Finally, the coating enters the solidification stage, and argon cooling on the surface is involved, which accelerates the solidification speed of the molten pool. The latent heat of crystallization of the liquid metal can be obtained from both the lower substrate and the upper cold argon, resulting in the smallest G/R ratio. The crystals are formed without directionality and become more fragmented, and the crystal size is smaller, which is clearly shown in Figure 4b. It is worth noting that there is only a small amount of TiC present in the middle and top regions of the coating with 10% TiC addition, as shown in Figure 4c,d. This also explains why no TiC phase was found in the XRD results. After the TiC addition amount reaches 20%, there are more slightly melted TiC particles in the top and middle regions of the coating, as seen in Figure 4e,f. The presence of more TiC hard phases in the top and middle regions of the S2 coating will improve the wear resistance of the coating.

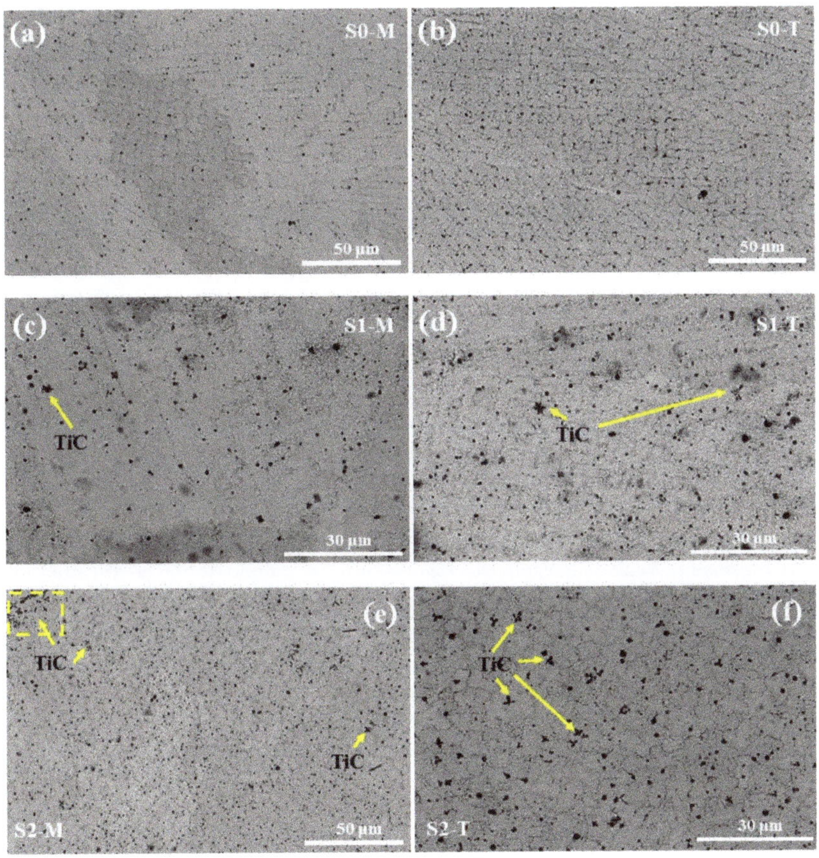

Figure 4. Cross-sectional SEM images of different samples. (**a**,**b**) S0 sample; (**c**,**d**) S1 sample; (**e**,**f**) S2 sample.

3.3. Wear Resistance

Figure 5a,b show the COF curves of the CoCrFeNiMn-$(TiC)_x$ HEA composite coatings when rubbed against Si_3N_4 steel balls. The COF first fluctuates significantly and then tends to be stable. It is generally believed that this effect is based on two factors: one is surface wearing, and the other is indenter sinking [31]. It can be seen that before the addition of TiC, the COF of the CoCrFeNiMn coating clearly fluctuates. After adding TiC, the COF of the coating begins to decrease and tends to be stable. When the amount of added TiC reaches 20%, the COF of the coating drops to its lowest at 0.699. By observing the wear profile of each coating in Figure 5c,d, it can be found that after adding TiC, the volume of the wear profile of the coating first increases and then decreases. Among them, the wear profile of sample S2 is the smallest, and the wear volume is 0.0398 mm^3.

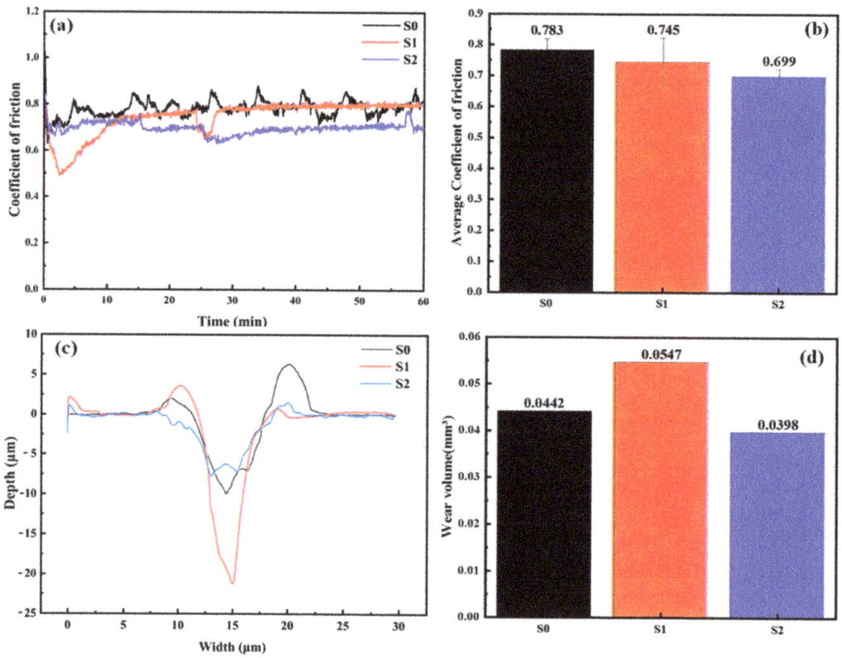

Figure 5. (**a**) COF; (**b**) average COF; (**c**) two-dimensional wear patterns; (**d**) wear volume.

Combined with the COF coating results, it can be determined that sample S2 has the best wear resistance. The increase in the wear volume of sample S1 can be attributed to the fact that the small amount of hard phases leads to the lack of stress concentration in the coating during the wear process. During the wear process, a small amount of TiC hard phases accumulate in the middle and lower regions of the coating, causing the top of the coating to be worn away first. Subsequently, after the steel ball contacts the TiC hard phases in the middle and lower parts, it will wear away the hard phases from the surface of the coating and wear the coating together with the steel ball, resulting in an increase in the wear volume. The distribution of TiC in the middle and lower parts of the coating is also the main reason for the fluctuation of the COF of sample S1 at about 22 min. The main reason for the decrease in the wear volume of sample S2 is the increase in the TiC hard phases in the coating. There are more and more evenly distributed TiC hard phases in the middle part of the coating of sample S2 compared with sample S1. When the wear reaches the middle area of the coating, the stress imposed by the steel ball spreads uniformly among the TiC hard phases, reducing the wear loss of the coating. The decrease in the COF of sample S2 at 22 min indicates that the wear has reached the distribution area of TiC.

Figure 6 shows the SEM images of the wear tracks of the worn samples S0, S1, and S2. The low-magnification images on the top allow us to observe the complete wear profiles, while the images below are locally magnified high-magnification ones, enabling us to clearly see the details of the wear. These samples exhibit both abrasive wear and adhesive wear. As can be seen from Figure 6a, the wear of sample S0 is extremely severe, with deep furrows existing and some wear debris emerging simultaneously. In terms of the nature of the wear, this can be attributed to abrasive wear. The furrows of S1 and S2 are much shallower, and an adhesive layer appears at the same time, indicating that the addition of TiC can reduce abrasive wear while adhesive wear occurs. In addition, after adding TiC particles, the wear profiles of the coatings are significantly narrowed, as shown in Figure 6b,c. Among them, the wear profile of sample S2 is the narrowest and shallowest, and the delamination of the adhesive layer is quite obvious. This is consistent with the result shown in Figure 6c. It is worth noting that there is no TiC present in the wear profile of sample S1, which indicates that during the wear process, TiC is detached due to wear, aggravating the wear situation. However, TiC is distributed within the wear profile of sample S2, which reduces the stress concentration and enhances the wear properties of the coating. This is in line with the above analysis.

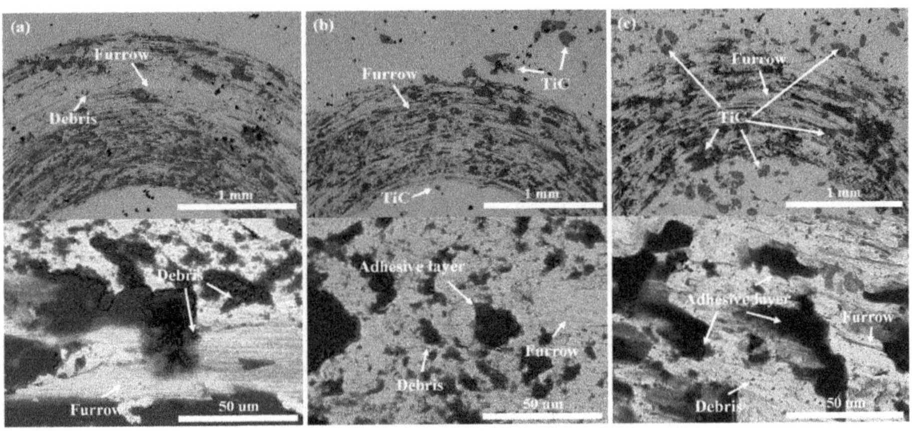

Figure 6. The wear morphologies of the samples at high and low magnifications. (**a**) S0 sample; (**b**) S1 sample; (**c**) S2 sample.

3.4. Corrosion Resistance

An OCP test for 3600 s was carried out on the samples that had been immersed in a 3.5% NaCl solution for 3 h to determine whether the samples formed a stable passivation film. Figure 7a presents the open-circuit potential curves of all the coatings. It can be observed that the open-circuit potential curves of the S0 and S2 coatings remained stable from the initial stage, which means that after 3 h of immersion treatment, one can successfully obtain stable passivation films. In contrast, the open-circuit potential curve of the S1 coating continued to decline. This indicates that samples S0 and S2 obtain more protective passivation films than sample S1. Among them, the test results of the open-circuit potential show that the S2 coating exhibits a higher open-circuit voltage, indicating its better steady-state response characteristics and stronger passivation tendency. It is capable of producing a passivation film that is more protective and stable, which effectively decreases the susceptibility to corrosion and decelerates the rate of the corrosion reaction [32,33]. In the

10 wt% TiC coating (S1), due to gravitational sedimentation and laser decomposition, the reinforcing phases, such as TiC, are unevenly distributed. This uneven distribution will lead to differences in composition and structure at different parts of the coating surface. As a result, in a corrosive environment, the reaction activities of various parts of the coating surface are different, making it easy to form local corrosion cells, accelerating the corrosion process, and further affecting the formation and stability of the passive film, thus resulting in poor passivation behavior.

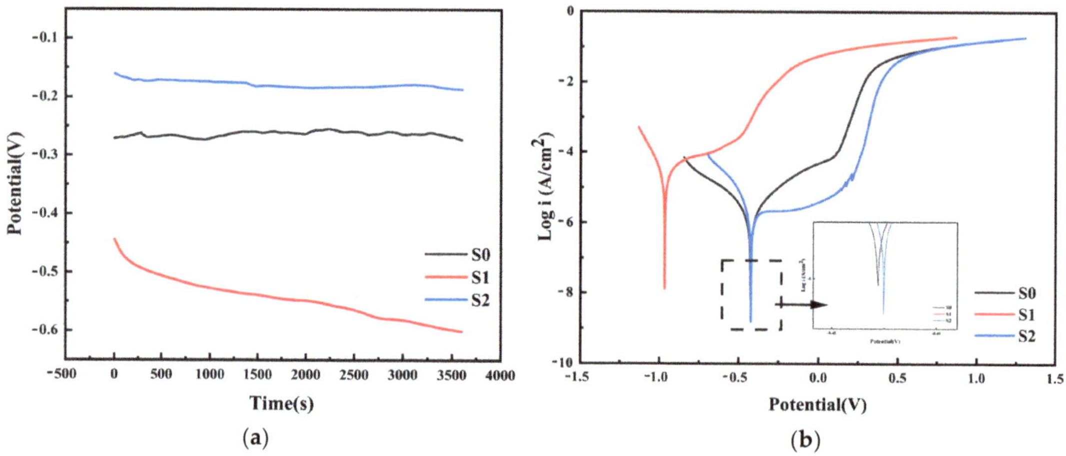

Figure 7. (**a**) Open circuit potential curves of HEA coatings; (**b**) Tafel polarization plots of HEA coatings.

Figure 7b shows the Tafel polarization curves of all the coatings. The anodic passivation region is a key indicator for evaluating the stability of the passivation film [34]. A wider and more stable passivation region means that there is a stable passivation film in the corrosion environment of the coating, indicating that these coatings have excellent corrosion resistance. In the anodic region, all three coatings enter the passivation region. This phenomenon indicates that it is possible for oxides and hydroxides to form on the surfaces of these three coatings, and then a passivation film is generated [33]. In the passivation region, with the increase in the voltage, the corrosion current of the S1 and S2 coatings remains almost unchanged, which indicates that the surfaces of these coatings are in a stable passivation state. In contrast, the corrosion current of the S0 coating keeps rising in the passivation region, which indicates that its passivation state is less stable compared with the other two coatings. In addition, the S2 coating has the lowest passivation current density and a larger passivation region, highlighting its strong surface passivation ability [35].

The self-corrosion potential (E_{corr}) and self-corrosion current density (I_{corr}) obtained by the Tafel extrapolation method are significant parameters in assessing the corrosion resistance of the coating, with the outcomes depicted in Table 3. E_{corr} is a concept within thermodynamics, signifying the probability of corrosion occurring, and the I_{corr} is a corrosion kinetic parameter, clearly reflecting the corrosion rate [36]. The S2 coating has the largest E_{corr} and the smallest I_{corr}, followed by the S0 coating. The S1 coating has the smallest E_{corr} and the largest I_{corr}. This indicates that, from the perspective of the thermodynamic and kinetic parameters of corrosion, the S2 coating has the best corrosion resistance, the S0 coating comes second, and the S1 coating has the worst corrosion resistance.

Table 3. Electrochemical parameters derived from Tafel curves.

Sample \ Parameter	E_{corr} (V)	I_{corr} (A/cm^2)
S0	−0.428	2.669×10^{-6}
S1	−0.968	2.920×10^{-5}
S2	−0.425	1.119×10^{-7}

In order to more thoroughly explore the performance of the surface passivation films of the three coatings (S0, S1, and S2), as well as their kinetic characteristics during the corrosion process, EIS tests are carried out on all these samples. The Nyquist diagram is presented in Figure 8a. Scrutiny of this figure reveals that all the samples exhibit a nearly identical sunken capacitive semi-circular shape. This shape generally implies the existence of a charge transfer mechanism on the non-uniform surface, thus fully confirming the presence of the passivation film.

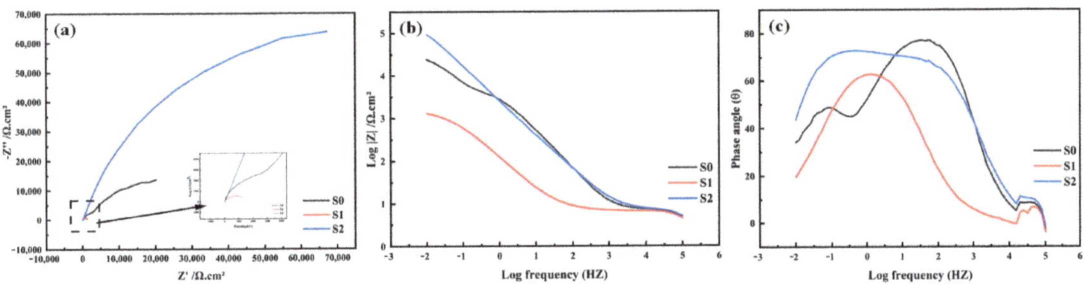

Figure 8. EIS spectra for samples. (**a**) Nyquist diagram; (**b**,**c**) Bode diagram.

By analyzing the situation in the high-frequency region in depth, it can be found that the radii of the capacitive arcs have the following order: the radius of the capacitive arc of the S2 coating is larger than that of the S0 coating, and the S0 coating is larger than the S1 coating. The larger the radius of the capacitive loop, the higher the charge transfer resistance of the coating, the less likely the electron transfer is, the weaker the electrical conductivity is, the higher the impedance is, and correspondingly, the stronger the corrosion resistance is. Therefore, it can be concluded that among these three samples, the S2 coating has the best corrosion resistance, while the S1 coating has the worst corrosion resistance.

Figure 8b,c show the Bode diagrams of the three coatings. In the low-frequency region, the value of the impedance modulus can reflect the situation of the charge transfer resistance. Consequently, the overall corrosion resistance of the coating is able to be evaluated in an intuitive manner through the parameter of $|Z|_{0.01Hz}$. Through analysis, it can be known that the order of the values of $|Z|_{0.01Hz}$ is S2 coating > S0 coating > S1 coating, and this order is consistent with the change trend of the radius of the capacitive arc in the Nyquist plot. Meanwhile, in the low-frequency region, the phase angle of S2 is also the largest. The increase in both the $|Z|_{0.01Hz}$ value and the phase angle further confirms that a thicker and denser passivation film has been formed in the S2 coating. Therefore, the protective performance of this coating has been significantly improved.

Above all, adding a small amount of TiC is not conducive to the improvement of the corrosion resistance of the CoCrFeNiMn coating. However, when the amount of TiC added reaches 20%, the corrosion resistance of the CoCrFeNiMn coating can be greatly improved.

4. Conclusions

In this work, CoCrFeNiMn high-entropy alloy-based composite coatings with diverse TiC contents (0, 10, 20 wt%) were successfully prepared using LC technology, and the synergistic regulation mechanism of TiC addition on the wear properties and corrosion resistance of the coatings was revealed. Experiments show that the 20 wt% TiC composite coating (S2) exhibits comprehensive performance advantages due to the uniformly distributed TiC hard phases, optimized stress dispersion ability, and stable passivation behavior. The cardinal research inferences are presented as follows:

(1) In the coating (S2) with 20 wt% TiC addition, the TiC and $M_{23}C_6$ phases are uniformly distributed in the high-entropy alloy matrix, and lattice distortion occurs in the coating (broadening and leftward shift of the XRD peaks). In contrast, in the 10 wt% TiC coating (S1), due to gravitational sedimentation and laser decomposition, the reinforcing phase is locally enriched and unevenly distributed.

(2) The S2 coating has the lowest coefficient of friction and wear volume because the uniformly distributed hard phases can disperse the stress. However, in the S1 coating, due to the detachment of TiC and stress concentration, the wear is aggravated.

(3) In a 3.5% NaCl solution, compared with samples S0 and S1, sample S2 showed the most outstanding corrosion resistance because of the protective effect of the dense passivation film.

Author Contributions: Conceptualization, Q.Z. and C.L.; Methodology, F.L.; Software, J.H.; Investigation, Q.Z.; Resources, J.H., Z.W., B.M. and C.L.; Writing—original draft, Q.Z.; Writing—review & editing, F.L., Z.W., B.M. and C.L.; Supervision, Z.W., B.M. and C.L.; Funding acquisition, C.L. All authors have read and agreed to the published version of the manuscript.

Funding: This research was funded by the National Natural Science Foundation of China (12074398).

Data Availability Statement: The original contributions presented in this study are included in the article. Further inquiries can be directed to the corresponding author. The raw data supporting the conclusions of this article will be made available by the authors on request.

Conflicts of Interest: The authors declare no conflict of interest.

References

1. Li, P.; Zhang, J.; Yang, T.; Zhang, T.; Zhang, J.; Lin, J.; Yan, Y.; Li, C.; Si, X.; Cao, J.; et al. Characteristics, applications and perspective of high entropy alloys for interfacial joining: A review. *J. Manuf. Process.* **2024**, *110*, 303–317. [CrossRef]
2. Li, D.; Liaw, P.K.; Xie, L.; Zhang, Y.; Wang, W. Advanced high-entropy alloys breaking the property limits of current materials. *J. Mater. Sci. Technol.* **2024**, *186*, 219–230. [CrossRef]
3. Abdullah, M.R.; Peng, Z. Review and perspective on additive manufacturing of refractory high entropy alloys. *Mater. Today Adv.* **2024**, *22*, 100497. [CrossRef]
4. Liang, G.; Jin, G.; Cui, X.; Qiu, Z.; Wang, J. Synthesis and characterization of directional array TiN-reinforced AlCoCrCuNiTi high-entropy alloy coating by magnetic-field-assisted LC. *J. Mater. Eng. Perform.* **2021**, *30*, 3568–3576. [CrossRef]
5. Sun, D.; Zhu, L.S.; Cai, Y.C.; Yan, Y.N.; Ge, F.Y.; Shan, M.D.; Tian, Y.B.; Han, J.; Jiang, Z.Y. Tribology comparison of laser-cladded CrMnFeCoNi coatings reinforced by three types of ceramic (TiC/NbC/B4C). *Surf. Coat. Technol.* **2022**, *450*, 129013. [CrossRef]
6. Sun, D.; Cai, Y.C.; Zhu, L.S.; Gao, F.F.; Shan, M.D.; Manladan, S.M.; Geng, K.P.; Han, J.; Jiang, Z.Y. High-temperature oxidation and wear properties of TiC-reinforced CrMnFeCoNi high entropy alloy composite coatings produced by LC. *Surf. Coat. Technol.* **2022**, *438*, 128407. [CrossRef]
7. Wei, Y.N.; Chen, Y.; Guo, B.B.; Zhu, L.H. The microstructure and mechanical properties of TiC-reinforced W-matrix composites prepared by spark plasma sintering. *Int. J. Refract. Met. Hard Mater.* **2023**, *112*, 106158. [CrossRef]
8. Zhuang, D.D.; Tao, W.W.; Du, B.; Zhang, S.H.; Lian, X.L.; Wang, F. Microstructure and properties of TiC-enhanced CrMnFeCoNi high-entropy alloy coatings prepared by LC. *Tribol. Int.* **2023**, *180*, 108246. [CrossRef]
9. Xiong, J.K.; Sun, D.; Dong, N.; Cai, Y.C.; Wang, D.Y.; Luo, Z. Oxidation mechanism of TiC reinforced CrMnFeCoNi composite coatings by LC at 600 °C/900 °C. *Mater. Lett.* **2022**, *324*, 132710. [CrossRef]

10. Zhou, J.L.; Cheng, Y.H.; He, B.; Wan, Y.X.; Chen, H.; Wang, Y.F.; Yang, J.Y. Enhancement of high-entropy alloy coatings with multi-scale TiC ceramic particles via high-speed LC: Microstructure, wear and corrosion. *Appl. Surf. Sci.* **2025**, *685*, 162061. [CrossRef]
11. Zhao, W.; Kedong, Y.; Ma, Q.I.; Song, C.; Xiao, G.; Zhang, H.; Lv, Y.; Guo, N.; Li, Z. Synergistic effects of Mo and in-situ TiC on the microstructure and wear resistance of AlCoCrFeNi high entropy alloy fabricated by LC. *Tribol. Int.* **2023**, *188*, 108827. [CrossRef]
12. Zhang, X.R.; Cui, X.F.; Qi, M.; Zhang, Q.; Qi, Y.P.; Jin, G. Corrosion and passivation behavior of in-situ TiC reinforced Al 0.1 CrNbSi 0.1 TaTiV refractory high entropy alloy coatings via doping C. *Corros. Sci.* **2024**, *227*, 111736. [CrossRef]
13. Shang, X.; Liu, Q.; Guo, Y.; Ding, K.; Liao, T.; Wang, F. Nano-TiC reinforced [Cr–Fe 4 4 4]Cr 3 high-entropy-alloy composite coating fabricated by LC. *J. Mater. Res. Technol.* **2022**, *21*, 2076–2088. [CrossRef]
14. Wu, H.; Wang, L.; Zhang, S.; Wu, C.L.; Zhang, C.H.; Sun, X.Y. Corrosion and cavitation erosion behaviors of laser clad FeNiCoCr high-entropy alloy coatings with different types of TiC reinforcement. *Surf. Coat. Technol.* **2023**, *471*, 129910. [CrossRef]
15. Zhang, J.; Jia, T.; Zhu, H.G.; Xie, Z.H. Microstructure and mechanical properties of in-situ TiC reinforced FeCoNiCu2.0 high entropy alloy matrix composites. *Mater. Sci. Eng. A* **2021**, *822*, 141671. [CrossRef]
16. Wilson, J.M.; Shin, Y.C. Microstructure and wear properties of laser-deposited functionally graded Inconel 690 reinforced with TiC. *Surf. Coat. Technol.* **2012**, *207*, 517–522. [CrossRef]
17. Yu, K.D.; Zhao, W.; Li, Z.; Guo, N.; Xiao, G.C.; Zhang, H. High–temperature oxidation behavior and corrosion resistance of in–situ TiC and Mo reinforced AlCoCrFeNi–based high entropy alloy coatings by LC. *Ceram Int.* **2023**, *49*, 10151–10164. [CrossRef]
18. Wu, C.L.; Xu, T.Z.; Wang, Z.Y.; Zhang, C.H.; Zhang, S.; Ni, C.L. Laser surface alloying of FeCoCrAlNiTi high entropy alloy composite coatings reinforced with TiC on 304 stainless steel to enhance wear behavior. *Ceram Int.* **2022**, *48*, 20690–20698. [CrossRef]
19. Liu, H.; Liu, J.; Chen, P.J.; Yang, H.F. Microstructure and high temperature wear behaviour of in–situ TiC reinforced AlCoCrFeNi–based high–entropy alloy composite coatings fabricated by LC. *Opt. Laser Technol.* **2019**, *118*, 140–150. [CrossRef]
20. Jiang, D.; Hong, C.; Zhong, M.; Alkhayat, M.; Weisheit, A.; Gasser, A.; Zhang, H.; Kelbassa, I.; Poprawe, R. Fabrication of nano-TiCp reinforced Inconel 625 composite coatings by partial dissolution of micro-TiCp through LC energy input control. *Surf. Coat. Technol.* **2014**, *249*, 125–131. [CrossRef]
21. Luo, F.Y.; Wang, S.S.; Shi, W.Q.; Xiong, Z.Y.; Huang, J. Wear behavior of single-layer graphene oxide reinforced CoCrFeNiMn HEA coating by LC. *Intermetallics* **2024**, *175*, 108512. [CrossRef]
22. Yeh, J.; Chen, S.; Lin, S.; Gan, J.; Chin, T.; Shun, T.; Tsau, C.; Chang, S. Nanostructured high-entropy alloys with multiple principal elements: Novel alloy design concepts and outcomes. *Adv. Eng. Mater.* **2004**, *6*, 299–303. [CrossRef]
23. Cantor, B.; Chang, I.T.H.; Knight, P.; Vincent, A.J.B. Microstructural development in equiatomic multicomponent alloys. *Mater. Sci. Eng. A* **2003**, *3571*, 257–264. [CrossRef]
24. Wang, T.; Zhu, L.; Song, H.Y.; Wang, H. Effect of WC-17Co content on microstructure and properties of IN718 composites prepared by laser cladding. *Opt. Laser Technol.* **2022**, *148*, 107780. [CrossRef]
25. Luo, F.Y.; Shi, W.Q.; Xiong, Z.Y.; Huang, J. Microstructure and properties analysis of AlCoCrFeNi high-entropy alloy/iron-based amorphous composite coatings prepared by LC. *J. Non-Cryst. Solids* **2024**, *624*, 122732. [CrossRef]
26. Luo, F.Y.; Wang, S.S.; Shi, W.Q.; Zhao, Y.; Huang, J. Wear behavior and corrosion resistance of laser-clad Ni60-1% carbon nanotubes coating. *Surf. Coat. Technol.* **2024**, *482*, 130686. [CrossRef]
27. Gu, Z.; Xi, S.Q.; Sun, C.F. Microstructure and properties of laser cladding and $CoCr_{2.5}FeNi_2Tix$ high-entropy alloy composite coatings. *J. Alloys Compd.* **2020**, *819*, 152986. [CrossRef]
28. Li, Y.T.; Fu, H.G.; Ma, T.J.; Wang, K.M.; Yang, X.J.; Lin, J. Microstructure and wear resistance of AlCoCrFeNi-WC/TiC composite coating by laser cladding. *Mater. Charact.* **2022**, *194*, 112479. [CrossRef]
29. Ismail, H.; Vaclav, O.; Kornel, C.; Jeffthm, D. Microstructure and Phase Formation in a Rapidly Solidified Laser-Deposited Ni-Cr-B-Si-C Hardfacing Alloy. *Metall. Mater. Trans. A* **2014**, *45A2*, 878–892.
30. Liu, X.M. Mechanism and control of the nick-based alloy WC laser cladded layer microstructure. *Appl. Laser* **2006**, *5*, 299–302.
31. Li, Y.H.L.J.; Tao, Y.F.; Hu, L.F. High-temperature wear and oxidation behaviors of $TiNi/Ti_2Ni$ matrix composite coatings with TaC addition prepared on Ti6Al4V by laser cladding. *Appl. Surf. Sci.* **2017**, *402*, 478–494.
32. Yuan, W.Y.; Li, R.F.; Zhu, Y.Y.; Zhao, Y.; Zhang, X.Q.; Liu, B.; Zhang, B.S. Structure and properties of nickel-plated CNTs/Fe-based amorphous composite coatings fabricated by high-speed LC. *Surf. Coat. Technol.* **2022**, *438*, 128363. [CrossRef]
33. Shahriari, A.; Ghaffari, M.; Khaksar, L.; Nasiri, A.; Hadadzadeh, A.; Amirkhiz, B.S.; Mohammadi, M. Corrosion resistance of 13wt% Cr martensitic stainless steels: Additively manufactured CX versus wrought ni-containing AISI 420. *Corros. Sci.* **2021**, *184*, 109362. [CrossRef]
34. Muangtong, P.; Rodchanarowan, A.; Chaysuwan, D.; Chanlek, N.; Goodall, R. The corrosion behaviour of CoCrFeNix x = Cu, Al, Sn) high entropy alloy systems in chloride solution. *Corros. Sci.* **2020**, *172*, 108740. [CrossRef]

35. Zhou, Y.Q.; Kong, D.C.; Wang, L.; Li, R.X.; Ni, X.Q.; Cheng, M.; Dong, C.F.; Engelberg, D. Pit growth kinetics of additively manufactured MoNi over-alloyed type 316L stainless steel. *J. Mater. Res. Technol.* **2023**, *27*, 7532–7547. [CrossRef]
36. Li, J.X.; Liu, Z.D.; Ma, H.R.; Liu, Q.B.; Mao, J.; Zhang, J.; Kong, Y. Comparative study on the corrosion behaviour of 1.4529 super austenitic stainless steel and laser-cladding 1.4529 coating in simulated desulfurized flue gas condensates. *Corros. Sci.* **2022**, *209*, 110794. [CrossRef]

lubricants

MDPI

Article

Effect of Quenching Temperature on Microstructure and Wear Resistant Properties of Mo_2FeB_2 Cermet Coating

Hao Zhang [1], Yongqi Hu [1] and Yang Zhang [2,*]

[1] College of Optoelectronic Manufacturing, Zhejiang Industry & Trade Vocational College, Wenzhou 325003, China; zhanghao573@163.com (H.Z.)
[2] School of Engineering + Technology, Western Carolina University, Cullowhee, NC 28723, USA
* Correspondence: yzhang@wcu.edu

Abstract: H13 steel, a widely used material in hot work tooling, faces premature failure due to insufficient hardness and wear resistance. To address this limitation, Mo_2FeB_2 cermet coatings were fabricated on H13 alloy steel via plasma spray welding, and subsequently quenched at 850 °C, 1000 °C, and 1150 °C. The effects of the quenching temperature on the microstructure and wear resistance were investigated using optical microscopy (OM) for cross-sectional morphology, scanning electron microscopy (SEM) for microstructural and wear surface analyses, energy-dispersive spectroscopy (EDS) for elemental composition analysis, and X-ray diffraction (XRD) for phase identification. The coating primarily consisted of α-Fe, Mo_2FeB_2, $(Mo,Fe,Cr)_3B_2$, and $Fe_{23}(B,C)_6$ phases. Increasing the temperature to 1150 °C increased the Mo_2FeB_2 hard phase and elevated microhardness by 32.04% (from 827 $HV_{0.5}$ to 1092 $HV_{0.5}$). Wear resistance improved by 46.38% (mass loss reduced from 6.9 mg to 3.7 mg). The main wear mechanism was identified as abrasive wear due to the spalling of hard phase particles. These results demonstrate that optimizing quenching temperature enhances the hardness and wear resistance in Mo_2FeB_2 coatings, offering a viable strategy to extend H13 steel service life in high-temperature industrial applications.

Keywords: plasma spray welding; Mo_2FeB_2; quenching temperature; microstructure; wear resistance

1. Introduction

With the rapid development of the molding industry, mold materials have advanced significantly [1]. Among these, H13 steel has emerged as one of the most widely used materials due to its strong carbide-forming elements (Cr, Mo, and V), excellent toughness [2], high hardenability [3], and resistance to heat softening [4]. These properties make it suitable for applications such as non-ferrous metal die casting, hot extrusion, hot forging, and plastic molding [2,5]. However, H13 steel often fails to meet the demands of harsh working conditions due to its softness (~202–241 HV) [6]; limited wear resistance; and vulnerability to thermal fatigue [5], erosion, and stress corrosion [2].

Many large-scale engineering components operate in harsh environments involving water vapor, liquids, and solid particles, which leads to severe corrosion and wear, ultimately resulting in premature failure. The surface remanufacturing techniques offer an effective solution by restoring and enhancing the performance of damaged or worn regions. To extend H13's service life, various surface modification techniques have been explored. Meng et al. [7] studied the mechanical properties of TiC-reinforced H13 steel fabricated via bionic laser treatment. The results showed that with an increasing TiC fraction, the microstructure of the laser alloying zone was refined, and the microhardness was

https://doi.org/10.3390/lubricants13060233

improved to 1156 HV when the TiC fraction was 70%. However, tensile strength and wear resistance initially increased before declining at higher TiC fractions. Narvan et al. [8] investigated the feasibility of fabricating defect-free functionally graded bi-materials (FGMs) by incorporating vanadium carbide (VC) into H13 tool steel to enhance wear resistance. They found that the microhardness values of 1, 3 and 5 wt% VC were 700, 785, and 840 HV, respectively. The nanoindentation maximum penetration depth decreased to 515, 489, and 470 nm, respectively. Ref. [9] investigated the microhardness, wear properties, and microstructure of Fe-based coatings containing various WC contents deposited on H13 die steel via laser cladding. The results demonstrated that the coating microhardness increased with higher WC particle mass fractions. The average COFs of the coatings with 0%, 3%, 6%, and 9% were 0.349, 0.342, 0.336, and 0.313, respectively. The wear rates were 4.3326, 3.9476, 3.3240, and 2.7194×10^{-6} $mm^3 \cdot N^{-1} \cdot m^{-1}$, respectively.

Among the surface modification methods, plasma spray welding offers distinct advantages: a fine heat-affected zone, strong metallurgical bonding at the coating–substrate interface, minimal porosity, and high coating thickness [10,11]. In this work, plasma spray welding was employed to deposit in situ Mo_2FeB_2-based cermet coatings on H13 steel, followed by quenching at varying temperatures to further enhance the performance. Compared to alternatives like Mo_2NiB_2 [12,13] and WCoB [14,15], Mo_2FeB_2-based cermets are cost-effective, while retaining hardness, chemical stability, and wear resistance, making them suitable for tools, cladding materials, and industrial components [16–18]. However, prior research on Mo_2FeB_2 cermets has primarily focused on sintering [19,20] or non-heat-treated cladding [21], leaving their post-deposition heat treatment underexplored.

Quenching is a rapid heat and mass transfer process where a high-temperature surface is rapidly cooled by a fluid below its boiling point [22,23]. Appropriate heat treatment not only defines the coating's microstructure, relieving residual stresses, promoting crystallization, and enhancing mechanical properties, but can also alter its phase composition and morphology [24–26].

In this study, coatings containing Mo_2FeB_2 hard phase were deposited on H13 tool steel by plasma spray welding, and then quenched at various temperatures. The effects of the quenching temperature on the coating morphology, phase composition (XRD), microhardness, microstructure, and wear resistance were systematically investigated. The primary objective is to demonstrate that the combined use of plasma spray welding and optimized quenching can substantially improve the hardness and wear resistance of H13 steel, thereby extending its service life and reducing the overall tooling costs. By depositing a Mo_2FeB_2 coating on the material surface, the limitations of metallic materials in terms of insufficient hardness and poor wear resistance can be effectively addressed. At the same time, the brittleness of ceramic materials and the interfacial incompatibility between ceramics and metallic substrates can be mitigated.

2. Materials and Methods

2.1. Materials

An H13 alloy steel substrate ($50 \times 30 \times 10$ mm^3) was used in this study, with its elemental composition provided by the manufacturer shown in Table 1. Prior to the experiments, the H13 substrates were prepared by polishing the surface to a roughness of Ra < 1.0 μm, followed by ultrasonic cleaning in alcohol for 20 min. The cleaned substrates were then preheated in a heat treatment furnace at 300 °C for 10 min to remove residual moisture. Plasma spray welding powder consisted of Mo (3–5 μm), FeB (8–10 μm, 19.7 wt.% B), carbonyl Fe (2–3 μm), Cr (5–8 μm), Ni (1–5 μm), and SiC (1–2 μm) powders (chemical composition provided by the manufacturer shown in Table 2). Before deposition, the powders were premixed and ball-milled using a QM-1SP4 planetary ball milling machine. Milling was performed with cemented carbide vials and balls (10 mm diameter) at a ball-to-powder weight ratio of 5:1. The mill ran at 300 rpm for 24 h, with absolute ethanol added as a process control agent. After milling, the powder was dried at 200 °C for 2 h in a drying oven.

Table 1. Elemental composition (wt.%) of H13 alloy steel.

C	Si	Mn	Cr	Mo	V	P	S	Fe
0.35~0.40	1.13~1.20	0.40~0.50	4.50~4.80	1.30~1.50	0.80~1.00	≤0.03	≤0.03	Rest

Table 2. Elemental composition (wt.%) of plasma spray welding powder.

Mo	FeB	Cr	SiC	Ni	Fe
18~25	10~15	5~8	0.3~0.6	1~1.5	Rest

2.2. Experimental Procedure

Plasma spray welding was performed using a DPT100 welding gun, 400A plasma current, and an MDS-3 powder feeder (United Coatings Technologies Co., Ltd., Beijing, China). Detailed process parameters are summarized in Table 3.

Table 3. Process parameters of plasma spray welding.

Nozzle Height (mm)	Voltage (V)	Non-Transferred Arc (A)	Transferred Arc (A)	Plasma Ar Gas Flow (L/min)	Powder Ar Gas Flow (L/min)	Powder Feeding (g/min)	Surfacing Speed (mm/min)
10	30	11	90	1.0	4	10	80

After spray welding, the samples were heat-treated in an SX-4-10 furnace at a rate ≤ 400 °C/h, followed by a 30 min hold. The samples were categorized based on heat treatment: C0 (no heat treatment), C1 (heated to 850 °C), C2 (heated to 1000 °C), and C3 (heated to 1150 °C). Quenching temperatures of 850 °C, 1000 °C, and 1150 °C were selected based on the high-temperature characteristics of the primary hard phase Mo_2FeB_2. These elevated temperatures were chosen to promote the re-nucleation and refinement of new Mo_2FeB_2 particles during the heat treatment process. Samples C1, C2, and C3 were quenched in mineral oil after reaching their target temperatures. For characterization, all the samples were cut to 10 mm × 10 mm × 5 mm, ground, and polished. The polished surfaces were then etched for 20 s in a solution containing $K_3[Fe(CN)_6]$, NaOH, and H_2O (volume ratio 1:1:10), followed by ultrasonic cleaning in alcohol.

2.3. Characterization

The cross-sectional microstructure of the coatings was examined using a ZEISS Axio Plan 2 optical microscope (OM) (Zeiss, Shanghai, China). Phase composition analysis was performed via X-ray diffraction (XRD) (PANalytical, Almelo, The Netherlands) on an X'Pert Pro MPD system with Cu Kα radiation (λ = 0.15418 nm) operated at 40 kV and 20 mA. Scans were conducted over a 2θ range of 10–90° with a step size of 0.05° and a dwell time of 10 s per step. Wear morphology and microstructure were characterized using a field-emission scanning electron microscope (FE-SEM, Nova 400 Nano, FEI, Hillsboro, OR, USA) equipped with an energy-dispersive X-ray spectroscopy (EDS) (INCA IE 350 PentaFETX-3, Oxford, UK) system for elemental analysis. Microhardness was measured on polished cross-sections using an HX-500 microhardness tester (Laizhou Yutong Test Instrument Co. Ltd., Laizhou, Shandong, China) under 500 g force load applied for 10 s. To ensure the accurate characterization of the transition region, microhardness measurements were conducted more densely near the interface, specifically at 50 μm intervals above and below the fusion line. Wear resistance was evaluated under dry sliding conditions with a UMT-TriboLab tribometer (Bruker, Billerica, MA, USA). A Si_3N_4 ball (cemented carbide, 2200 $HV_{0.5}$) served as the counterface. The tests were run at a normal load of 100 N, a reciprocating frequency of 5 Hz, and a total duration of 1 h. For each condition, the average coefficient of friction and mass loss were determined from three replicate tests; mass loss was measured with an electronic balance.

3. Results and Discussion

3.1. Coating Morphology

Figure 1 shows the cross-sectional microstructure of the coating, where the thickness of the coating is approximately 575 μm. In Figure 1a, a large number of hard phase particles (black) are embedded in the binder phase (white). Some regions show the noticeable agglomeration of these hard phase particles. This agglomeration is attributed to the short duration of the plasma spray welding process, which limited the time available for the formation of a liquid molten pool and resulted in a rapid solidification rate. As a consequence, the hard phase particles were distributed unevenly within the binder; the middle region of the coating exhibited more pronounced agglomeration, whereas such clustering was almost absent in the fusion zone between the coating and the substrate. In the fusion zone, the high-energy thermal cycle extended the high-temperature duration near the molten pool, allowing for the hard phase particles sufficient time to react and diffuse, thereby reducing the tendency for agglomeration.

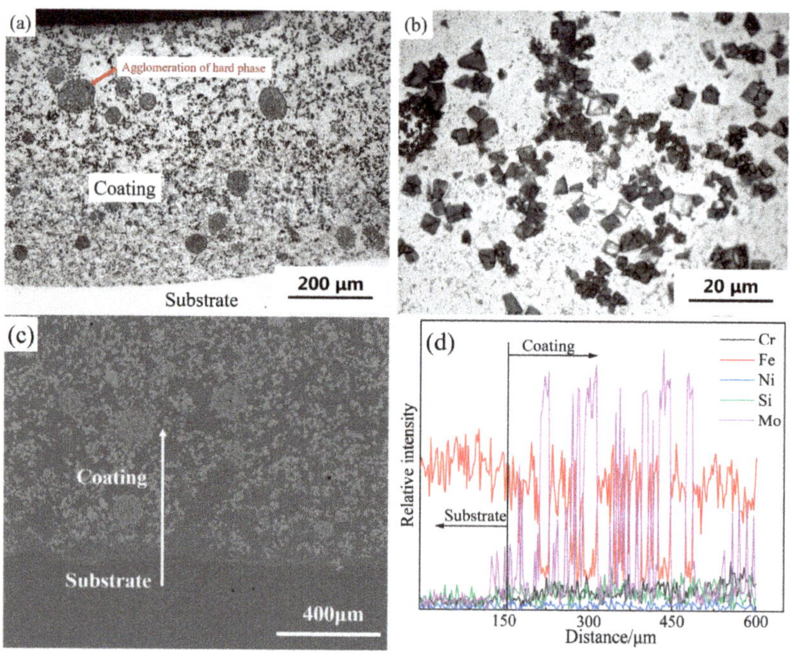

Figure 1. The cross-section of the coating. (**a**) The OM morphology of the bonding zone; (**b**) the OM morphology of the coating; (**c**) the SEM morphology of the bonding zone; (**d**) an EDS line scan in the marked area of (**c**).

Figure 1b reveals that the coating is composed of three distinct phases: a binder phase, a hard phase, and a eutectic structure. Interfacial bonding between the coating and the substrate is further examined in Figure 1c, which demonstrates a strong metallurgical bond with no visible defects, such as pores, inclusions, and cracks.

The EDS line scan result in Figure 1c is shown in Figure 1d, indicating significant changes in the contents of Mo, Fe, and Cr from the substrate to the coating. Notably, there is considerable fluctuation in the Mo and Fe concentrations on both sides of the interface. This variation is attributed to the interdiffusion of elements during the high-energy welding process, with Mo from the coating diffusing into the substrate and Fe from the substrate diffusing into the coating. Overall, the Mo and Cr contents are considerably lower in the substrate compared to the coating, while the Fe content is higher in the substrate. These EDS results confirm the existence of strong metallurgical bonding between the coating and the substrate.

3.2. XRD Analysis

Figure 2 shows the XRD patterns of the coatings subjected to different quenching temperatures. The C0 coating consists of α-Fe, Mo_2FeB_2, $(Mo,Fe,Cr)_3B_2$, and $Fe_{23}(C,B)_6$. After high-temperature quenching (C2: 1000 °C; C3: 1150 °C), the $Fe_{23}(C,B)_6$ phase disappears, while the diffraction peak intensities of Mo_2FeB_2 and $(Mo,Fe,Cr)_3B_2$ increase significantly. This indicates that elevated temperatures (≥1000 °C) promote the dissolution of $Fe_{23}(C,B)_6$ and enhance the formation of Mo_2FeB_2 and $(Mo,Fe,Cr)_3B_2$.

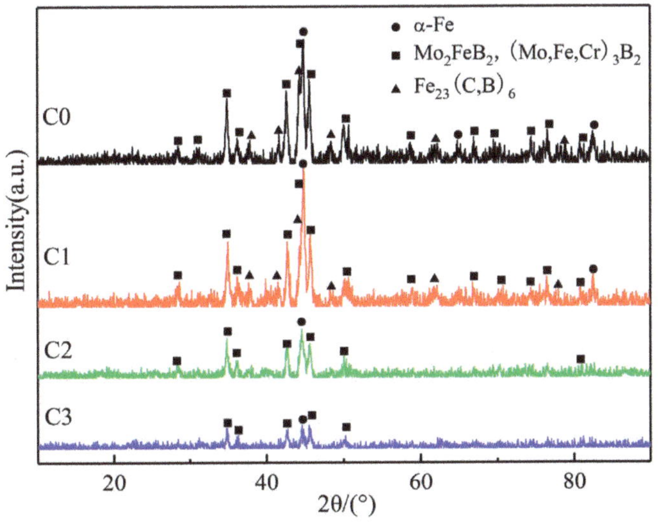

Figure 2. XRD patterns of boride coating with different quenching temperatures.

During the formation of the Mo_2FeB_2 hard phase, a portion of the Cr atoms replaces the Mo or Fe atoms [21], leading to the formation of $(Mo,Fe,Cr)_3B_2$. This substitution induces lattice distortion in the Mo_2FeB_2 crystals, which is beneficial for enhancing the coating's hardness [21]. In addition, the presence of the α-Fe binder phase effectively improves the plasticity of the coating. The eutectic structure consists of (α-Fe + $Fe_{23}(B,C)_6$). During the rapid cooling of the molten pool, grain boundaries serve as channels for rapid atomic diffusion, promoting the segregation of impurity atoms. Elements such as C and B diffuse and precipitate along these boundaries, forming $Fe_{23}(B,C)_6$, which then reacts with α-Fe to produce the eutectic structure.

3.3. Microstructure

Figure 3 shows the SEM morphologies of the hard phases in the coatings subjected to different quenching temperatures. The in situ Mo_2FeB_2 and $(Mo,Fe,Cr)_3B_2$ hard phases retained similar morphologies across all the samples, though their volume fraction increased with higher quenching temperatures. As the quench temperature rose, finer white hard phase particles gradually appeared within the coating. The average grain size of the coatings was measured using Image-Pro Plus, and the values are approximately 18.78 μm, 9.88 μm, 9.70 μm, and 9.48 μm for C0, C1, C2, and C3, respectively. In contrast, the binder phase and the reticulated eutectic network undergo pronounced morphological changes, in agreement with the XRD findings (Figure 2). To examine these transformations more closely, Figure 4 and Table 4 summarize the EDS point analyses of representative regions.

In the C0 coating, reticulated eutectic structures (point 2) were observed. At 850 °C (C1), these structures break up, yielding elongated or rod-like boride phases (point 4). Further increasing the temperature to 1000 °C (C2) and 1150 °C (C3) promoted the formation of fine hard phase particles (points 6 and 8) within the binder. EDS analysis (Table 4) revealed a Mo/Fe/Cr atomic ratio of ~(5–5.5):3:1, confirming the hard phases as Mo_2FeB_2 and $(Mo,Fe,Cr)_3B_2$. The disappearance of the eutectic structure at high quench temperatures (1000 °C and 1150 °C) indicates that Mo from the binder reacts with B from the eutectic structure to form new boride particles, which then grow into granular or spherical shapes. With increasing temperature, these particles coarsen and become more evenly distributed.

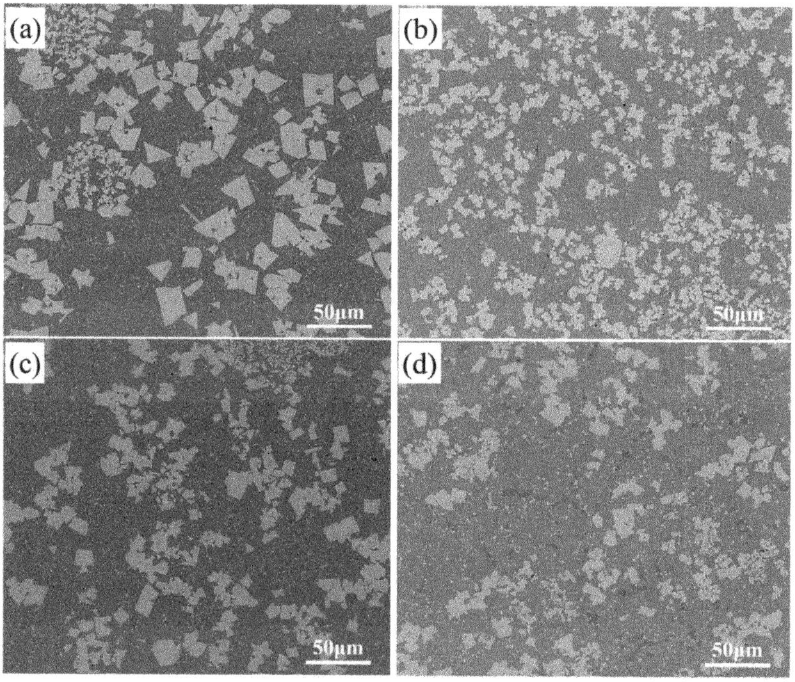

Figure 3. SEM morphologies of hard phases in coating at different quenching temperatures (**a**) C0; (**b**) C1; (**c**) C2; and (**d**) C3.

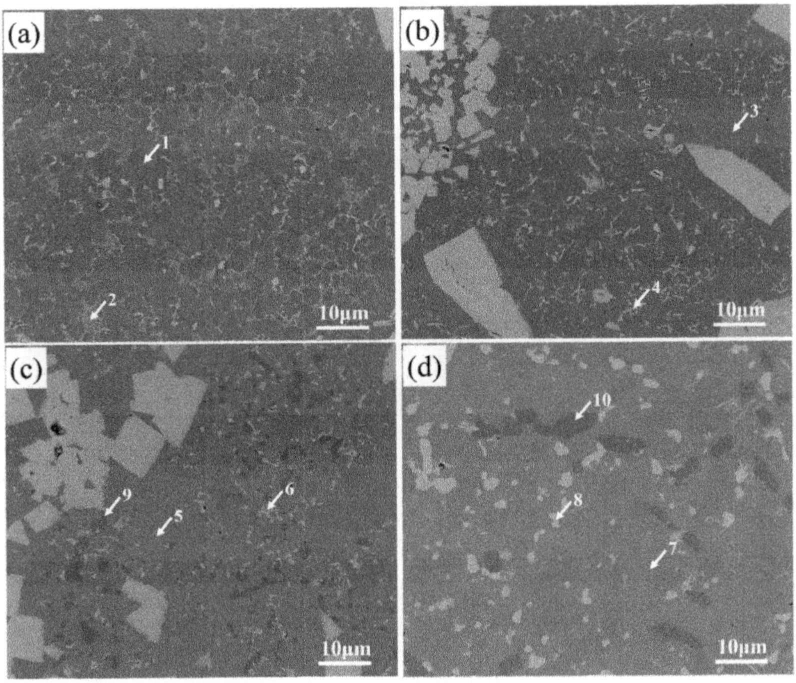

Figure 4. SEM morphologies of binder phases in coatings at different quenching temperatures (**a**) C0; (**b**) C1; (**c**) C2; and (**d**) C3.

Table 4. EDS analysis results of marked points in Figure 4.

Point	Weight%					Atomic%				
	Mo	Fe	Cr	Ni	Si	Mo	Fe	Cr	Ni	Si
1	1.13	41.03	1.97	3.10	1.51	0.96	83.10	3.87	5.97	6.10
2	13.00	7.64	2.57	0.78	0.27	39.30	39.67	14.35	3.88	2.80
3	1.41	57.20	2.28	4.01	1.94	1.19	83.24	3.92	5.56	6.08
4	24.93	32.26	3.05	1.67	1.01	26.97	59.91	6.09	2.94	4.09
5	-	52.01	1.69	3.87	3.41	-	80.88	2.83	5.73	10.56
6	38.95	15.34	5.65	-	-	51.43	34.81	13.76	-	-
7	-	53.68	2.04	4.01	2.56	-	82.57	3.73	5.87	7.83
8	41.47	14.68	5.42	-	-	54.07	32.88	13.05	-	-
9	-	45.92	7.95	-	-	-	84.32	15.68	-	-
10	-	46.61	7.34	-	-	-	85.53	14.47	-	-

The EDS points in the binder phase (points 1, 3, 5, and 7) represent Fe along with Cr, Ni, and Si, elements known to stabilize ferrite and provide solid-solution strengthening. In C1, the binder composition is similar to C0, indicating limited diffusion at 850 °C. In C2 and C3, however, Mo originally dissolved in the binder reacts fully to form additional Mo_2FeB_2, while Cr incorporates into Mo_2FeB_2 lattices to form $(Mo,Fe,Cr)_3B_2$ [19]. Notably, the darker binder regions (points 9 and 10) appear in C2 and C3, composed exclusively of Fe and Cr. At quench temperatures $\geq$ 1000 °C, Ni and Si (which do not participate in Mo_2FeB_2 formation) were excluded from the reaction front. Rapid quenching trapped these elements in a depleted area, leaving localized Fe-Cr binder domains upon cooling.

3.4. Microhardness

Figure 5 shows the microhardness distributions of the coatings quenched at different temperatures. The microhardness values across each coating were relatively uniform, with no significant fluctuations. The average microhardness values of samples C0, C1, C2, and C3 were 827 $HV_{0.5}$, 905 $HV_{0.5}$, 1061 $HV_{0.5}$, and 1092 $HV_{0.5}$, respectively. Compared with C0, the microhardness of C1, C2, and C3 increased by 9.43%, 28.29%, and 32.04%, respectively.

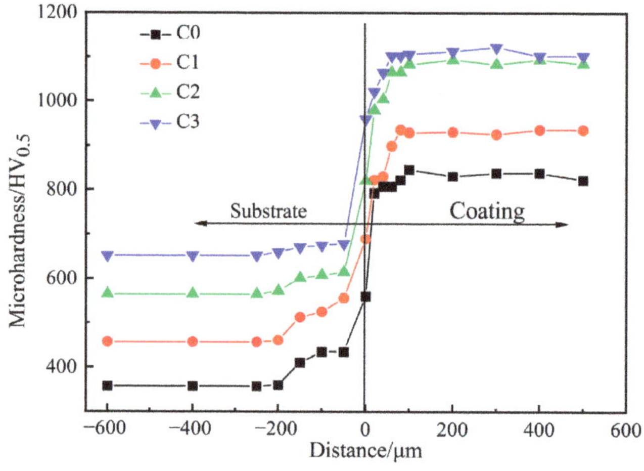

Figure 5. Microhardness distribution at different quenching temperatures.

The observed hardness enhancement is closely related to the microstructural evolution during quenching. After high-temperature quenching (C2 and C3), the reticulated eutectic structure disappeared, and the coating had sufficient thermal energy and time to undergo in situ phase transformations. This promoted the formation of a larger quantity of Mo_2FeB_2 hard phase particles. These boride particles significantly contributed to increased hardness. Therefore, it can be inferred that both the disappearance of the eutectic structure and the generation of hard phase particles have a strong influence on the coating's microhardness.

3.5. Wear Resistance

Figure 6 shows the variation in the coefficient of friction (COF) of the coatings at different quenching temperatures. The COF curves of C0 and C1 exhibit significant fluctuations as wear time increased, indicating poor wear resistance. This can be attributed to the rapid heating and cooling cycles inherent in plasma spray welding, which result in uneven residual stress distribution within the coating. For the C0 sample, which did not undergo any heat treatment, residual stress was not relieved. This led to stress concentrations and the formation of microcracks during the wear process, thereby causing large fluctuations in the COF. In addition, due to the relatively low microhardness of the C0 coating compared to the friction counterpart, substantial wear debris was generated, further accelerating wear and increasing the COF over time. For the C1 sample, although heat-treated at 850 °C with 30 min of holding time followed by quenching in mineral oil (a medium with good cooling efficiency), residual stress was only partially released [27]. The resulting decrease in lattice distortion contributed to a slight reduction in the COF compared with that of C0. However, because 850 °C was insufficient to fully relieve internal stresses, some fluctuation in the COF curve still remains.

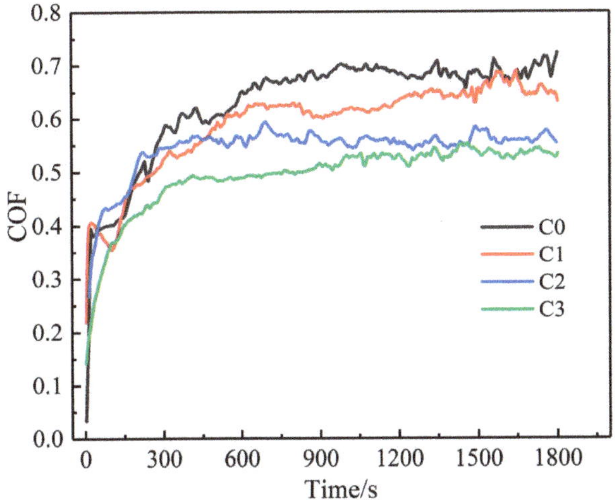

Figure 6. Variation in coefficient of friction of coatings at different quenching temperatures.

In contrast, after high-temperature quenching (C2 and C3), the COF curves showed a much flatter trend with increasing wear time, indicating improved wear stability. The enhanced wear resistance at higher quenching temperatures was associated with the progressive disappearance of eutectic structures (as seen in Figure 4) and the increased formation of Mo_2FeB_2 hard phase particles. These microstructural changes strengthened the coating and reduced its wear rate. Furthermore, residual stress was more effectively relieved at higher quenching temperatures, contributing to the coating's enhanced wear performance. Overall, the C3 coating demonstrated the best wear resistance, as evidenced by the most stable COF curve during the wear process.

Figure 7 shows the variation in the average COF and mass loss of the coatings at different quenching temperatures. The average COF decreased progressively from C0 to C3. The wear mass loss data represent the average values obtained from three repeated measurements to ensure reliability. The mass losses of samples C0, C1, C2, and C3 were 6.9 mg, 5.7 mg, 4.0 mg, and 3.7 mg, respectively. Under similar wear test conditions, the wear mass losses of C1, C2, and C3 decreased by 17.4%, 42.03%, and 46.38% compared with that of C0. Greater mass loss indicates poorer wear resistance [28–30].

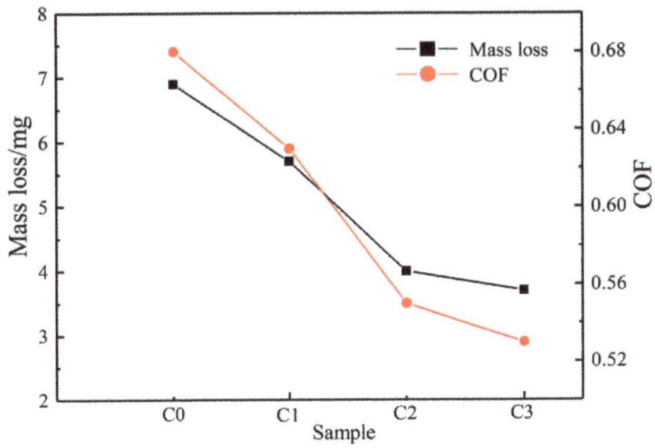

Figure 7. Variation in average COF and mass loss of coatings at different quenching temperatures.

These results indicate that increasing the quenching temperature significantly improved the wear resistance of the coatings. The improvements are attributed to the disappearance of eutectic structures, the increased formation of Mo_2FeB_2 hard phase particles, and the strengthening of the binder phase.

Figure 8 shows the wear morphologies of the coatings at different quenching temperatures. In Figure 8a, the wear surface of the C0 coating exhibits deep, "granular" spalling pits (point I), along with shallower, "flaky", peeling pits (point II), which are associated with the fracture and detachment of eutectic structures. After quenching at 850 °C (Figure 8b), the number of spalling pits slightly decreased, and some evolved into elongated or short bone-shaped pits (point III). These morphological changes suggest that the eutectic structure plays a critical role in determining the coating's wear resistance. At higher quenching temperatures (C2 and C3), the number of spalling pits decreased significantly. In Figure 8c, for the C2 coating, only a few spalling pits and some black oxides (point IV) can be observed. The C3 coating in Figure 8d exhibits the fewest spalling features, indicating an excellent wear performance.

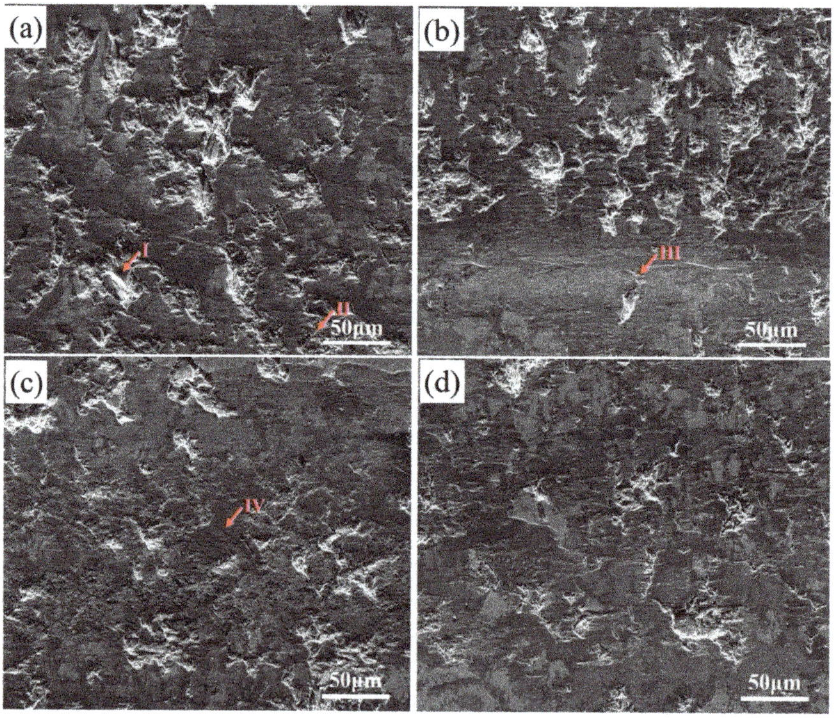

Figure 8. Wear morphologies of coating at different quenching temperatures (**a**) C0; (**b**) C1; (**c**) C2; and (**d**) C3.

In plasma spray-welded coatings, the Mo_2FeB_2 hard phase particles and the eutectic structures possess higher hardness than the α-Fe binder phase, creating a distinct hardness contrast within the microstructure. While the binder phase provides mechanical support, it is more susceptible to wear and tends to generate abrasive debris. During wear, Mo_2FeB_2 hard phases may protrude above the worn surface, and under the shear stress of the Si_3N_4 counter body, these bulging particles can become detached, forming deeper spalling pits. Additionally, the eutectic structure, being brittle due to boron carbide components, is easily fractured and peeled off during wear. However, because boron carbide exhibits hardness and thermal stability, it does not adhere to the Si_3N_4 ball. Instead, the fractured boron carbide and Mo_2FeB_2 particles are gradually removed from the surface by reciprocating motion, without forming pronounced grooves or plowing tracks.

On the one hand, the disappearance of eutectic structures and the increased formation of Mo_2FeB_2 hard phases after high-temperature quenching significantly enhance wear resistance. The binder phase in the C2 and C3 coatings primarily consists of Fe and Cr and lacks high-temperature-resistant elements such as Ni or Si. As a result, the binder phase is prone to oxidation during wear, leading to the formation of an oxide layer on the worn surface. This oxide layer, gradually formed and distributed by the reciprocating Si_3N_4 friction pair, acts as a lubricant [31], thereby improving wear resistance [32]. Once this oxide layer is completely worn off, broader plowing marks may appear, leveling the worn surface and causing the limited spalling of the hard phases. On the other hand, the increased and evenly distributed Mo_2FeB_2 hard phases in quenched coatings contribute to dispersion strengthening and serve as an anti-wear skeleton. These hard phases mitigate the plowing action of wear debris and abrasives, reduce the fracture of hard phase particles, and protect the worn surface, thereby further enhancing the wear resistance of the coating [21]. In summary, the primary wear mechanism of the coating is abrasive wear resulting from the detachment of hard phase particles.

4. Conclusions

This study aimed to enhance the hardness and wear resistance of H13 alloy steel by fabricating Mo_2FeB_2-based ternary boride coatings through plasma spray welding, followed by quenching at different temperatures. The coating was successfully prepared with a defect-free surface without pores, inclusions, and cracks and a good metallurgical bond with the H13 alloy steel substrate. The main phases in the coating were identified as α-Fe, Mo_2FeB_2, $(Mo,Fe,Cr)_3B_2$, and $Fe_{23}(B,C)_6$.

The morphology of the Mo_2FeB_2 hard phase particles remained relatively unchanged across different quenching temperatures. However, high-temperature quenching (1000 °C and 1150 °C) led to the formation of more Mo_2FeB_2 particles and the gradual disappearance of the reticulated eutectic structure. Both the eutectic structure and the Mo_2FeB_2 hard phase were the key factors influencing the microhardness and wear resistance of the coating.

Microhardness increased progressively with quenching temperature, reaching 1092 $HV_{0.5}$ at 1150 °C (32% higher than that of the as-sprayed coating). Concurrently, the average coefficient of friction (COF) and wear mass loss decreased by 46.4%, demonstrating superior wear resistance in the quenched coatings.

The wear mechanism was primarily abrasive wear caused by the detachment of Mo_2FeB_2 hard phase particles from the coating. The overall wear resistance was significantly enhanced due to the hardness of the Mo_2FeB_2 phase, the increased strength of the binder phase, and the lubricating effect of oxides formed from the binder during wear.

Author Contributions: Conceptualization, H.Z. and Y.Z.; methodology, H.Z.; formal analysis, H.Z., Y.H. and Y.Z.; investigation, H.Z.; writing—original draft preparation, H.Z., Y.H. and Y.Z.; writing—review and editing, Y.Z.; funding acquisition, H.Z. All authors have read and agreed to the published version of the manuscript.

Funding: This research was funded by the PhD research startup foundation of Zhejiang Industry & Trade Vocational College, grant number YJRC202304.

Data Availability Statement: Data will be available upon reasonable request.

Conflicts of Interest: The authors declare no conflicts of interest.

References

1. Hao, J.; Hu, F.; Le, X.; Liu, H.; Yang, H.; Han, J. Microstructure and high-temperature wear behaviour of Inconel 625 multi-layer cladding prepared on H13 mould steel by a hybrid additive manufacturing method. *J. Mater. Process. Technol.* **2021**, *291*, 117036. [CrossRef]
2. Zhang, J.; Yu, M.; Li, Z.; Liu, Y.; Zhang, Q.; Jiang, R.; Sun, S. The effect of laser energy density on the microstructure, residual stress and phase composition of H13 steel treated by laser surface melting. *J. Alloys Compd.* **2021**, *856*, 158168. [CrossRef]
3. Yang, C.; Jing, C.; Fu, T.; Lin, T.; Guo, W.; Liu, N. Effect of TiC addition on the microstructure and properties of laser cladding Ni60A coatings on H13 steel surface. *Mater. Today Commun.* **2024**, *38*, 107904. [CrossRef]
4. Zong, H.; Kang, N.; El Mansori, M. Impact of applied loads on wear mechanisms in H13 steel at various preheating temperatures during laser powder bed fusion additive manufacturing. *Wear* **2024**, *556–557*, 205538. [CrossRef]
5. Lu, H.F.; Xue, K.N.; Xu, X.; Luo, K.Y.; Xing, F.; Yao, J.H.; Lu, J.Z. Effects of laser shock peening on microstructural evolution and wear property of laser hybrid remanufactured Ni25/Fe104 coating on H13 tool steel. *J. Mater. Process. Technol.* **2021**, *291*, 117016. [CrossRef]
6. Pedroso da Silva, S.; Abrão, A.M.; Rodrigues da Silva, E.; Câmara, M.A. Surface modification of AISI H13 steel by die-sinking electrical discharge machining and TiAlN coating: A promising hybrid technique to improve wear resistance. *Wear* **2020**, *462–463*, 203509. [CrossRef]
7. Meng, C.; Cao, R.; Li, J.; Geng, F.; Zhang, Y.; Wu, C.; Wang, X.; Zhuang, W. Mechanical properties of TiC-reinforced H13 steel by bionic laser treatment. *Opt. Laser Technol.* **2021**, *136*, 106815. [CrossRef]
8. Narvan, M.; Ghasemi, A.; Fereiduni, E.; Elbestawi, M. Laser powder bed fusion of functionally graded bi-materials: Role of VC on functionalizing AISI H13 tool steel. *Mater. Des.* **2021**, *201*, 109503. [CrossRef]
9. Lu, J.Z.; Cao, J.; Lu, H.F.; Zhang, L.Y.; Luo, K.Y. Wear properties and microstructural analyses of Fe-based coatings with various WC contents on H13 die steel by laser cladding. *Surf. Coat. Technol.* **2019**, *369*, 228–237. [CrossRef]
10. Mao, X.; Zhu, P.; Sun, X.; Huang, S.; Xu, C.; He, H.; Chen, Y.; Cheng, Z. Nucleation mechanism, particle shape and strengthening behavior of (Ti, Nb) (C, N) particles in Fe-based composite coatings by plasma spray welding. *Ceram. Int.* **2024**, *50*, 27296–27304. [CrossRef]
11. Mao, X.; Zhu, P.; Sun, X.; Xu, C.; Huang, S.; He, H.; Cheng, Z. Effect of N on microstructure, nucleation mechanism and mechanical performance of (Ti, Nb) C particles in Fe-based composite coating via plasma spray welding. *Surf. Coat. Technol.* **2024**, *490*, 131177. [CrossRef]
12. Zhang, L.; Gao, Y.; Huang, Z.; Ma, H.; Liu, Y.; Jie, W.; Zhou, L. Effect of Ni content on the corrosion behavior of Mo2NiB2–Ni cermet in H2SO4 solution. *Ceram. Int.* **2023**, *49*, 36289–36298. [CrossRef]
13. Zhang, L.; Huang, Z.; Jie, W.; Cao, Z.; Liu, Y.; Zhou, L. Effect of impact angle on wear behavior of Mo2NiB2–Ni cermets with different Ni content. *Ceram. Int.* **2022**, *48*, 16944–16955. [CrossRef]
14. Zhang, H.; Pan, Y.; Zhang, Y.; Lian, G.; Cao, Q.; Yang, J. Influence of laser power on the microstructure and properties of in-situ NbC/WCoB–TiC coating by laser cladding. *Mater. Chem. Phys.* **2022**, *290*, 126636. [CrossRef]
15. Zhu, X.; Pan, Y.; Zhang, H.; Ke, D. Effect of Ni content on the microstructure and properties of WCoB-TiC cermets. *Int. J. Refract. Met. Hard Mater.* **2023**, *117*, 106381. [CrossRef]
16. Zhang, H.; Zhang, Y. In-situ synthesized WC reinforcement phase on microstructural evolution, toughness and tribological properties of Mo2FeB2-based composite coatings fabricated by laser cladding. *J. Mater. Res. Technol.* **2025**, *34*, 677–690. [CrossRef]
17. Zhang, Y.; Zhang, H. Influence of laser remelting on the microstructure and properties of Mo2FeB2 ternary boride coatings prepared by laser cladding. *Mater. Today Commun.* **2024**, *41*, 110548. [CrossRef]
18. Lang, X.-Y.; Zhang, G.-H. Microstructure and mechanical properties of Mo2FeB2-316SS cermets prepared by vacuum sintering. *J. Alloys Compd.* **2025**, *1027*, 180626. [CrossRef]
19. Li, B.; Zheng, Y.; Shao, X.; Li, K.; Xu, X.; Chen, M.; Liang, H.; Yang, M. Growth kinetics of ceramic grains during liquid phase sintering in Mo2FeB2-based cermets. *Int. J. Refract. Met. Hard Mater.* **2023**, *110*, 105991. [CrossRef]
20. Cao, Z.; Jian, Y.; Zhao, Z.; Xiao, P.; Xu, L.; Huang, Z. Effect of V doping on the microstructure and mechanical properties of Mo2FeB2-based cermets and investigation on the enhancement mechanism. *J. Alloys Compd.* **2023**, *931*, 167545. [CrossRef]
21. Wei, X.; Chen, Z.; Zhong, J.; Xiang, Y. Feasibility of preparing Mo2FeB2-based cermet coating by electrospark deposition on high speed steel. *Surf. Coat. Technol.* **2016**, *296*, 58–64. [CrossRef]
22. Duan, S.; Ren, W.; Lei, W.; Wang, Y. Study on the microstructure and properties of rail cladding layer after laser quenching. *J. Manuf. Process.* **2023**, *108*, 180–193. [CrossRef]
23. Chen, C.; Feng, A.; Liu, B.; Wei, Y.; Song, X. Effect of quench-tempering and laser quenching on the microstructure and properties of high-chromium cast iron. *J. Mater. Res. Technol.* **2022**, *19*, 2759–2773. [CrossRef]
24. Sun, H.; Ding, H.; Liu, T. Unveiling the effect of vacuum heat treatment on HVOF-sprayed high entropy cantor alloy coatings: Microstructure, diffusion behavior and mechanical property. *J. Mater. Res. Technol.* **2024**, *33*, 9033–9043. [CrossRef]

25. Li, B.; Tian, M.; Feng, L.; Wu, D.; Guo, H.; Gao, Y.; Xu, Y.; An, C.; Liang, C. Effect of heat treatment on the microstructure and mechanical properties of TiAl alloys. *J. Mater. Res. Technol.* **2024**, *33*, 8545–8554. [CrossRef]
26. Xiao, G.; Jiang, J.; Wang, Y.; Liu, Y.; Zhang, Y.; Huang, M. Microstructure and mechanical properties of thixoformed GH4037 parts before and after heat treatment. *Mater. Sci. Eng. A* **2021**, *815*, 141304. [CrossRef]
27. Kuroda, P.A.B.; de Mattos, F.N.; Grandini, C.R.; Afonso, C.R.M. Influence of the heat treatment temperature on the MAO coating produced in the Ti–25Ta–25Zr alloy. *J. Mater. Res. Technol.* **2023**, *26*, 3881–3892. [CrossRef]
28. Zhang, G.; Liu, W.; Bian, H.; Xi, W.; Zhang, K.; Wang, H. A novel approach to enhance the wear resistance of laser-cladded Tribaloy T-800 coatings on DD5 single crystal alloys by addition of Si. *Mater. Charact.* **2025**, *221*, 114728. [CrossRef]
29. Qin, J.; Shu, L.; Wang, J.; Li, P.; Li, A.; Huang, T.; Shi, J.; Qiu, Y.; Dong, H. Study on the preparation and wear properties of laser cladded Incoel718/WC gradient composite coatings. *Mater. Today Commun.* **2024**, *41*, 111080. [CrossRef]
30. Zhang, G.; Liu, W.; Bian, H.; Wang, H.; Wang, W.; Xu, X.; Liu, J. Effect of laser energy density on cracking susceptibility and wear resistance of laser-clad tribaloy T-800 coatings on DD5 single-crystal alloys. *Mater. Today Commun.* **2024**, *40*, 109604. [CrossRef]
31. Belov, D.S.; Blinkov, I.V.; Sergevnin, V.S.; Smirnov, N.I.; Volkhonskii, A.O.; Bondarev, A.V.; Lobova, T.A. Abrasive, hydroabrasive, and erosion wear behaviour of nanostructured (Ti,Al)N-Cu and (Ti,Al)N-Ni coatings. *Surf. Coat. Technol.* **2018**, *338*, 1–13. [CrossRef]
32. Yang, F.; Su, B.; Zhang, A.; Meng, J.; Han, J.; Wu, Y. Tribological properties and wear mechanisms of Mo2FeB2 based cermets at high temperatures. *Tribol. Int.* **2018**, *120*, 391–397. [CrossRef]

lubricants

MDPI

Article

Investigation of the Friction and Wear Behavior of Cr-Mo-V Steel with Different Surface Treatment Processes

Wei Zhang [1,2,†], Jian Zhang [3,4,†], Shizhong Wei [1,*], Liuliang Chen [4], Wei Zhang [4], Zhenhuan Sun [4], Chong Chen [1,2], Feng Mao [1,2], Xiaodong Wang [1], Caihong Dou [1] and Cheng Zhang [1,2,*]

1 National Joint Engineering Research Center for Abrasion Control and Molding of Metal Materials and Longmen Laboratory, School of Materials Science and Engineering, Henan University of Science and Technology, Luoyang 471000, China; 230320020303@stu.haust.edu.cn (W.Z.); chenchong8812@163.com (C.C.); maofeng718@163.com (F.M.); nmxdwang@163.com (X.W.); dch1805@163.com (C.D.)
2 Longmen Laboratory, Luoyang 471003, China
3 State Key Laboratory for Advanced Metals and Materials, University of Science and Technology Beijing, Beijing 10003, China; zhj85401@126.com
4 Henan Zhongyuan Special Steel Equipment Manufacturing Co., Ltd., Jiyuan 454685, China; 8127yyxyyx@163.com (L.C.); zjzycso@163.com (W.Z.); szhzysc@163.com (Z.S.)
* Correspondence: wsz@haust.edu.cn (S.W.); zhangch06@126.com (C.Z.)
† These authors contributed equally to this work.

Abstract: Hot work die steel is an alloy steel with good high-temperature performance, which is widely used in mechanical manufacturing, aerospace, and other fields. During the working process of hot working mold steel, it is subjected to high temperature, wear, and other effects, which can lead to a decrease in the surface hardness of the mold, accelerate surface damage, shorten the service life, and reduce the quality of the workpiece. In order to improve the wear resistance of the mold, this paper conducts two surface treatments, chrome plating and nitriding, on the surface of hot work mold steel, and compares the high-temperature wear behavior of the materials after the two surface treatments. The results indicate that the hot work die steel obtained higher surface hardness and wear resistance after nitriding surface modification. After nitriding treatment, the surface of hot work die steel contains ε phase ($Fe_{2-3}N$), which improves its surface hardness and wear resistance, thus exhibiting better surface hardness and wear resistance than the chrome-plated sample. In this study, the high-temperature wear behavior of hot work die steel after two kinds of surface strengthening treatments was deeply discussed, and the high-temperature wear mechanism of steel after surface strengthening was revealed. It provides a theoretical basis and experimental basis for the surface modification of hot working die steel, and also provides new ideas and methods for improving the service life and workpiece quality of hot working die steel in industrial production. In this study, the advantages and disadvantages of high-temperature wear resistance of hot working die steel after chromium plating and nitriding were systematically compared for the first time, which provided a scientific basis for the selection of surface strengthening technology of hot working die steel and had important academic value and practical application significance.

Keywords: hot work die steel; surface treatment; nitriding; chrome plating; high-temperature wear mechanism

1. Introduction

In modern industrial production, the surface properties of materials play a decisive role in their service life and reliability, especially under severe operating conditions such as high temperatures and high loads. In high-temperature working environments, the surface hardness of the mold decreases [1] and the elevated temperatures during operation can accelerate wear [2–4], thereby shortening its service life. To enhance the hardness and wear resistance of molds, surface strengthening treatments are commonly employed [5–8], which can effectively extend the service life of the molds [9,10]. Commonly used surface modification techniques for hot work die steel currently include thermal spraying, chemical heat treatment, electroplating deposition, laser surface modification, magnetron sputtering, etc. Among them, surface chromium plating and nitriding treatment are surface strengthening treatment methods used more in the industrial field. It is of great practical significance and representativeness to choose these two surface modification technologies for comparative study. At the same time, it provides a reference for the application of industrial manufacturing.

Nitriding is a chemical heat treatment process. It involves diffusing nitrogen atoms into the surface of a component at high temperatures to form a high-nitrogen surface layer. The surface of nitrided components exhibits high hardness, wear resistance, fatigue strength, hot hardness, and anti-scuffing properties [11–13]. Chen [14] carried out plasma nitriding on Ti6Al4V alloy, which enhanced the tribological properties and prolonged the service life of the substrate. These improvements in mechanical and tribological properties effectively reduce wear compared to materials that are not nitrided [15,16]. Ze [17] compared aluminum-modified plasma nitriding with conventional plasma nitriding and found that the wear resistance was doubled after aluminum-modified plasma nitriding. Chromium plating is an electrochemical process. It involves depositing a layer of chromium on the material surface, which not only enhances surface hardness and wear resistance but also improves its high-temperature resistance and oxidation resistance. The chromium plating layer, with its excellent wear and corrosion resistance, is most widely used in the manufacturing, automotive, and aerospace industries [18–22]. König [23] studied how atmospheric conditions influence the wear behavior between cast iron and chromium-plated steel under high-temperature conditions reaching 800 °C. Below 200 °C, the adhesion of cast iron to the chromium-plated layer results in high wear and friction. Between 200 °C and 400 °C, wear and friction slightly decrease due to the oxidation and agglomeration of debris. Above 400 °C, a wear-resistant glaze layer primarily composed of Fe_2O_3 forms. At 700 °C, the sheet exhibits slight wear. Wang [24] investigated the failure mechanisms of electroplated chromium layers using repetitive scratch tests under sub-critical loads. The proportion of micro-cutting increased with the increase in load and number of cycles. Compared with progressive scratch tests, repetitive scratch tests under sub-critical loads can more comprehensively evaluate failure under different loads and provide a clearer understanding of the failure formation process of the coating.

Although a large number of researchers have paid attention to the effect of chromium plating and nitriding on the wear resistance of materials, most of the research focuses on the performance analysis at room temperature, and there are relatively few comparative studies at high temperature. There is a lack of systematic comparative analysis of these two surface modification technologies in terms of high-temperature wear performance and wear mechanism in the existing literature. In this study, for the first time, two kinds of surface treatment methods, chromium deposition and nitriding, were carried out on the same matrix material Cr-Mo-V hot working die steel. The wear properties and wear mechanism of the two coatings at room temperature and high temperature were compared and studied, which provided theoretical and data support for production practice.

2. Sample Preparation and Testing

2.1. Coating Preparation

The base material is 2Cr3Mo3VA mold steel. The detailed chemical composition can be found in Table 1. These compositional data were supplied by the material supplier. Prior to surface treatment, the base material was subjected to heat treatment. Normalization: the base material, hot-work die steel, was heated to 960 °C, held at this temperature for 3 h, and then cooled in air. Quenching: the hot-work die steel was heated to 1040 °C, held at this temperature for 1 h, and then quenched in oil. Tempering: the steel was tempered at 580 °C for 4 h, followed by a second tempering at 620 °C for 4 h to achieve optimal mechanical properties. The heat treatment process is illustrated in Figure 1. The material was cut into disc-shaped specimens with the dimensions shown in Figure 2 using an electrical discharge machining (EDM) wire cutter. The specimens were then ground with 200# to 1200# sandpaper, rinsed with deionized water, ultrasonically cleaned in an acetone solution at room temperature for 15 min, and subsequently dried with cold air.

Table 1. Chemical composition of the experimental steel (mass fraction, %).

Steel	C	Si	Mn	Cr	Mo	V	P	S	Fe
2Cr3Mo3VA	0.22	0.50	0.25	2.79	2.80	0.15	<0.015	<0.015	Bal.

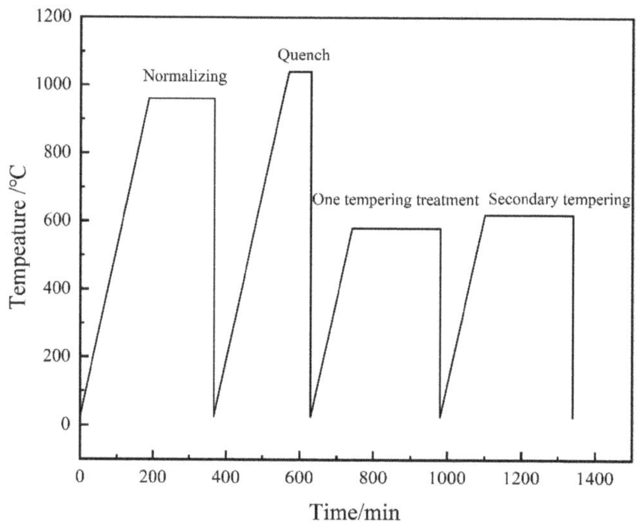

Figure 1. Heat treatment process flowchart of 2Cr3Mo3VA steel.

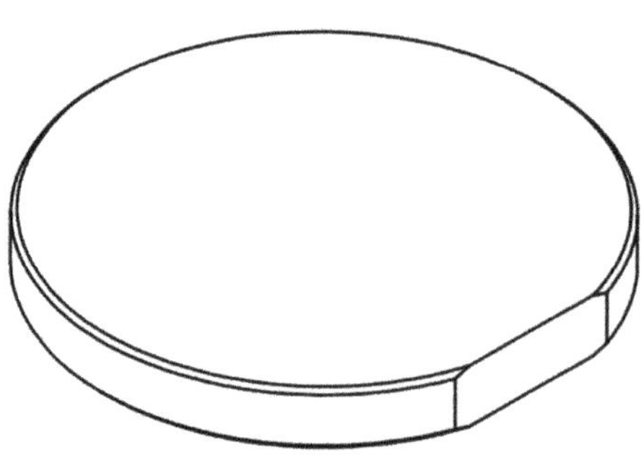

Figure 2. Wire Electrical Discharge Machining (WEDM) diagram of the specimen.

2.1.1. Surface Treatments

Cut the heat-treated sample into the size shown in Figure 2, and then proceed with surface chrome plating treatment. The preparation process parameters for electroplated chromium are as follows: CrO_3 concentration of 250 g/L, H_2SO_4 concentration of 2.5 g/L, current density of 20 A/dm^2, and temperature of 60 °C. For another set of samples, the heat-treated specimens were subjected to nitriding surface treatment. The nitriding process is as follows: First, the samples were cleaned. Then, they were preheated at a constant temperature of 350 °C for 40 min. Subsequently, the samples were nitrided for 2 h in a nitriding salt bath (consisting of base salt and adjusting salt) at 500 °C in a box resistance furnace under nitrogen protection. Next, the samples were oxidized for 40 min in an oxidizing salt bath at 380 °C. Finally, the samples were cleaned, dried, and then immersed in oil.

2.1.2. Microstructural and Property Characterization of Chromium Plating/Nitriding

The surface coating and cross-sectional morphology of the samples before and after wear were examined using a field-emission scanning electron microscope (JSM-IT800, JEOL, Tokyo, Japan). Elemental distribution analysis of the coating surface and wear tracks was performed using Energy Dispersive Spectroscopy (EDS) techniques. The crystal structure was characterized using an X-ray diffractometer (Bruker D8, Bruker, Billerica, MA, USA) with Cu-Kα radiation at 40 kV and 40 mA. The X-ray diffractometer was operated with a Cu target, a scanning range of 20° to 90°, and a scanning speed of 6°/min. Microhardness tests were conducted using a microhardness tester (HVT-1000, HYDAC, Sulzbach/Saar, Germany). This instrument is used to measure surface microhardness at a certain depth, with a load of 50 g and a dwell time of 15 s. After polishing the surface of the samples, microhardness tests were performed at five uniformly distributed points on the structure, and the average value of the data points and the standard deviation of the data set were calculated. The wear track's cross-sectional hardness was evaluated via nanoindentation (Keysight G200, Keysight, Santa Rosa, CA, USA).

2.1.3. Tribological Properties

Wear tests were conducted using the MFT-5000 high-temperature friction and wear tester (Rtec-Instruments, San Jose, CA, USA). The tribological evaluations employed a ball-on-disk setup, utilizing a 9.525 mm diameter Si_3N_4 ball as the upper specimen. The specific schematic diagram is shown in Figure 3. The prepared samples were cleaned with ethanol and then air-dried. The tribological test parameters are as follows: the test temperatures were room temperature and 700 °C, the test load was 50 N, the sliding radius was 15 mm, the sliding linear velocity was 0.628 m/s, and the total sliding duration was 30 min. The testing rotational speed was 400 rpm. Both the chromium-plated samples and the nitrided samples were tested for friction and wear at room temperature and at 700 °C. The specific parameters of the above experiments are determined based on the working conditions in actual industrial production. In the following article, different samples were numbered and named, as shown in Table 2.

During the tests, automatic heating was achieved using a resistor, and automatic temperature regulation and control were realized using a thermocouple and a temperature controller. The friction coefficient was measured using a microcomputer module. The wear surface morphology was modeled and analyzed using a laser confocal microscope (LCM) (OLS5100, Evident Scientific, Waltham, MA, USA). The wear rate was calculated through 3D surface profiling. Five cross-sections at equal distances from the wear track were selected, and the average cross-sectional area was calculated. The wear track volume

V was determined by multiplying the average cross-sectional area by the length of the wear track. The wear rate calculation formula is as follows:

$$\text{Wear rate} = \frac{V}{NL}\left(\text{mm}^3/\text{Nm}\right), \quad (1)$$

In the formula, *V* represents the wear volume (mm^3), *N* represents the load (N), and *L* represents the total sliding distance (m).

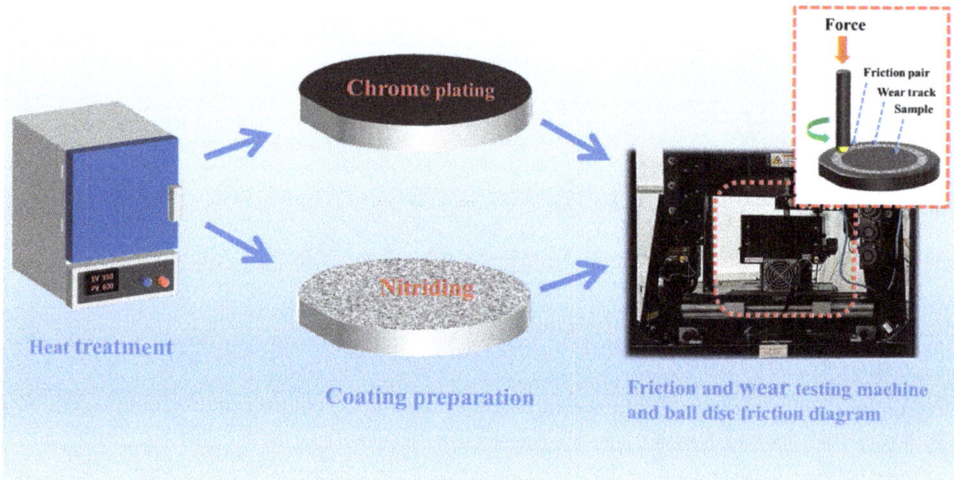

Figure 3. Test flowchart and ball-on-disc friction diagram.

Table 2. Comparison table of sample number and wear test conditions.

Sample Number	Surface Treatment	Wear Test Temperature
1-1	Chromium plating	Room
1-2	Chromium plating	700 °C
2-1	Nitride	Room
2-2	Nitride	700 °C

3. Results and Discussion

3.1. Microstructure of the Coating

Figure 4 is the cross-sectional morphology of the chromium plating layer. There are micro-cracks in the cross section of the coating, as shown in the arrow in Figure 4a. From the cross-sectional morphology of the chromium plating layer in Figure 4b, it can be seen that the cracks are distributed inside the chromium plating layer and do not extend to the inside of the substrate. This is due to the weak plastic deformation ability of metal chromium in the electroplating process. Under the combined action of electrochemical reaction and stress, the stress release in the chromium plating layer will produce cracks [25]. Although these micro-cracks do not directly penetrate into the matrix, the cracks are easy to expand and connect under the action of thermal stress and other factors, extending from the surface to the matrix. Then, the micro-cracks expand horizontally along the interface between the coating and the substrate, resulting in a large area of peeling off of the coating and losing the protective effect on the substrate [26].

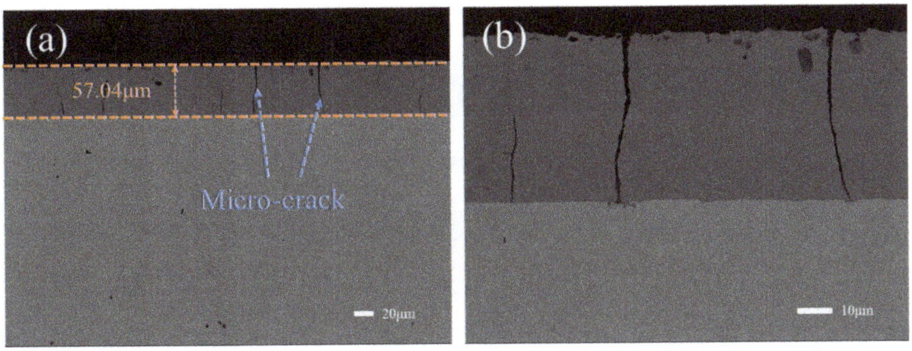

Figure 4. SED diagrams of chromium plating coating cross-section: (**a**) low SED diagram, (**b**) high SED diagram.

In Figure 5a,b, it can be clearly seen that the nitriding layer is loose and there are many veins extending into the matrix. According to relevant research, due to the coarse original structure and high nitrogen potential, the permeable layer after nitriding will have defects such as loose compound layer and vein nitride [27]. Compared with the obvious interface between the chrome plating layer and the substrate, the existence of the nitriding diffusion layer can buffer the stress [26].

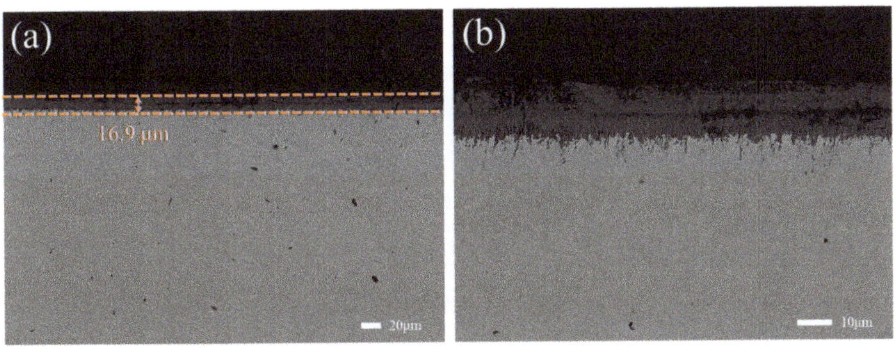

Figure 5. Nitriding section SED diagrams: (**a**) low SED diagram, (**b**) high SED diagram.

3.2. Friction Coefficient and Wear Rate of the Coating

It can be seen from Figure 6 that the surface hardness of the sample after nitriding is the highest, and the average hardness is 1087.9 Hv. The surface hardness of the sample after chromium plating is 975.5 Hv. It can be seen from Figure 7 that the surface hardness of 1-1 and 2-1 samples is almost the same as that of the samples without wear test. This also shows that the friction and wear experiments at room temperature have little effect on the hardness of the sample surface. It can be seen from Figure 7 that the surface hardness of the sample after high-temperature wear is lower than that of the sample after room-temperature wear. The average hardness of sample 1-2 was 401.0 Hv, while that of sample 2-2 was 566.1 Hv. This shows that the surface hardness of the two samples decreases with the increase in temperature. At the same time, it can be seen that the surface hardness of the chromium plating layer is lower than that of the nitriding layer. It can be seen from the wear rate trend in Figure 8 that the wear rate of the 2-1 sample is the lowest (10.67×10^{-6} $mm^3/N{\cdot}m$), while the wear rate of the 1-2 sample is the highest (42.57×10^{-6} $mm^3/N{\cdot}m$). The data show that the wear rate of the nitrided layer at 700 °C is about 38% lower than that of the chrome-plated layer, and the wear rate of the nitrided layer at room temperature is about 50% lower than that of the chrome-plated layer, indicating that the nitrided layer has better wear resistance than the chrome-plated layer.

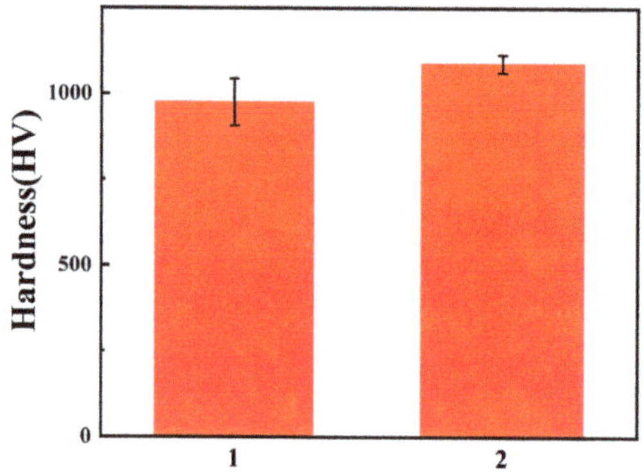

Figure 6. Surface hardness of different samples: no. 1 is the surface hardness of the sample after chromium plating; no. 2 is the surface hardness of the sample after nitriding.

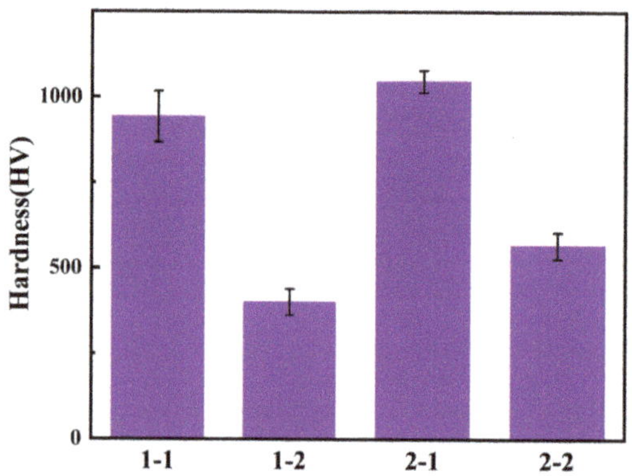

Figure 7. Surface hardness of different samples after wear.

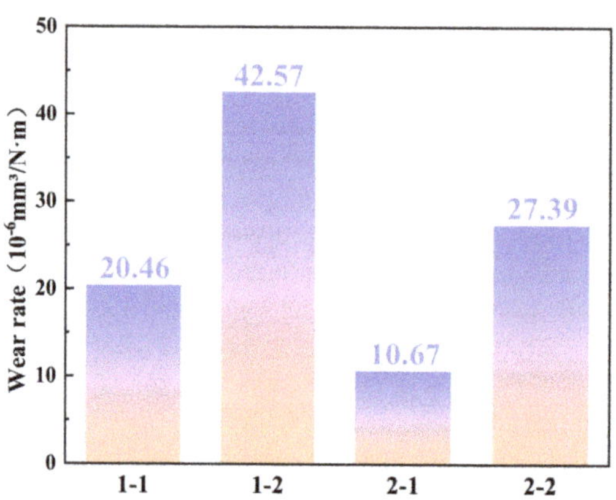

Figure 8. Wear rates of different specimens.

Figure 9 is the curve of the friction coefficient of 1-1, 1-2, 2-1, and 2-2 specimens with the change of sliding time. The sample experienced two stages in the friction process at different temperatures: running stage and stable stage. When the Si_3N_4 balls of the friction pair first came into contact with the surface of the sample, the friction resistance increased rapidly under the action of the asperities on the surface; that is, the friction coefficient of the running and stage increased rapidly. In the stable wear stage, the friction coefficient basically fluctuates in a constant value. The average friction coefficients of 1-1, 1-2, 2-1, and 2-2 samples are 0.6084, 0.4442, 0.7063, and 0.2740, respectively. The data show that the friction coefficient of the two samples gradually decreases with the increase in wear temperature. This is due to the softening of the material at high temperatures. It can be seen from the average friction coefficient of the two samples at different friction temperatures that the average friction coefficient at 700 °C drops sharply, which is due to the intensification of material surface softening and the formation of a continuous oxide film on the surface. The results show that the nitriding treatment has a certain friction drag reduction effect, which is because the nitriding layer with higher hardness effectively enhances the plastic deformation resistance of the alloy compared with the matrix.

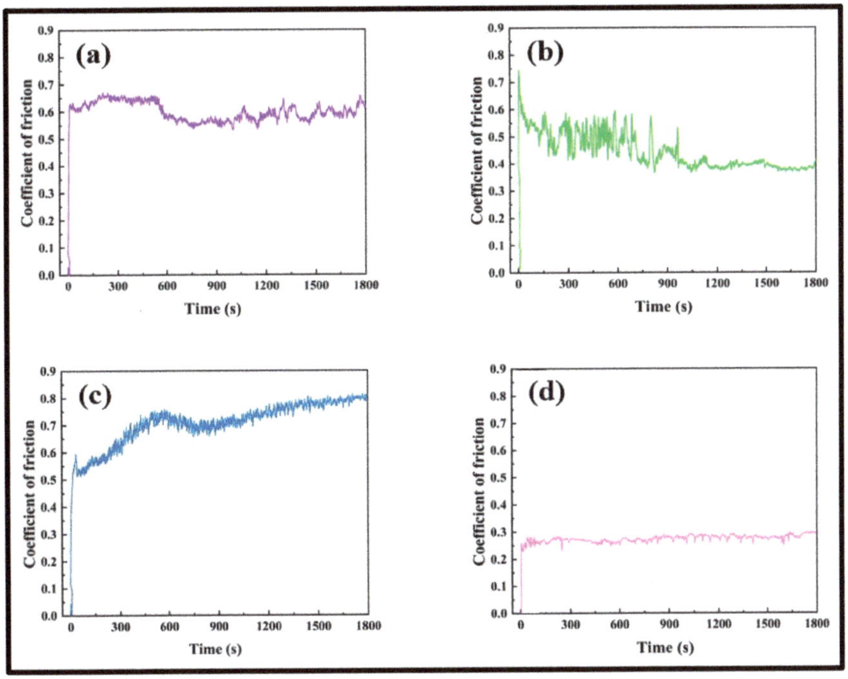

Figure 9. Variation of friction coefficient with time (**a**–**d**): 1-1 chromium plating sample at room-temperature wear, 1-2 chromium plating sample at 700 °C high-temperature wear, 2-1 nitriding sample at room-temperature wear, and 2-2 nitriding sample at 700 °C high-temperature wear.

In the process of high-temperature wear, friction heat, ambient temperature, and load will cause plastic deformation, dynamic recovery, and recrystallization of the sample, resulting in the change of the performance of the substructure, thus affecting the wear behavior [28]. Figure 10 shows the hardness distribution of the wear cross section. The cross-section hardness curve divides the sample into three parts: coating area, coating/substrate bonding area, and substrate area. It was found that the hardness of the coating was higher than that of the substrate after wear at room temperature and 700 °C. It can be seen from

Figure 10a,b that the coating hardness of 1-2 and 2-2 samples after wear at high temperature 700 °C is lower than that of 1-1 and 2-1 samples after wear at room temperature. This can be attributed to the increased softening of the sample caused by the increase in temperature. It can also be seen from Figure 10 that the hardness of the coating/substrate bonding zone is higher than that of the substrate. The hardness of the matrix is also affected by the temperature, resulting in a decrease in hardness. This softening of the sample is intensified due to the increase in temperature, which leads to the delamination and spalling of the generated oxide layer, resulting in severe oxidative wear and ultimately a decrease in hardness. This delamination phenomenon may further aggravate the damage degree of the sample surface.

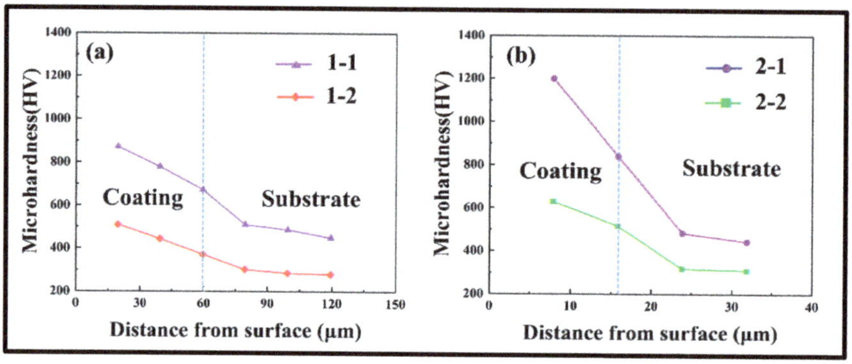

Figure 10. Hardness distribution of the worn cross-section: (**a**) chromium plating; (**b**) nitriding.

3.3. Morphology and Compositional Analysis of the Worn Surface

Figure 11a shows the XRD pattern of chromium-plated samples after friction at different temperatures for 30 min. It can be seen that the surface phase of the sample is mainly a Cr-containing phase when the 1-1 sample is worn at room temperature. The phase of Cr can also be observed in the X-ray diffraction pattern of 1-2 samples at a high temperature of 700 °C. These are mainly the phases of the coating. At the same time, it can be found that the oxide peak intensity of Cr gradually increases in the XRD pattern of 1-2 samples. It can be known that the XRD phase has a significant change with the increase in temperature. It is mainly related to the increase in temperature oxidation. Figure 11b shows the XRD pattern of the nitrided sample after friction at different temperatures for 30 min. It can be seen that the nitrided layer on the surface of the 2-1 sample is mainly ε phase ($Fe_{2-3}N$), which is also consistent with the results of other scholars [29,30]. In the salt bath nitriding process, the cyanate ion (CNO^-) in the salt bath decomposes to produce active nitrogen atoms. These active nitrogen atoms gradually aggregate on the surface of the material, and then diffuse to the lattice gap of the material under the impetus of the nitrogen potential. With the extension of nitriding time, the content of solid solution nitrogen atoms in the material increases gradually. Finally, a layer of iron nitride (ε-phase) is formed on the surface of the material, which has high hardness and wear resistance. In the XRD pattern of 2-2 samples in Figure 11b, there is not only ε phase ($Fe_{2-3}N$), but also Fe oxide phase. The change of the above phase is mainly due to the increase in oxidation caused by the increase in friction and wear temperature, which makes the oxide of Fe appear.

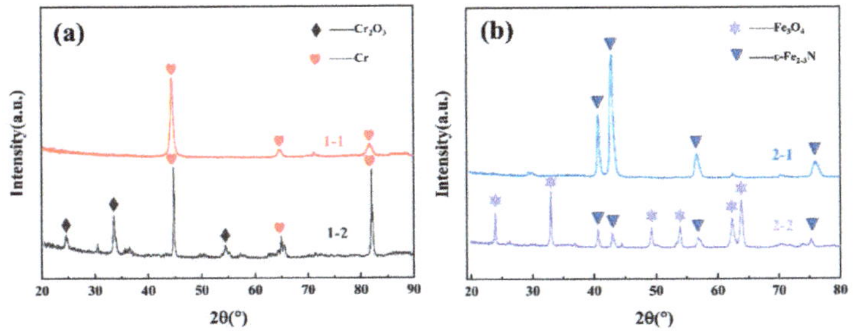

Figure 11. XRD patterns of surface phases after wear for two types of specimens: (**a**) XRD patterns of chromium-plated specimens under room-temperature wear and 700 °C high-temperature wear; (**b**) XRD patterns of nitrided specimens under room-temperature wear and 700 °C high-temperature wear.

Figures 12 and 13 show the microstructure image and the corresponding element distribution of the wear scar surface of the chrome-plated sample and the nitrided sample, respectively. Figure 12a is the wear scar microstructure diagram and element distribution diagram corresponding to the 1-1 sample. It can be seen from Figure 12a that there are a large number of flake thin layers, fish scale thin layers, furrows, and scratches. From the corresponding element distribution map, it can be seen that these flake thin layers and fish scale thin layers are oxides of Cr. At room temperature, the surface of the chromium plating layer will undergo oxidation reaction due to flash heat phenomenon, forming a thin layer of oxide. During the wear process at room temperature, the oxide film is weak and easy to peel off. The flake thin layer formed granular debris after sliding friction. This wear debris will cause furrows and scratches on the material when sliding, thereby accelerating the wear of the material. It can be seen that the wear mechanism of the chromium plating layer at room temperature is dominated by abrasive wear and accompanied by slight oxidative wear. Figure 12b is the wear scar microstructure diagram and element distribution diagram corresponding to the 1-2 sample. It can be observed in Figure 12b that there are spalling pits, and there are also a large number of flake thin layers, gullies, and scratches. It can be seen from the corresponding element distribution map that the surface of the wear scar has been completely covered by the oxide film. Compared with the wear surface at room temperature, the wear surface after sliding wear at 700 °C has wider and deeper gully traces, and the ploughing traces basically disappear. In addition, the correlation between the appearance of the gully phenomenon and the wear mechanism is significant. Under the condition of high-temperature wear, the surface of the sample is continuously worn and peeled off under the action of high temperature and friction, and gradually forms a gully-like morphology. The evolution of this microstructure will lead to a significant increase in surface roughness, thereby reducing the stability of the friction coefficient and weakening the wear resistance of the material. It can be seen that the wear mechanism of chrome-plated samples at 700 °C is the combined effect of oxidation wear, adhesive wear, and abrasive wear [31]. Figure 13a is the micro-graph of the wear scar and the element distribution map corresponding to the 2-1 sample. It can be observed in Figure 13a that there are some flake thin layers and furrows. From the corresponding element distribution map, it can be seen that the flake thin layer in the wear scar should be the oxide of Fe. It is shown that the wear form of the 2-1 sample is mainly adhesive wear at room temperature, accompanied by slight oxidative wear. Figure 13b is the wear scar microstructure diagram and element distribution diagram corresponding to the 2-2 sample. It can be seen from the corresponding element distribution map that the surface of the wear scar has been completely covered by the oxide film. The wear process generates an oxide film on the contact surface, where wear debris becomes thermally compacted and strongly bonded to the track under elevated temperatures and material softening. This results in the formation of a smooth, wear-resistant tribolayer that functions as a solid lubricant, effectively lowering both the coefficient of friction and wear rate [32–34]. The dominant wear mechanisms are identified as oxidative and abrasive wear.

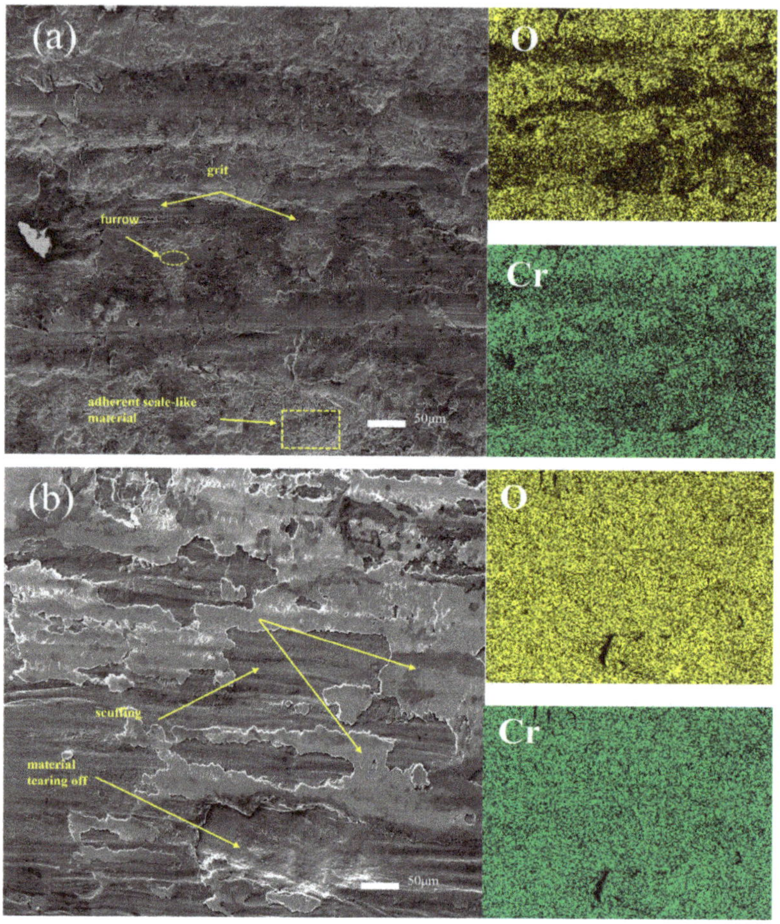

Figure 12. SEM images of wear scar surface and corresponding element distribution map: (**a**) room-temperature wear scar map and corresponding element distribution map of chromium-plated 1-1 sample; (**b**) wear scar diagram and corresponding element distribution diagram of chromium-plated 1-2 samples at 700 °C.

Figure 14a–d shows the three-dimensional topography and corresponding cross-section curves of the wear traces of samples 1-1, 1-2, 2-1, and 2-1 (Figure 14(a1)–(d1)). It can be seen from Figure 14(a1,b1) that the width and depth of the wear scar of the chrome-plated sample at high temperature are 1956.93 μm and 82.33 μm, respectively. The width and depth of the wear scar at room temperature are 2036.73 μm and 52.33 μm, respectively. The wear depth and wear volume at high temperature are larger than those at room temperature. This can be attributed to the softening of the sample caused by the increase in temperature, so that the wear is more serious. This also corresponds to the wear rate of Figure 8. It can be seen from Figure 14(c1,d1) that the width and depth of the wear scar of the nitrided sample at high temperature are 1479.76 μm and 45.35 μm, respectively. The width and depth of the wear scar at room temperature are 1829.06 μm and 11.13 μm, respectively. This also shows the same rule as the chrome-plated sample. The results show that high temperature will soften the material and make the wear resistance worse. The continuous Fe oxide film formed by the nitriding layer at 700 °C has a lubricating effect, which significantly reduces the friction coefficient and wear rate. However, due to the micro-crack propagation and softening effect of the chromium plating layer, the oxide film is easy to peel off, resulting in increased abrasive wear [26]. Thus, the nitrided layer exhibits better wear resistance under high-temperature wear. The wear resistance of the nitriding layer at 700 °C and room temperature is better than that of the chromium plating layer.

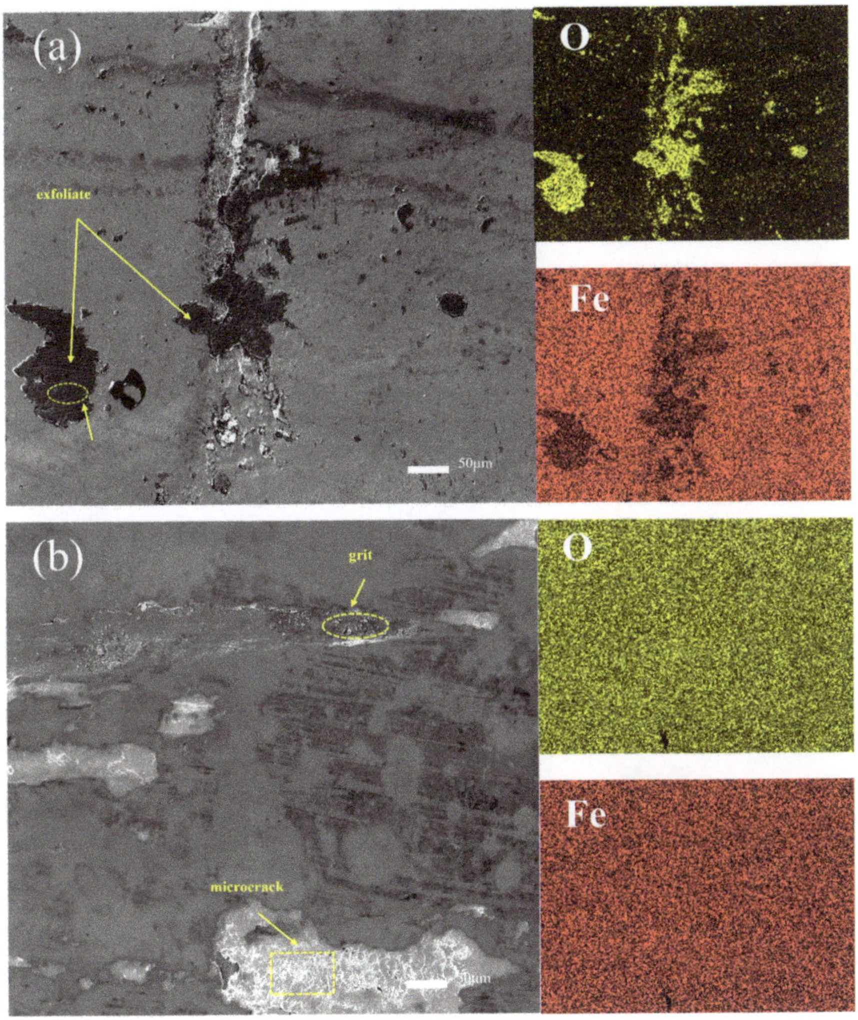

Figure 13. SEM images of wear scar surface and corresponding element distribution map: (**a**) room temperature wear scar map and corresponding element distribution map of nitriding 2-1 sample; (**b**) wear scar diagram and corresponding element distribution diagram of nitrided 2-2 samples at 700 °C.

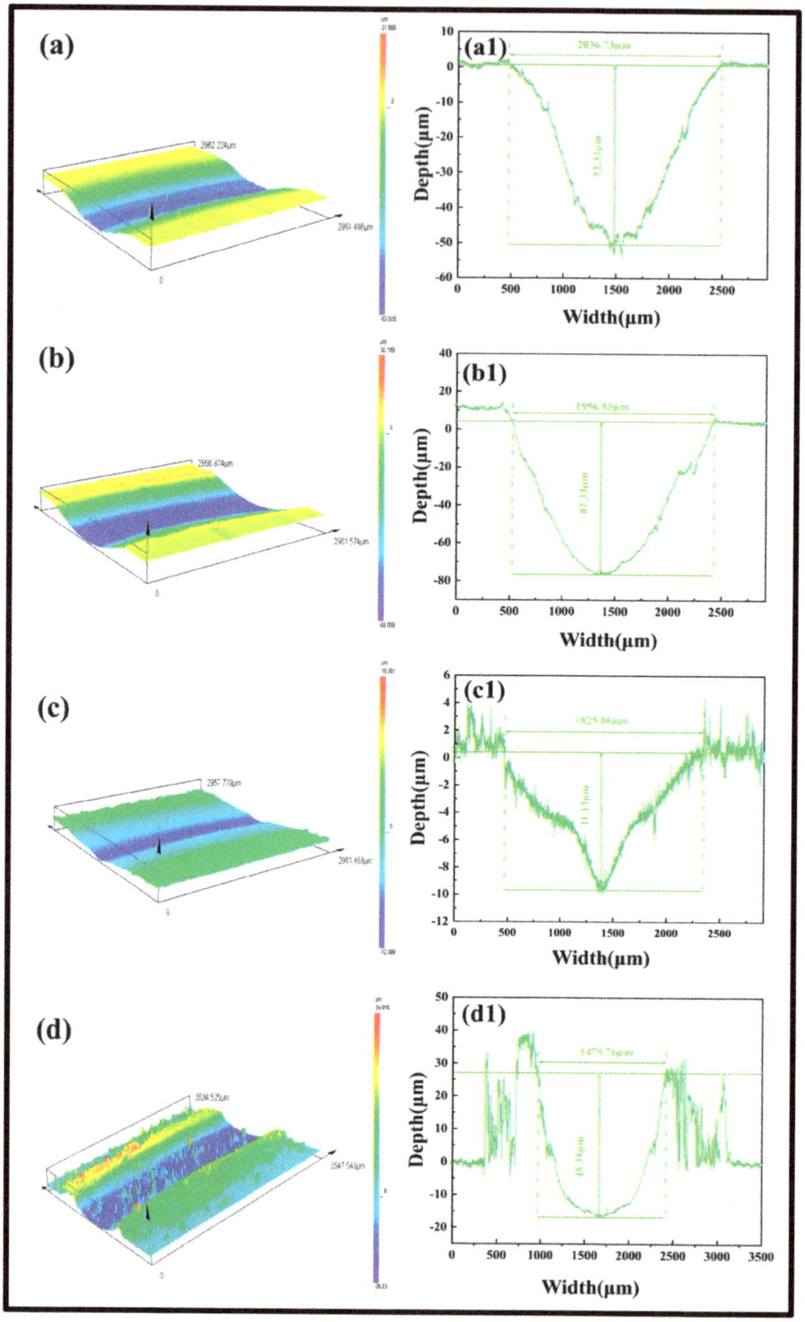

Figure 14. Three-dimensional topography (**a–d**) and corresponding cross-sectional curves (**a1–d1**): (**a**,**a1**) 1-1 chrome-plated specimen under room-temperature wear, (**b**,**b1**) 1-2 chrome-plated specimen under 700 °C high-temperature wear, (**c**,**c1**) 2-1 nitrided specimen under room-temperature wear, and (**d**,**d1**) 2-2 nitrided specimen under 700 °C high-temperature wear.

4. Conclusions

In this study, the chromium plating layer and nitriding treatment were prepared on the surface of the matrix material 2Cr3Mo3VA hot work die steel. The tribological properties and wear mechanism of two different surface treatments at room temperature and 700 °C were studied. The following conclusions were obtained:

1. The chromium plating layer and nitriding treatment on the surface of 2Cr3Mo3VA hot work die steel were effectively modified. The surface hardness and wear resistance of 2Cr3Mo3VA hot work die steel were improved due to the presence of ε phase ($Fe_{2-3}N$) in the 2Cr3Mo3VA hot work die steel after nitriding surface treatment. As a result, the surface hardness of the nitrided layer is better than that of the chrome-plated layer.
2. The surface hardness of the nitrided sample is higher than that of the chrome-plated sample after room-temperature wear or 700 °C high-temperature wear. At the same time, the wear rate of the nitrided layer at 700 °C is about 38% lower than that of the chromium plating layer, and the wear rate of the nitrided layer at room temperature is about 50% lower than that of the chromium plating layer, indicating that the nitrided layer has better wear resistance than the chromium plating layer.
3. The wear mechanism of chromium coating at room temperature is mainly abrasive wear accompanied by slight oxidation wear. The wear mechanism at 700 °C is the combined effect of oxidation wear, adhesive wear, and abrasive wear. The wear mechanism of the nitrided layer at room temperature is mainly adhesive wear, accompanied by slight oxidative wear. The wear mechanism at 700 °C is mainly oxidation wear and abrasive wear.
4. The research results provide important theoretical and experimental basis for surface modification of hot work die steel, especially in improving die life and workpiece quality in high-temperature environment. Future work should focus on further research on the wear mechanism of the coating under more temperature parameters, and explore the high-temperature wear mechanism of more different surface modification technologies, so as to provide a more comprehensive solution for the surface modification technology of hot work die and aerospace parts, and promote technological progress and application expansion in related fields.

Author Contributions: Conceptualization, Methodology, Investigation, and Data Curation, W.Z. (Wei Zhang 1) and J.Z.; Writing—Original Draft, W.Z. (Wei Zhang 2); Software, S.W. and C.Z.; Conceptualization and Writing—Review and Editing L.C., Z.S., X.W. and C.D.; Methodology, Study Design, and Writing—Review and Editing, F.M., C.C. and C.Z. All authors have read and agreed to the published version of the manuscript.

Funding: This project was supported by the Frontier Exploration Projects of Longmen Laboratory (No. LMQYTSKT011), the Natural Science Foundation of Henan Province (No. 242300420020), the Program for Science & Technology Innovation Talents in Universities of Henan Province (No. 25HASTIT007), and the Postdoctoral Research Project of Henan Province (No. HN2025145 and No.HN2025146).

Data Availability Statement: The raw data supporting the conclusions of this article will be made available by the authors on request.

Acknowledgments: We wish to take this opportunity to thank the Provincial and Ministerial Co-construction of Collaborative Innovation Center for Non-ferrous Metal New Materials.

Conflicts of Interest: Jian Zhang, Liuliang Chen, Wei Zhang, and Zhenhuan Sun are employed by Henan Zhongyuan Special Steel Equipment Manufacturing Co., Ltd. The remaining authors declare that the research was conducted in the absence of any commercial or financial relationships that could be construed as a potential conflict of interest.

References

1. Kashani, H.; Amadeh, A.; Ghasemi, H.M. Room and high temperature wear behaviors of nickel and cobalt base weld overlay coatings on hot forging dies. *Wear* **2007**, *262*, 800–806. [CrossRef]
2. Kumar, S.; Maity, S.R.; Patnaik, L. Friction and tribological behavior of bare nitrided, TiAlN and AlCrN coated MDC-K hot work tool steel. *Ceram. Int.* **2020**, *46*, 17280–17294. [CrossRef]
3. Mellouli, D.; Haddar, N.; Köster, A.; Ayedi, H.F. Hardness effect on thermal fatigue damage of hot-working tool steel. *Eng. Fail. Anal.* **2014**, *45*, 85–95. [CrossRef]
4. Yang, C.; Jing, C.; Fu, T.; Lin, T.; Guo, W.; Liu, N. Effect of TiC addition on the microstructure and properties of laser cladding Ni60A coatings on H13 steel surface. *Mater. Today Commun.* **2024**, *38*, 107904. [CrossRef]
5. Yang, X.; Li, C.; Zhang, Z.; Zhang, X.; Gu, J. Effect of cobalt-based coating microstructure on the thermal fatigue performance of AISI H13 hot work die steel. *Appl. Surf. Sci.* **2020**, *521*, 146360. [CrossRef]
6. Lu, J.Z.; Cao, J.; Lu, H.F.; Zhang, L.Y.; Luo, K.Y. Wear properties and microstructural analyses of Fe-based coatings with various WC contents on H13 die steel by laser cladding. *Surf. Coat. Technol.* **2019**, *369*, 228–237. [CrossRef]
7. Liu, B.; Wang, B.; Yang, X.; Zhao, X.; Qin, M.; Gu, J. Thermal fatigue evaluation of AISI H13 steels surface modified by gas nitriding with pre-and post-shot peening. *Appl. Surf. Sci.* **2019**, *483*, 45–51. [CrossRef]
8. Beake, B.; Ning, L.; Gey, C.; Veldhuis, S.; Kornberg, A.; Weaver, A.; Khanna, M.; Fox-Rabinovich, G. Wear performance of different PVD coatings during hard wet end milling of H13 tool steel. *Surf. Coat. Technol.* **2015**, *279*, 118–125. [CrossRef]
9. Xue, S.; Yang, T.; Guo, R.; Deng, A.; Liu, X.; Zheng, L. Crack analysis of Cr-Mo-V-Si medium-carbon alloy steel in casting die. *Eng. Fail. Anal.* **2021**, *120*, 105083. [CrossRef]
10. Klobčar, D.; Tušek, J.; Taljat, B. Thermal fatigue of materials for die-casting tooling. *Mater. Sci. Eng. A* **2008**, *472*, 198–207. [CrossRef]
11. Gu, Y.; Xia, K.; Wu, D.; Mou, J.; Zheng, S. Technical Characteristics and Wear-Resistant Mechanism of Nano Coatings: A Review. *Coatings* **2020**, *10*, 233. [CrossRef]
12. Qian, H.; Chen, S.; Wang, T.; Cheng, G.; Chen, X.; Xu, Z.; Zeng, Q.; Liu, Y.; Yan, D. Silicon nitride modified enamel coatings enable high thermal shock and corrosion resistances for steel protection. *Surf. Coat. Technol.* **2021**, *421*, 127474. [CrossRef]
13. Zhang, J.; Wang, H. Development of offshore wind power and foundation technology for offshore wind turbines in China. *Ocean Eng.* **2022**, *266*, 113256. [CrossRef]
14. Chen, H.; Wang, W.; Le, K.; Liu, Y.; Gao, X.; Luo, Y.; Zhao, X.; Liu, X.; Xu, S.; Liu, W. Effects of substrate roughness on the tribological properties of duplex plasma nitrided and MoS2 coated Ti6Al4V alloy. *Tribol. Int.* **2024**, *191*, 109123. [CrossRef]
15. Liu, S.; Liu, C.; Yang, Z.; He, L.; Zeng, G.; Zhang, W.; Long, J.; Chang, H. Microstructure, high-temperature corrosion resistance and oxidation properties of (TiVZrCrAl)N high entropy nitride coatings with different N2/Ar ratios. *Surf. Coat. Technol.* **2024**, *476*, 130226. [CrossRef]
16. Mirhosseini, S.S.; Mahboubi, F.; Azadfalah, M. Effect of different plasma nitriding durations on the tribological characteristics of nickel-boron-nanodiamond electroless nanocomposite coatings. *Surf. Coat. Technol.* **2024**, *476*, 130181. [CrossRef]
17. He, Z.; Wei, W.; Hu, J.; Gu, J. Aluminum-Modified Plasma Nitriding with High Efficiency and Enhanced Performance. *Coatings* **2024**, *14*, 1373. [CrossRef]
18. Lausmann, G.A. Electrolytically deposited hardchrome. *Surf. Coat. Technol.* **1996**, *86*, 814–820. [CrossRef]
19. Pan, P.; Gao, J.; Si, C.; Yao, Q.; Guo, Z.; Zhang, Y. Tribological properties of TIN coating on cotton picker spindle. *Coatings* **2023**, *13*, 959. [CrossRef]
20. Sun, Y.; Liang, Y.; Zhao, O. Local—Flexural interactive buckling behaviour and resistances of high-chromium stainless steel slender welded I-section columns. *Eng. Struct.* **2020**, *220*, 111022. [CrossRef]
21. Skripnyak, N.V.; Emelyanova, E.S.; Skripnyak, V.A.; Skripnyak, E.G. Damage of high-chromium steels under deformation in a wide temperature range. In *AIP Conference Proceedings*; AIP Publishing: Melville, NY, USA, 2017; p. 1909. [CrossRef]
22. Liu, Y.; Sun, Y.; Wu, H. Effects of chromium on the microstructure and hot ductility of Nb-microalloyed steel. *Int. J. Miner. Metall. Mater.* **2021**, *28*, 1011–1021. [CrossRef]
23. König, T.; Kimpel, T.; Kürten, D.; Kailer, A.; Dienwiebel, M. Influence of atmospheres on the friction and wear of cast iron against chromium plated steel at high temperatures. *Wear* **2023**, *522*, 204695. [CrossRef]
24. Wang, Y.; Hu, X.; Shi, X.; Yuan, Y.; Gao, J.; Zhang, Y. Failure mechanism analysis of electroplated chromium coating in repeated scratch tests under subcritical loads. *Wear* **2025**, *564*, 205708. [CrossRef]
25. Beake, B.D.; Ogwu, A.A.; Wagner, T. Influence of experimental factors and film thickness on the measured critical load in the nanoscratch test. *Mater. Sci. Eng. A* **2006**, *423*, 70–73. [CrossRef]
26. Fu, H.T.; Zhang, J.; Huang, J.F.; Lian, Y.; Zhang, C.; Gao, W. Comparison of nitrided hot work tool steel and chromium coated 30SiMn2MoVA. *China Surf. Eng.* **2015**, *28*, 1–6. [CrossRef]
27. Kumar, A.; Keerti, S.; Jain, J.; Sinha, S.; Tekumalla, S.; Gupta, M. Investigations of Wear Response of Pure Mg and Mg-0.4 Ce-Y_2O_3/ZnO Nanocomposites Using a Single and Repeated Scratch Tests. *Tribol. Trans.* **2018**, *61*, 951–959. [CrossRef]

28. Zhen-Yu, Z.; Zhi-Guo, J.; Qiu-Yang, Z.; Yu, L.; Zhi-Peng, Y.; Cong, D.; Zhong-Yu, P. Research on the construction of gradient nanostructure and anti-tribocorrosion behavior of aluminum alloy surface. *Tribol. Int.* **2024**, *194*, 109448. [CrossRef]
29. King, P.C.; Reynoldson, R.W.; Brownrigg, A.; Long, J.M. Cr (N, C) diffusion coating formation on pre-nitrocarburised H13 tool steel. *Surf. Coat. Technol.* **2004**, *179*, 18–26. [CrossRef]
30. Feng, L.; Hu, J.; Yan, S.; He, Z.; Shi, J.; Li, J. Effect of high-speed steel surface nitriding treatment on adhesion and wear resistance properties of nitrogen-doped diamond-like carbon coatings. *Diamond Relat. Mater.* **2023**, *136*, 110006. [CrossRef]
31. Tima, R.; Mahboubi, F. Ability of plasma nitriding to improve tribological behavior of medium and high boron electroless nickel coatings. *Tribol. Int.* **2021**, *156*, 106822. [CrossRef]
32. Jiang, W.; Wang, S.; Deng, Y.; Guo, X. Microstructure stability and high temperature wear behavior of an austenite aging steel coating by laser cladding. *Mater. Charact.* **2022**, *184*, 111700. [CrossRef]
33. Derelizade, K.; Rincon, A.; Venturi, F.; Wellman, R.; Kholobystov, A.; Hussain, T. High temperature (900 °C) sliding wear of CrNiAlCY coatings deposited by high velocity oxy fuel thermal spray. *Surf. Coat. Technol.* **2022**, *432*, 128063. [CrossRef]
34. Xing, Q.; Feltrin, A.C.; Akhtar, F. Processing, microstructure and high temperature dry sliding wear of a Cr-Fe-Hf-Mn-Ti-Ta-V high-entropy alloybased composite. *Mater. Today Commun.* **2021**, *28*, 102657. [CrossRef]

MDPI AG
Grosspeteranlage 5
4052 Basel
Switzerland
Tel.: +41 61 683 77 34

Lubricants Editorial Office
E-mail: lubricants@mdpi.com
www.mdpi.com/journal/lubricants

www.ingramcontent.com/pod-product-compliance
Lightning Source LLC
Chambersburg PA
CBHW061546210926
51850CB00009B/2014
9783725851799